FINITE SUPERSTRINGS

FINITE SUPERSTRINGS

J. G. Taylor
Department of Mathematics, King's College London, UK

P. C. Bressloff
Long Range Research, Hirst Research Laboratory
GEC, Wembley, UK

A. Restuccia
Department of Physics, Universidad Simon Bolivar
Caracas, Venezuela

World Scientific
Singapore • New Jersey • London • Hong Kong

Published by

World Scientific Publishing Co. Pte. Ltd.

P O Box 128, Farrer Road, Singapore 9128

USA office: Suite 1B, 1060 Main Street, River Edge, NJ 07661

UK office: 73 Lynton Mead, Totteridge, London N20 8DH

FINITE SUPERSTRINGS

ISBN 981-02-0969X

Printed in Singapore by General Printing Services (Pte) Ltd

FOREWORD

The writing of this book was begun in 1989, and is only now, two years later, completed. The reason for this rather lengthy period, in a field in which new results come out every few days and new fashions can emerge in a few months, is that the authors were geographically distant from each other for most of the period of writing. Yet the delay so resulting has, we think, led to a clearer picture of where the subject has reached and what further steps might be taken. We can do that with a certain degree of tranquillity since superstring research is no longer the frenzied field of activity it was in the mid-80's. In fact there are few important new results in the field over the last two years, and what results there have been appear as consolidation of old ideas.

The book is a contribution to that consolidation. Superstring fanatics (wherever they are now) certainly believe (without proof having been given) in the finiteness of the physical results of superstrings. On the other hand critics of the theories will dismiss them as either (a) too complicated, (b) two remote from the physical world or (c) unproven as being finite. A combination of any two or all three of such attitudes also occurs. We are trying to present a case for the incorrectness of a dismissal of superstrings under item (c). This book is devoted to showing that superstrings, and in particular the heterotic superstring, do provide a finite quantum theory of gravity. This, we feel, is an important step, and that is why it was felt necessary to try to present the details of the proof in a connected manner. That also may help to reduce the criticism of superstrings under (a), although certain parts of this book are difficult (especially Chapters 6 and 7). However we hope that the liberal supply of Appendices can augment missing detail from the text where necessary. We do not presume to attempt to answer criticism (b); that is for others to develop completely new methods or continue where we have left off.

This book is not meant to be an introduction to superstrings, although we have tried to make it reasonably self-contained. The first few chapters go over the construction of superstring field theory in the light-cone gauge; we recommend the books of Green, Schwarz and Witten[48] as a companion text of flesh out features not discussed by us. The meat of the book is Chapters 5, 6 and 7. The first of these presents our explicit construction of multi-loop amplitudes, the second their finiteness properties, and Chapter 7 the proof of closure of the ten-dimensional supersymmetry algebra. We give careful references to published material to aid the reader, as well as help in Appendices, as we indicated. In particular, the Appendices I to VII area brief survey of the theory of Riemann surfaces, whilst VIII to XVIII give more detailed derivations of some of the material used in Chapters 5 and 6. We recommend reading of Chapters 6 and 7 if it is felt to be desired.

Finally we would like to thank all of those in the subject, and our Departments, who have helped in this work and made the book possible. We would also like to thank the British Council and the International Centre for Theoretical Physics, Trieste for support allowing the authors to meet, and the Department of Mathematics at King's College London where the bulk of this book was written. Especially we would like to thank Marianne Fuller for her heroic work in laying out and typing the material so effectively.

King's College London
16.10.91

TABLE OF CONTENTS

FINITE SUPERSTRINGS

CHAPTER 1: INTRODUCTION

Section 1.1: Towards a Finite Quantum Gravity

The problem of obtaining a sensible theory of quantum gravity has been a central one in the development of theoretical physics in the latter half of this century. This arose in particular from the desire to unify all of the forces of nature. In this book we present a solution to the problem of quantizing gravity by means of the 'hetorotic' superstring. A heterotic superstring is first of all a string (a one-dimensional object which may indeed be visualised as a length of ordinary string, or as a closed loop of such material, in the case of open and closed strings respectively) which is moving in a space-time of ordinary co-ordinate values (as is the space in which we live) and in addition a set of anticommuting co-ordinates. It is these latter which are the additional fermionic variables which may begin to allow the string to possess a super-symmetry in which the fermionic (anti-commuting) and bosonic (commuting) co-ordinates are rotated into each other at each point of the string in a manner which will be delineated later in the book. Supersymmetry is regarded as one of the important ingredients for any ultimate theory for reasons to be explained later in this section; in brief it grants less divergent high energy behaviour to a theory possessing it. A superstring theory would seem to be more effective than a theory of point interactions in that the former has an infinite number of supersymmetries, one for each point of the superstring.

The epithet 'heterotic' has Greek origin, meaning 'being stronger by bringing together disparate elements'. The two different elements for the heterotic superstring are a closed bosonic string (involving right-going modes along the string) and a closed fermionic string with only left-going modes. These can be so combined as to lead to a string with suitable supersymmetry invariances along with extra symmetries which may be used to generate the internal symmetries corresponding to those forces of nature other than gravity.

It is our purpose in this book to show that the heterotic superstring is aptly named. We claim that our analysis gives the first complete construction of a finite version of quantum gravity in ten dimensions. That is achieved by developing a proof of the finiteness of the multi-loop amplitudes for the scattering of the massless modes (gravitons, etc) of the string. This proof is only for each order of the amplitudes in the strength of the string interaction; it may be that the total series, when summed over the loop number, is not well defined. The massless modes are the transverse excitations of the string and correspond to the particles we observe around us at our present (low) energies; they are the only stable particles described by the string. It is only possible to give proof of the finiteness of the scattering amplitudes for these massless modes by expanding the amplitudes in powers of the interaction strength between a pair of strings (corresponding to the strength with which a pair of strings fuse together, at one point, to generate a third). This expansion is a pertubation series in powers of the interaction strength.

In this book we give a self-contained account of the construction, as well as the finiteness analysis, of the multiloop amplitudes. As such the work presented here is only a first tentative step forward down to distances of order of 10^{-33}cms (the so-called Planck length). The existence of a natural length, energy, or time arises in any theory attempting to combine G (Newton's gravitational constant), h (Plank's constant) and c (the velocity of light). These constants are contained respectively in the theories we are trying to combine, the latter being gravity, quantum mechanics and special relativity. One is then able to construct the so-called Planck length (having dimension of length) and given by

$$L_p = (hG/C^3)^{1/2}$$

and with value about 10^{-33}cms. The corresponding Planck energy is 10^{19}GeV, and the Planck time 10^{-43}sec, being that taken by light to travel the Planck length. It is at such short lengths or times or at such high energies (seventeen orders of magnitude greater than available at man-made particle accelerators), that superstrings would appear to play an essential role in physics. The very early Universe, or deep inside a black hole where quantum gravitational effects may become important can thus only begin to be properly described in terms of a superstring. For similar reasons it seems only possible to construct a theory of everything by means of superstrings. The arena of space-time (or even its supersymmetric extension to so-called superspace) is not big enough. What is required is to consider the set of all loops in such a space; loop space is the proper arena for such discussions.

We do not attempt in this book to explore fully the physical implications of our results. Such physical features are, in any case, being investigated as part of an active research programme in their own right, with too many details for such inclusion. Our aim is to give chapter and verse to the finiteness proof, so as to justify our claim that the heterotic superstring gives a finite quantum gravity. That appears a worthy enough exercise in its own right, as well as an essential prerequisite for believing that superstrings can have any physical implications at all. For to try to discover the physics of superstrings (or any other claimed theory of quantum gravity) when physical results are infinite, and not removable in any sensible manner, would be like trying to build castles in the sand. Finiteness is a prerequisite.

How does such a result fit in with previous attempts to quantise gravity? We do not propose to write a history of the subject of quantum gravity to answer that here, but suffice it to say that all attempts to unify quantum mechanics and gravity have, up till the appearance of superstrings in the early 80's, proved in vain. However in the process of unravelling, two different strands can be seen apparent. Each of these has proven crucial in the final construction of finite superstrings. They are therefore of value to describe briefly.

The first strand is that of higher dimensional physics, which had already been suggested as a way of unifying gravity with other forces in the early part of the century [1]. Curvature in extra dimensions allowed for additional (non-gravitational) forces, such as electromagnetion or Yang-Mills gauge theories (at the basis of modern understanding of the unification of electromagnetism and radioactivity and of the nuclear-force) to emerge. Such a possibility was explored extensively in the last three decades, and leads both to viable classical unified theories (including gravity) and also to interesting quantum implications [2].

That higher dimensional theories were included in strings was known already in the early 70's. Ordinary bosonic strings were found only to avoid an inconsistent quantum theory (due to a conformal anomaly) if they were quantised in twenty-six dimensions. The problem of removing the extra twenty-two dimensions was immediately apparent, but is still unsolved, both for the string and superstring. The most popular present approach is to take as ground state of the string one in which the extra dimensions are compactified into small circles. More exotic compactifications are also possible, and indeed desirable to fit the known symmetries of particles. The inverse of the radii of these circles gives the mass gap separating massless from massive modes. Thus for radii less then 10^{-17}cms such massive states will be presently unobservable. However a situation in which one chooses the ground state of the string from experiment appears unsatisfactory from the theoretical point of view; some further discussion of the problem of superstring compactification will be given at the end of the book.

The other strand which has driven development is that of supersymmetry (SUSY), the symmetry between bosons and fermions, and was introduced in the early seventies [3]. It was shown early on that this symmetry could ameliorate the ultra-violet divergences which had plagued local quantum field theories ever since their introduction. For example, the vacuum energy had a negative contribution from the fermions. These U.V. infinites were ultimately removed completely from certain super-symmetric Yang-Mills theories, although the resulting theories have not yet proved of value in helping guide experimental predictions.

Applied to gravity, supersymmetry leads to supergravity [4], in which the graviton has its super partner, termed the gravitino. Supergravity can be extended by the addition of up to eight gravitini in that if more (identical) gravitini were allowed then particles of spin $^5/_2$ or higher would be required. Such particles proved very hard to include in a consistent manner. The superstring later resolved this problem in a very special manner, but keeping to ordinary space-time did not allow more than eight gravitini. The resulting theory was called N=8 supergravity (N being the number of gravitini). This theory was linked up to higher dimensional theories in that the N=8 theory was found to arise from an N=1 supergravity in eleven dimensions by dimensional reduction (neglecting dependence of the field variables on the extra dimensions beyond four).

There were hopes that the amelioration of infinites discovered for supersymmetric Yang-Mills theories could extend to supergravities, and particularly to the N=8 version. However, it was soon realised that such an extension is not possible and it is generally accepted that the infinite number of divergences which still plague supergravities could never be removed satisfactorily. Something different from a local quantum field theory had to be considered.

The source of the infinite set of U.V. infinites in quantum gravity is the existence of arbitrary high powers of fields at the same point, of the form $[g(x)]^n$ (where g is the field variable at the space-time point x). Such powers arise from the fact that gravity couples universally to the energy-momentum tensor, which itself has a gravitational contribution. The interaction has to be local, from causality criteria, so that the interaction must always involve many fields interacting at the same point, together with some derivatives of fields at that point.

Objects 'distributed' on a point (i.e. fields), must always interact pointwise to preserve locality. The relative size of the interaction region in comparison to the support of the object itself is thus 1:1. To reduce the importance of the interaction region one may increase the size of the support of the field whilst preserving the point-like nature of the interaction region. This leads naturally to the consideration of string-like objects, in which the string is a one-dimensional (open or closed) contour. Open strings can interact at their ends, closed ones by fusing together at one point. The size of the interacting region is now of measure zero on the string, so that there is still hope of amelioration of the ultra-violet divergences of the resulting interacting string theory, especially because as remarked earlier, there could now be an infinite number of supersymmetries, one for each point of the string.

The construction of string theory was begun in the early 70's. The existence of an unphysical tachyon in the spectrum was removed by the late 70's by the introduction of fermions on the string, and the reduction of divergences down to at most quadratic conjectured. The free superstring, in the light-cone gauge, was then constructed in 1981 [5)] and in covariant form three years later [6)]. Since that time there have been many attempts to construct and analyse for finiteness the multi-loop amplitudes for the superstring; they will be discussed briefly in the next section.

It is clear that the proof of finiteness of the heterotic superstring multi-loop amplitudes does not begin to touch on some of the deep problems dicussed over the decades by those trying to quantise gravity [7], especially of a non-perturbative character. However recent developments in the non-perturbative physics of two-dimensional quantum gravity [8] may be a hint of a framework for strings in which some of these question may be analysed. We return to such features briefly at the end of the book.

Section 1.2: Why the Light-Cone Gauge?

All fundamental physical theories of present interest-electroweak chromodynamics, or gravity - are gauge theories, involving unitary gauge groups (SU(2)xU(1) for E.W., SU(3) for QCD, or the orthogonal group SO(3,1) for gravity). Unifications of some of these theories, involve higher rank unitary or orthogonal groups (SU(5) or SO(10)), whilst for the more recent superstring theories $E_8 x E_8$ or SO(32) are possibilities, as well as many others. These gauge symmetries generate constraints (which we denote $\phi_\alpha=0$ for α belonging to some set of parameters) between the co-ordinates and their canonical momenta which render the quantisation of these theories non-trivial.

There are two equivalent ways of dealing with the constraints. One is to impose them directly on states, in an analogous manner to the Gupta-Bleuler method in quantum electrodynamics. For this approach the positive frequency part of $\partial\mu A^\mu$ (where A^μ is the four-vector potential of the electromagnetic field) is required to annihilate all physical states. One may also introduce a further set of gauge fixing constraints (which we denote λ_β) so that the total set of contraints $\phi_\alpha,\lambda_\beta$ have non-singular Poisson bracket ($\{\phi_\alpha,\lambda_\beta\}$). In general this second way leads to a more explicit representation of the symmetries of the system under study although it still involves constraints on the states. The original theory has now been modified by the addition of the new constraints $\lambda_\beta=0$ (included, say, in the action, by Lagrange multipliers). However there still exists a global symmetry - the BRST symmetry [9] - for the modified theory. This invariant is a relic of the original local symmetry, and allows proof of the independence of physcial quantities (S-matrix elements) on the particular choice of the constraints λ_β. In order to incorporate the BRST symmetry it is well-known that it is necessary to introduce ghost degrees of freedom into the theory, as Grassmann-valued gauge parameters. The ghosts may then be shown to cancel, in physical quantities, the gauge degrees of freedom in the original theory (this is well reviewed in [10], for example).

The other way to deal with constraints is to solve them classically. In addition, in gauge theories, one may choose λ_β in a non-covariant manner so that the ghosts decouple completely, and the non-physical modes in the original fields may be expressed directly in terms solely of the physical ones. One such gauge choice is the light-cone gauge (L.C. for short), originally introduced by Dirac [11]. In a space-time of D dimensions, this uses the light-cone variables

$$x^\pm = \frac{1}{\sqrt{2}}(x^0 \pm x^{D-1})$$

$$\underline{x} = (x^1, x^2, ---, x^{D-2})$$

where $\underline{x}$ denotes the transverse vector part of x to the light-cone directions $x^\pm$. For electrodynamics [12], with x^+ the L.C. 'time', it is then possible to choose the gauge $A^+=0$ and solve $\partial_\mu A^\mu=0$ as $A^-=(\partial_-)^{-1}$ $\underline{\partial}.\underline{A}$. It is necessary to be careful in the definition of the integral operator $(\partial_-)^{-1}$ [12], but by this gauge choice one has reduced the electromagnetic potential to the physical, transverse, photon modes.

A similar gauge choice allows the elimination of half of the degrees of freedom for the electron in electrodynamics [13], and of A^- in non-Abelian Yang-Mills theories. In the case of supersymmetric Yang-Mills theories, auxiliary fields are avoided completely, so that No-Go theorems [14] no longer apply. This was particularly the case in Mandelstam's construction [15] of a L.C. superfield version of N=4 supersymmetric Yang-Mills theory, leading to his proof of perturbative finiteness of the theory by explicit construction of the general superfield Feynman diagram. Similar attempts have more recently been made for supergravity [16], based on the fact [17] that the non-linear constraints of general relativity are also exactly soluble in this gauge. However there seems no indication from this approach that the resulting formulation of supergravity could be finite at each order of perturbation theory.

Superstring theory is presently the best candidate for a perturbatively finite quantum theory of gravity. This became manifest by the important proof of anomaly cancellation for the type I superstring, with gauge group SO(32) by Green and Schwarz in 1984 [18], and the relationship then proven between this and 1-loop finiteness [19]. One loop finiteness has also been demonstrated [20] for the more recently constructed heterotic string [21].

Since that time, serious attempts have been made to extend the proof of perturbative finiteness of massless superstring amplitudes to all loop orders (in string perturbative theory), as well as to analyse the dependence of this result on the background (massless) fields in which the strings move. The main thrust of this analysis has been through the use of the covariant approach first indicated by Polyakov [22], and since developed more fully using the concepts and techniques of Riemann surface theory (for a review of this see, for example, [23]). In order for this to be applied to the spinning (NRS) [24] or superstring [5],[6] it is necessary to extend the notion of a Riemann surface to that of a super-Riemann surface (SRS) [25], in which the complex variable z has a companion Grassmann-valued variable θ. Since multi-loop string amplitudes require integration over the class of conformally inequivalent Riemann surfaces, ie over the space of the moduli, it is to be expected that superstring multi-loop amplitudes require integration over super-moduli space. This latter space is still in the process of being defined [26], and moreover, requires a satisfactory and unambigious definition of integration. This specification has not yet been given in detail, except for the partition function at genus 2. [27].

In spite of the apparently greater elegance of the covariant (Polyakov) approach to strings, there is, besides the above difficulty for superstrings, the question of unitarity of the resulting multi-loop amplitudes. This has been proven for bosonic string amplitudes by reducing the covariant amplitudes to those of the light-cone gauge [28]. These latter are unitary, having been constructed from a LC field theory of strings [29]. A similar equality has not yet been proven for 'superstrings, mainly because both sides of this identity are apparently, as yet, incompletely known. However, there is supposedly a complete LC field theory for types I and II superstrings [29] and a more recently constructed one for heterotic strings [30],[31]. The latter version of that field theory has allowed construction of multi-loop amplitudes for the heterotic string [31], which were developed from earlier constructions of the type II multi-loop amplitudes [32].

Since there has already been construction of multi-loop superstring amplitudes in the LC gauge, it is possible to determine whether or not these are finite. This has indeed been analysed [33], with finiteness obtained for the heterotic, and the same, to within boundary terms, for type II superstrings. However the boundary terms found in [34] give a hint of something incomplete in the LC string field theory of [29]. The latter theory, for closed superstrings, was claimed to be only at most cubic in the closed superstring field ψ in the Hamiltonian. Such an interaction would lead to an unstable theory, as has recently been forcefully pointed out [35]. Both the need for cancellation of the boundary terms, in amplitudes, and for stability of the field theory required a quartic addition to the Hamiltonian, and very likely to the space-time supersymmetry generators, which are non-linearly realised. Various groups have analysed that situation, [36],[37]. Regularisation problems are considered in particular in [36], together with analysis of closure of the quartic generators so constructed. This latter feature is particularly important, since non-polynomiality of the completed LC superstring field theory may arise from introduction of higher order terms: the process may never cease. That would make sensible calculations very difficult to perform. A review of this situation up to the middle of 1988 was presented in [38], and a final complete proof of closure of the full super-Poincaré algebra sP_{10} for the heterotic string given in [39]. The Witten covariant superstring field theory [40] appears to evade the above difficulties completely. The action was only cubic in the superstring field in the original version, yet the algebra is closed. Moreover it has been suggested that the extra terms in the LC string field theory, such as considered above, may arise only on gauge fixing to the LC gauge. However that need not be the case, since there are also at least quartic if not higher terms required similar to those in the LC gauge [41]. That is a natural state of affairs to expect if one follows the Siegel-Zwiebach programme of covariantizing the LC string field theory [42]; or that of the Japanese [43]; the new terms in the LC field theory will also require the existence of their covariant counterparts.

After this somewhat lengthy preamble it is possible to say in more detail why it seems that the time is opportune to give a detailed and connected account of superstrings in the LC gauge. The reasons are:

(1) we can now present a complete superstring field theory in the LC gauge, including the so-called contact terms discussed

above;

(2) we can deduce, from field theory, explicit rules for multi-loop amplitudes, and even manipulate these expressions into explicitly covariant form;

(3) we can analyse these amplitudes for finiteness;

(4) in the process there are no ambiguities involved with integration over super-moduli;

(5) superstring LC gauge field theories can be constructed on general backgrounds so have physical implications;

(6) our field theories may lead to covariant string field theories with proper account of the contact terms ab initio;

(7) we may look for non-perturbative features from those field theories.

This book is devoted to an in depth analysis of results associated with (1)-(4) above; (5)-(7) will only be remarked on briefly at the end.

CHAPTER 2: FREE SUPERSTRINGS IN THE LIGHT CONE

Section 2.1: Free Bosonic Strings in the Light Cone

The light-cone was originally introduced by Goldstone, Goddard, Rebbi and Thorne in 1973 [44], in order to express the dynamics of the bosonic string purely in terms of the (D-2) oscillations transverse to the string's length. This followed the important step of Nambu [45] in giving the string a space-time interpretation, by writing the action as proportional to the area of the world sheet it swept out in the embedding space-time. The area a string delineates in its motion through space-time is thereby independent of the particular choice of co-ordinates (the parametrisation) of the world sheet. This leads to invariance of the action under reparametrisations of the world sheet (worldsheet diffeomorphisms). There also results (Virasoro) constraints, which may be seen to be the generators of these reparametrisations. As described in the first chapter, the constraints imply that there are fewer physical degrees of freedom than given by the original string co-ordinates. As we will see, there are really only those modes of the string corresponding to oscillations transverse to the worldsheet. The constraints cause similar difficulties to those arising in general relativity. The prime virtue of the L.C. gauge is the possibility of explicitly solving the Virasoro constraints for the two unphysical modes of the string in terms of the transverse oscillations.

To specify our notation, we will describe the manner in which the L.C. gauge may be introduced. The Nambu action S of the worldsheet W of fig. 1 is [45]

$$S = - T \int_W dA \tag{2.1}$$

where dA is an element of area W and T is the string tension with dimensions $(\text{mass})^2$. We will choose $T = \frac{1}{\pi}$, corresponding to the slope of Regge trajectories $\alpha' = \frac{1}{2}$, as will be discussed in section 2.3.

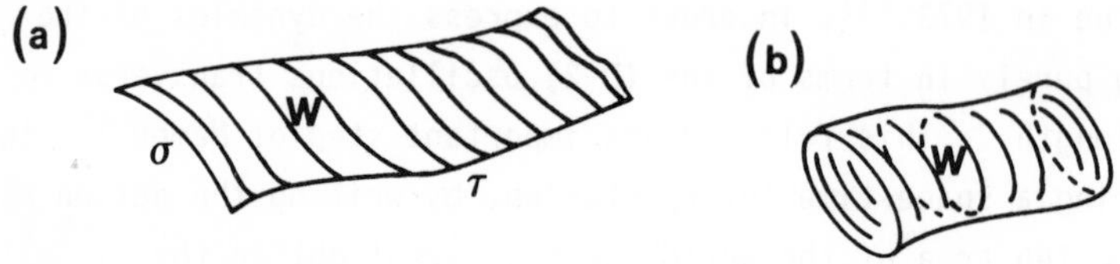

Fig 1:
(a) The world sheet W swept out by an open string
(b) W as swept out by a closed string

In terms of a particular set of co-ordinates (σ^0, σ^1) on W,

$$dA = (-\det g_{\alpha\beta})^{\frac{1}{2}} \, d\sigma^0 d\sigma^1 \tag{2.2}$$

where the induced metric $g_{\alpha\beta}$ $(\alpha,\beta=0,1)$ has value

$$g_{\alpha\beta} = \partial_\alpha X^\mu \partial_\beta X_\mu \tag{2.3}$$

The raising and lowering of indices on X^μ in (2.3) is performed by means of the Lorentz metric $\eta^{\mu\gamma} = \mathrm{diag}\,(1,-1,-1,\ldots,-1)$; in Chapter 8 we will consider the extension to the case of a general Riemannian embedding space time.

We use the notation $(\sigma^0, \sigma^1)=(t,\sigma)$ and $\partial X^\mu/\partial t = \dot{X}^\mu$, $\partial X^\mu/\partial\sigma = X^{\mu\prime}$ (which differs from the usual notation for σ^0 as τ, which we reserve for the Euclideanised world-sheet) so that, with $AB=A_\mu B^\mu$ and

$$\Delta^2 = (\dot{X}X')^2 - X'^2\dot{X}^2 \tag{2.4}$$

the lagrangian density in the action (2.2) has value

$$L = -\frac{1}{\pi}\Delta \tag{2.5}$$

On defining the canonical momenta

$$P_\mu = \partial L/\partial\dot{X}^\mu \tag{2.6}$$

with explicit form, from (2.4), (2.5)

$$P^\mu = -\pi^{-1}[(\dot{X}X')X^{\mu\prime} - X'^2\dot{X}^\mu]\Delta^{-1} \tag{2.7}$$

then a little algebra leads to the constraints

$$\phi_\sigma \equiv P.X' = 0 \tag{2.8a}$$

$$\phi_t \equiv P^2 + \frac{1}{\pi^2}X'^2 = 0 \tag{2.8b}$$

or

$$\phi_\pm \equiv (P \pm \frac{1}{\pi}X')^2 = 0 \tag{2.9}$$

The constraints ϕ_σ, ϕ_t, or $\phi_\pm$ are the Virasoro constraints corresponding to the parametrisation invariance of the action (2.1) mentioned earlier. We note that ϕ_σ, ϕ_t are the generators of σ and t reparametrisations on W. These latter parametrizations have to be fixed in the action S before quantisation can occur.

In order to specify one of the parameters σ, t, it turns out to be convenient to relate it to one component of the co-ordinates $X^\mu(\sigma,t)$ on the world-sheet. If we describe the string co-ordinates in the light-cone gauge, with co-ordinates

$$X^\pm = \frac{1}{\sqrt{2}}(X^0 \pm X^{D-1}),\ X^\mu Y_\mu = X^+Y^- + X^-Y^+ - X^iY^i \tag{2.10}$$

we realise that there may be an infinity of 'times' $X^+(\sigma,t)$, for each t, as t varies (where we initially take the parametrisation length $0 \leq \sigma \leq \pi$). There are still many unsolved problems in the development of many-time dynamics [46]. These may well be avoided by choosing the parametrisation gauge conditions, which reduce these many times to a single one, as

$$X^+(\sigma,t) = x^+ + p^+ t \tag{2.11a}$$

$$P^+(\sigma,t) = \pi^{-1}p^+ \tag{2.11b}$$

where x^+, p^+ are constants. If we now write the constraints (2.8) in L.C. co-ordinates as

$$P^+X^{-\prime} + P^-X^{+\prime} - \underline{P}.\underline{X}' = 0 \tag{2.12a}$$

$$2P^+P^- - \underline{P}^2 + \frac{1}{\pi^2}(2X^{+\prime}X^{-\prime} - \underline{X}'^2) = 0 \tag{2.12b}$$

then (2.11) allow (2.12) to be solved as

$$X^{-\prime}(\sigma)=(\pi/p^{+}).\underline{P}.\underline{X}'(\sigma) \tag{2.13a}$$

$$P^{-}(\sigma)=(\pi/2p^{+})\ [\underline{P}^{2}+\frac{1}{\pi^{2}}\underline{X}'^{2}](\sigma) \tag{2.13b}$$

(2.13a) may be integrated to give X^- in terms of the transverse modes $\underline{P}$,$\underline{X}$ and an additional zero mode x^-:

$$X^{-}(\sigma) = x^{-}+\pi(p^{+})^{-1}\int_{o}^{\pi} d\sigma^{1}\ \underline{P}(\sigma^{1}).\underline{X}'(\sigma^{1})\left[\frac{\sigma^{1}}{\pi}-\theta(\sigma^{1}-\sigma)\right] \tag{2.14}$$

where $\frac{1}{\pi}\int_{o}^{\pi} d\sigma X^{-}(\sigma)=x^{-}$. Equation (2.13b) expresses in particular the Hamiltonian, the conjugate to the 'time' t, from the action (2.1) as

$$H = p^{+}\int_{o}^{\pi} P^{-}(\sigma)d\sigma = \frac{1}{2}.\int_{o}^{\pi} d\sigma[\pi\underline{P}(\sigma)^{2}+\pi^{-1}\underline{X}'(\sigma)^{2}]=p^{+}P^{-} \tag{2.15}$$

(where $P^-=H/p^+$ is the bosonic part of the Hamiltonian, being the generator of translations in X^+). All possible gauge degrees of freedom are now fixed, since the lines of constant t and constant σ are completely specified on W (as pictured in fig 2).

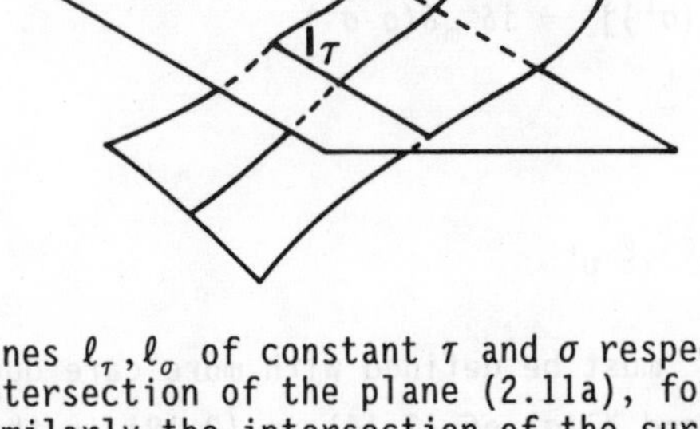

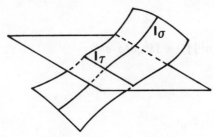

Fig. 2: The lines ℓ_τ,ℓ_σ of constant τ and σ respectively on W; ℓ_τ given by the intersection of the plane (2.11a), for given τ, with W, whilst ℓ_σ is similarly the intersection of the surface (2.14) for given σ, with W (where $x^\pm$ are kept fixed throughout).

We note in addition that the open bosonic string must also satisfy

$$X^{\mu\,\prime} = 0 \quad \text{at } t = 0,\pi \tag{2.16}$$

which corresponds to the parametrisation length of σ being kept constant under reparametrisations generated by ϕ_σ of (2.8a). On the other hand the closed bosonic string must satisfy the periodicity condition

$$X^\mu(\sigma+2\pi,t) = X^\mu(\sigma,t) \tag{2.17}$$

(where the range of σ is taken to be over an internal of length 2π for closed strings). An important aspect of any gauge-fixing proceedure is that it leads to covariant physical expressions. This may be achieved by constructing the generators of the full lorentz group in the L.C. gauge, and then ensuring that they satisfy the correct commutation relations [44]. The conserved lorentz generators j^μ have the formal expressions [44]

$$j^{\mu\nu} = \int_0^\pi d\sigma\, X^{[\mu}(\sigma)P^{\nu]}(\sigma) \tag{2.18}$$

For ℓ,m running over the transverse values 1,...,D-2,$j^{\ell m}$ satisfy the usual rotation algebra, as can be verified by direct calculation from (2.18) with

$$[X^\ell(\sigma),P_m(\sigma^1)]_- = i\delta^\ell{}_m\delta(\sigma-\sigma^1) \tag{2.19}$$

Moreover

$$j^{+\ell} = x^+p^\ell - x^\ell\, p^+ \tag{2.20}$$

The generators $j^{\ell -}$,j^{+-} must be defined with more care due to the need to use $P^-(\sigma)$ of (2.13b) and $X^-(\sigma)$ of (2.14) in (2.18) in the case μ or

$\upsilon = -$. There then results cubic expressions in the non-commuting variables $\underline{X}$, $\underline{P}$ of (2.19). These may be handled by normal ordering. We simply quote the result [44] here that the Lorentz algebra closes only for D=26, and with a suitable constant removed from (2.15), leading to the tachyon state. The resulting generators have been given in detail in [47] as, to within normal ordering,

$$j^{\ell -} = \int_o^\pi d\sigma \, [X^\ell(\sigma)P^-(\sigma) - X^-(\sigma)P^\ell(\sigma)] + ip^\ell \frac{\partial}{\partial p^+} \tag{2.21a}$$

$$j^{+-} = x^+ P^- + ip^+ \frac{\partial}{\partial p^+} \tag{2.21b}$$

where the second term on the right of (2.21) corresponds to the associated reparametrisation neccessary to preserve the gauge condition (2.11), p^+ being the conjugate variable to x^-.

Many of the physical features of strings are made more transparent by the use of mode expansions. It has also been attempted to use such expansions to give a better mathematical sense to some of the expressions we will meet later; however, that is not the method to be used in this book, where complex function techniques will prove more effective. We will discuss modes briefly in this and future sections so as to help develop the physical intuition.

In the case of open bosonic strings, the mode expansion is of the form

$$X^i(\sigma,t) = x^i + p^i t + i \sum_{n \neq o} \frac{1}{n} \alpha_n{}^i e^{int} \cos n\sigma \tag{2.22}$$

which satisfies the equation of notion

$$\ddot{X}^i = X''^i \tag{2.23}$$

together with the boundary condition, equation (2.16). The commutation relation (2.19) implies

$$[x^i, p^j] = i\delta^{ij}, \ [\alpha_m{}^i, \alpha_n{}^j] = m\,\delta_{m+n,o}\,\delta^{ij} \tag{2.24}$$

The string excitations are described by an infinite number of harmonic oscillators with $\alpha_n{}^i$ (n>0) and $\alpha_n{}^{i\dagger} = \alpha_{-n}{}^i$ (n>0) corresponding, respectively to raising and lowering operators. The vacuum state of the string (at the first quantised level) is defined by the condition

$$\alpha_n{}^i\,|0\rangle = 0, \quad n>0, \qquad i=1, \ldots D-2$$

The open string Hamiltonian has the mode expansion (after normal ordering)

$$H = \frac{1}{2}\sum_{i=1}^{D-2}\left[\left(p^i\right)^2 + 2\sum_{n>0}\alpha_{-n}^{\ i}\,\alpha_n^{\ i}\right] - a \tag{2.25}$$

where a is a zero-point contribution determined by the formula

$$a = \sum_{i=1}^{D-2}\left(-\frac{1}{2}\sum_{n=-\infty}^{\infty}\alpha_{-n}^{\ i}\,\alpha_n^{\ i} + \frac{1}{2}\sum_{n=-\infty}^{\infty}:\alpha_{-n}^{\ i}\,\alpha_n^{\ i}:\right)$$

$$= -\frac{1}{2}\,(D-2)\sum_{n=1}^{\infty} n$$

and repeated summation over the variable i labelling transverse vector components is always taken, with i=1,2,...,D-2. It is clear that a is not well-defined, but may be regularised by zeta function regularisation, with

$$\zeta (s) = \sum_{n\neq 0} n^{-s}, \quad \text{Res} > 1$$

$$\zeta (-1) = - 1/12$$

as may be obtained by analytic continuation. This gives the result

$$a = \frac{D-2}{24}$$

Each state of the string corresponds to a particle of specific mass and spin (leading to the Regge trajectories of the old duel-resonance model (see 48, which we refer to as GSW in the sequel, for a review). The spectrum of the states is given by the mass-shell condition

$$p^2 = 2p^+p^- - (p^i)^2$$

$$= 2\left[N - \frac{1}{24} (D-2)\right] \qquad (2.26)$$

where

$$N = \sum_{n>0} \alpha_{-n}^{\ i} \, \alpha_n^{\ i} \qquad (2.27)$$

Note that the state with lowest value of p^2 is the vacuum state |o> which corresponds to a tachyon (a particle with negative mass). One of the desirable features of the superstring, section 2.2, will be the elimination of any tachyonic states. The next larger value of p^2 occurs with states of the form $\alpha_{-1}^{\ i}\ |0)$ and p^2 value

$$\frac{1}{2} p^2 = 1 - \frac{D-2}{24} \tag{2.28}$$

Since there are exactly (D-2) such states this must correspond to a vector mode, so must be massless (a massive vector mode would have (D-1) physical states). Therefore, $p^2 = 0$ in equation (2.28) implying that D=26. This is the well-known quantum-mechanical condition that a bosonic string can only evolve consistently in 26 dimensions (in a flat background); in other (flat) dimensions non-physical modes will occur. A more convincing method of obtaining the critical dimension may be followed by requiring the closure of the total D-dimensional Lorentz algebra, as noted above.

Similarly, the mode expansion of the closed string, satisfying equations (2.23) and (2.17) is

$$X^i\ (\sigma,t) = x^i + \frac{1}{2} p^i t + i \sum_{n\neq 0} \frac{1}{2n} \left[\alpha_n{}^{\mu} e^{-in(t-\sigma)} + \tilde{\alpha}_n{}^{\mu} e^{-in(t+\sigma)}\right] \tag{2.29}$$

where α_n and $\tilde{\alpha}_n$ correspond, respectively to right- and left- moving modes. The commutation relations are

$$[x^i, p^j] = i\delta^{ij},\ [\alpha_m{}^i, \alpha_n{}^j] = [\tilde{\alpha}_m{}^i, \tilde{\alpha}_n{}^j] = m\delta_{m+n,0}\delta^{ij} \tag{2.30}$$

with $[\alpha, \tilde{\alpha}] = 0$. The mass-shell condition is now

$$p^2 = 4[N + \tilde{N} - (D-2)/12] \tag{2.31}$$

where

$$N = \sum_{n>0} \alpha_{-n}{}^{i}\ \alpha_{n}{}^{i}, \quad \tilde{N} = \sum_{n>0} \tilde{\alpha}_{-n}{}^{i}\ \tilde{\alpha}_{n}{}^{i} \tag{2.32}$$

However, in the closed string case there is an extra complication arising from the periodicity condition

$$X^{-}(\sigma + 2\pi) = X^{-}(\sigma) \tag{2.33}$$

Applying (2.33) to (2.13a) gives

$$\Lambda = \int_{-\pi}^{\pi} d\sigma\ X^{i\,\prime} P^{i} = 0 \tag{2.34}$$

Comparing with equation (2.8a) we see that in the LC gauge (2.34) is just $\int d\sigma \phi_{\sigma}$ and hence generates constant translations in σ. In other words, for the closed string there is an extra unsolved contraint expressing invariance under choice of origin of σ parametrisation. One immediate consequence of (2.34) is that for physical on-shell states

$$N = \tilde{N} \tag{2.35}$$

(also see section 4.5). The analysis of the closure of the closed string Lorentz algebra is similar to that of the open string and again requires D = 26. The first excited level ($N = \tilde{N} = 1$) then consists of the massless states $\alpha^{i}_{-1}\ \tilde{\alpha}^{j}_{-1}\,|0>$ which represent the graviton $g_{\mu\nu}$, the dilaton ϕ, and the antisymmetric tensor $B_{\mu\nu}$.

We end this section by discussing the orientation of the string which arises in certain models. This orientation may be imagined as a preferred direction along the length of the string and is equivalent to the possibility to distinguish between a string described by $X^\mu(\sigma)$ and one by $X^\mu(2\pi-\sigma)$. For a closed string the replacement $\sigma \to 2\pi-\sigma$ corresponds to the interchange $\alpha_n \leftrightarrow \tilde{\alpha}_n$. Type I (unoriented) closed strings only have states which are symmetric with respect to this interchange whereas type II (oriented) strings include both symmetric and anti-symmetric states. One consequence of the orientability of strings is the orientability of the world-sheets describing scattering processes. Hence non-orientable surfaces, such as the Klein bottle at one-loop, are excluded in type II theories.

In the case of open strings $\sigma \to \pi - \sigma$ corresponds to $\alpha_n \to (-1)^n\, \alpha_n$. Such a transformation is generated by the twist operator

$$\Omega = (-1)^N \ , \ N = \sum_n \alpha^i_{-n}\, \alpha^i_n \qquad (2.36)$$

The behaviour of an open string under Ω determines the sort of gauge group which can be attached to the ends of the string using the Chan Paton method. (GSWI, section 1.5) Unoriented open string states are symmetric with respect to Ω and carry SO(n) or Sp(2n). Oriented open string on the other hand have both symmetric and anti-symmetric states and carry U(n).

Finally note that open strings which are singlets of the gauge group can join their ends to form closed strings. Unoriented (oriented) open strings give rise to Type I (Type II) closed strings. However, oriented open strings are excluded from Type II theories at the superstring level. This is related to the inconsistency of N = 1 supersymmetry matter coupling to N = 2 supergravity. (open strings and Type I closed strings form N = 1, D = 10 supersymmetry multiplets; Type II closed strings form N = 2, D = 10 supergravity multiplets).

In our treatment of strings on tori, chapter 4, we shall only consider Type II closed strings. Therefore the separation of left- and right-moving modes in the compactified sector, described in section 4.3, will be in terms of oriented open strings. The above inconsistency argument does not apply to the supersymmetric extension of this construction as the compactified components are internal degrees of freedom.

Section 2.2: Free Superstrings in the Light-Cone

The superstring originally developed by Green, Schwarz and Brink [49] in the L.C. gauge, was achieved by adding to the bosonic vector $X^\mu(\sigma)$ two [8]-spinor variables $\theta_1(\sigma)$, $\theta_2(\sigma)$. The theory was later expressed in a Lorentz-covariant form [50], using [10]-spinors. Since we are only working in flat space-time the original L.C. version of [49] is all that is needed. If we were to turn to the problem of superstrings in an arbitrary background it will be necessary to use the covariant version to analyse the constraints.

The L.C. superstring version of [49] was originally used in an explicitly SO(8)-invariant version, SO(8) being the relic of SO(9,1) when L.C. gauge fixing has been done. The two eight-component spinors are those of SO(8), and have the same ten-dimensional chirality, so are in the same representation of SO(8), in types I (open and closed) and IIB (closed orientable), whilst for type IIA (closed, orientable) the spinors have opposite chirality and are in the different representations $\underline{8}_s$, $\underline{8}_c$ of SO(8). The corresponding spinors θ^ℓ, $\tilde{\theta}_\ell$ or $\theta^\ell, \tilde{\theta}^\ell$, respectively, satisfy Clifford algebra anticommutators at equal time (since θ is Majorana, and hence its own canonical conjugate momentum; the Dirac electron has a second class constraint!):

$$[\theta^a(\sigma,t),\theta^b(\sigma^1,t)]_+ = \delta^{ab}\delta(\sigma-\sigma^1) \tag{2.37a}$$

In order to use field theoretic methods it is necessary to construct fields as functions of commuting variables. This can be achieved by splitting θ^a into two parts, which will transform as a $\underline{4}$ and a $\underline{\bar{4}}$ under the decomposition SO(8)→SO(2)XSO(6)~U(1)XSU(4). For example, one can take new variables $\theta^{\bar{A}}$, λ^A, $\tilde{\theta}^{\bar{A}}$, $\tilde{\lambda}^A$ which transform as $\bar{4}$'s and 4's as specified and have zero equal-time anti-commutators except for

$$[\theta^{\bar{A}}(\sigma),\ \lambda^B(\sigma^1)]_+ = [\tilde{\theta}^{\bar{A}}(\sigma),\tilde{\lambda}^B(\sigma^1)]_+ = \delta^{\bar{A}B}\delta(\sigma-\sigma^1) \tag{2.37b}$$

Thus λ, $\tilde{\lambda}$ may be regarded as the conjugate momenta. It is convenient to take the parameter length of the string to be $2\pi|p^+|$ so that, on change of variables in (2.15) by $\sigma\rightarrow 2\,|p^+|\sigma$, the total free L.C. superstring Hamiltonian may be written as

$$H = p^+\int_0^{2\pi|p^+|}[\pi\underline{P}^2+\frac{1}{\pi}\underline{X}'^2-2i(\theta'\lambda-\tilde{\theta}'\tilde{\lambda})]\epsilon(p^+) \tag{2.38}$$

with $\epsilon(x)$ being the signature of x and

$$\underline{P} = (2\pi|p^+|)^{-1}\dot{\underline{X}}\ ,\ \pi_\theta = i\lambda, \pi_{\tilde{\theta}} = i\tilde{\lambda} \tag{2.39}$$

The resulting equations of motion are

$$\partial_+\partial_-\ X = \partial_+\theta = \partial_-\tilde{\theta} = 0 \tag{2.40}$$

where

$$\partial_\pm = \partial_t \pm \partial_\sigma \tag{2.41}$$

(Note that the time t has also been rescaled by $2|p^+|$). There are now additional terms in the Lorentz rotation generators (2.18), with the total generators $j^{\mu\nu} = \ell^{\mu\nu} + s^{\mu\nu}$, $\ell^{\mu\nu}$ being the orbital parts given by (2.18) and $s^{\mu\nu}$ the contributions from the spin expressed in terms of $\theta, \lambda, \tilde{\theta}, \tilde{\lambda}$. For the transverse SO(8) directions these spin terms can be written as [50]

$$s^{ij} = -\frac{1}{2} i \int_0^{2\pi|p^+|} [\lambda\rho^{ij}\tilde{\theta} + \rho^{ij}\theta]\, d\sigma \tag{2.42a}$$

$$s^{Ri} = \frac{i\epsilon(p^+)}{2\sqrt{2}}\,\pi \int_0^{2\pi|p^+|} [\lambda\rho^i\lambda + \tilde{\lambda}\rho^i\tilde{\lambda}]\, d\sigma \tag{2.42b}$$

$$s^{Li} = -i\frac{\epsilon(p^+)}{2\pi\sqrt{2}} \int_0^{2\pi|p^+|} [\theta\rho^i\theta + \tilde{\theta}\rho^i\tilde{\theta}]\, d\sigma \tag{2.42c}$$

$$s^{LR} = \frac{1}{2}\, i \int_0^{2\pi|p^+|} :[\lambda\theta + \tilde{\lambda}\tilde{\theta}]:\, d\sigma \tag{2.42d}$$

with $1 \le i,j \le 6$, $A^{R,L} = \frac{1}{\sqrt{2}}(A^7 \pm iA^8)$, ρ^i_{AB}, $\rho^i_{\bar{A}\bar{B}}$ are the Dirac matrices relating $(\underline{4}\times\underline{4})_S$ and $(\underline{\bar{4}}\times\underline{\bar{4}})_S$ to the $\underline{6}$ of SO(6) and

$\rho^{ij}_{A\bar{B}} = \frac{1}{2}(\rho^{i}_{AC}\,\rho^{j}_{\bar{C}\bar{B}} - \rho^{j}_{AC}\,\rho^{i}_{\bar{C}\bar{B}})$. Moreover the normal ordering in (2.42d)

for zero-modes (see below) is: $:Q_0^{\bar{A}}\,Q_0^{A}: = \frac{1}{2}(Q_0^{\bar{A}}\,Q_0^{A} - Q_0^{A}\,Q_0^{\bar{A}})$ containing the ususal minus sign on symmetrisation of anticommuting objects. There are also the generators $j^{\ell -}$ of the non-L.C. sub-algebra which have additional spin contributions from θ, etc of form, obtainable from the L.C. algebra commutators in Appendix AIX, [47]

$$S^{L-} = -(i/\surd 2)\,\rho^{jAB}\int_0^{2\pi|p^+|}(\theta_A\theta_B V^j + \tilde{\theta}_A\tilde{\theta}_B\tilde{V}^j)d\sigma \qquad (2.43a)$$

$$S^{R-} = (i\pi^2/\surd 2)\rho^{j\bar{A}B}\int_0^{2\pi|p^+|}(\lambda^A\lambda^B V^j + \tilde{\lambda}^A\tilde{\lambda}^B\tilde{V}^j)d\sigma$$

$$-\,2i\pi\epsilon(p^+):\int_0^{2\pi|p^+|}(\lambda^A\theta_A V^R + \tilde{\lambda}^A\tilde{\theta}_A\tilde{V}^R)d\sigma: \qquad (2.43b)$$

$$S^{i-} = -(i\pi^2/\surd 2)p^{i}_{AB}\int_0^{2\pi|p^+|}(\lambda^A\lambda^B V^L + \tilde{\lambda}^A\tilde{\lambda}^B\tilde{V}^L)d\sigma$$

$$+(i/\surd 2)\rho^{iAB}\int_0^{2\pi|p^+|}(\theta_A\theta_B V^R + \tilde{\theta}_A\tilde{\theta}_B\tilde{V}^R)d\sigma$$

$$- i\pi\epsilon(p^+)\rho^i_{AB}\ \rho^{jBC}\ :\int_0^{2\pi|p^+|} (\lambda^A\theta_C V^j + \tilde{\lambda}^A\tilde{\theta}_C\tilde{V}^j)d\sigma: \quad (2.43c)$$

$$S^{+-}=(i/2)\ :\int_0^{2\pi|p^+|} (\theta_A\lambda^A + \tilde{\theta}_A\tilde{\lambda}^A)d\sigma:+i \quad (2.43d)$$

where $\theta_A = \theta^{\bar{A}}$ etc.
There is no spin contribution to $j^{+\ell}$. The notation used in the above is that

$$V^\ell(\sigma) = p^\ell(\sigma) - \frac{1}{\pi} X^{\ell\prime}(\sigma) \quad (2.44a)$$

$$\tilde{V}^\ell(\sigma) = p^\ell(\sigma) + \frac{1}{\pi} X^{\ell\prime}(\sigma) \quad (2.44b)$$

The supersymmetry part of the ten-dimensional SUSY algebra splits into a L.C. part and that outside the L.C. sub-SUSY algebra. The sixteen L.C. generators, denoted by q^{+A}, $q^{+\bar{A}}$, $\tilde{q}^{+A}$, $\tilde{q}^{+\bar{A}}$ are just the zero modes of λ^A, $\theta^{\bar{A}}$, $\tilde{\lambda}^A$, $\tilde{\theta}^{\bar{A}}$ respectively:

$$q^{+\bar{A}} = \frac{1}{2\pi|p^+|}\int_0^{2\pi|p^+|} \theta^{\bar{A}}(\sigma)d\sigma \qquad , \text{ etc} \quad (2.45)$$

The remaining sixteen generators of the ten-dimensional SUSY algebra, outside the L.C. sub-algebra are denoted q_i^{-A}, $q_i^{-\bar{A}}$, $i=1,2$, and have the form [50]

$$q_1^{-A} = \int_0^{2\pi|p^+|} \left[\sqrt{2}\,\rho^i_{\bar{A}\bar{B}}\, V^i \theta^{\bar{B}} + 2\pi\epsilon(p^+) V^L \lambda^A\right] d\sigma \tag{2.46a}$$

$$q_2^{-A} = \int_0^{2\pi p^+|} \left[\sqrt{2}\,\rho^i_{\bar{A}\bar{B}}\, \tilde{V}^i \tilde{\theta}^{\bar{B}} + 2\pi\epsilon(p^+) \tilde{V}^L \tilde{\lambda}^A\right] d\sigma \tag{2.46b}$$

$$q_2^{-A} = \int_0^{2\pi p^+|} \left[\sqrt{2}\,\rho^i_{\bar{A}\bar{B}}\, \tilde{V}^i \tilde{\theta}^{\bar{B}} + 2\pi\epsilon(p^+) \tilde{V}^L \tilde{\lambda}^A\right] d\sigma \qquad (2.46b)$$

$$\bar{q}_1^{-A} = \int_0^{2\pi|p^+|} \left[2V^R \theta^{\bar{A}} - \sqrt{2}\pi\epsilon(p^+) \rho^i_{AB} V^i \lambda^B\right] d\sigma \tag{2.46c}$$

$$\bar{q}_2^{-A} = \int_0^{2\pi|p^+|} \left[2\tilde{V}^R \tilde{\theta}^{\bar{A}} - \sqrt{2}\pi\epsilon(p^+) \rho^i_{AB} V^i \tilde{\lambda}^B\right] d\sigma \tag{2.46d}$$

These satisfy the anticommutators

$$\left[q_i^{-A} , q_j^{-\bar{B}} \right]_+ = 2\delta^{A\bar{B}}\delta_{ij}h \tag{2.47}$$

We end this section by briefly discussing the mode expansions of the closed superstring. For the fermions these are of the form

$$\bar{\theta}^A(\sigma) = \frac{1}{p+} \sum_{n=-\infty}^{\infty} Q_n{}^{\bar{A}} \; e^{in\sigma/|p+|} \tag{2.48a}$$

$$\tilde{\bar{\theta}}^A(\sigma) = \frac{1}{p+} \sum_{n=-\infty}^{\infty} Q_n{}^{\bar{A}} \; e^{-in\sigma/|p+|} \tag{2.48b}$$

$$\lambda^A(\sigma) = \frac{1}{2\pi|p+|} \sum_{n=-\infty}^{\infty} Q_n{}^A \; e^{in\sigma/|p+|} \tag{2.48c}$$

$$\tilde{\lambda}^A(\sigma) = \frac{1}{2\pi|p^+|} \sum_{n=-\infty}^{\infty} \tilde{Q}_n^A \; e^{-in\sigma/|p^+|} \tag{2.48d}$$

Equations (2.37b) and (2.48) imply the anticommunication relations

$$[Q^A_m, \bar{Q}^B_n]_+ = [\bar{Q}^A_m, Q^B_n]_+ = p^+ \delta^{A\bar{B}} \delta_{m+n,0} \qquad (2.49)$$

all others zero. The bosonic modes are handled as in section 2.1.

The Hamiltonian for the closed superstring has the mode expansion

$$H = \{(p^i)^2/4 + N + \tilde{N}\} \qquad (2.50)$$

with

$$N = \sum_{n>0,i} \alpha^i_{-n} \alpha^i_n + \sum_{m>0} m\,(\bar{Q}^A_{-m} Q^A_m + Q^A_{-m} \bar{Q}^A_m)/p^+ \qquad (2.51)$$

and similarly for $\tilde{N}$. Note that there are no normal-ordering terms in (2.50) due to cancellation of the zero-point energies of the α's and the Q's; this leads to a tachyon-free theory. As with the bosonic string, invariance under choice of origin for σ results in the supplementary condition $N = \tilde{N}$. Hence the mass-shell condition is simply

$$p^2 = 4(N + \tilde{N}),\ N = \tilde{N} \qquad (2.52)$$

and consequently the ground state is massless. Moreover, the right-moving ground-state wavefunction can be represented by a superfield $\Phi(p^i,\theta)$ where $\bar{\theta}^A = \bar{Q}^A_o/p^+$. It has the expansion

$$\Phi(p^i,\theta) = u + u_{\bar{A}}\theta^{\bar{A}} + p^{+}u_{\bar{A}\bar{B}}\theta^{\bar{A}}\theta^{\bar{B}} + p^{+}\epsilon_{\bar{A}\bar{B}\bar{C}\bar{D}}\, u^{\bar{A}}\theta^{\bar{B}}\theta^{\bar{C}}\theta^{\bar{D}}\,/3$$

$$+ (p^{+})^2\epsilon_{\bar{A}\bar{B}\bar{C}\bar{D}}\, u\, \theta^{\bar{A}}\theta^{\bar{B}}\theta^{\bar{C}}\theta^{\bar{D}}\,/6 \qquad (2.53)$$

After imposing certain TCP self-conjugacy conditions the wavefunctions describes eight bosonic states and eight fermionic states.

The total ground state is given by the product $\Phi(p^i,\theta)\ \tilde{\Phi}\ (p^i,\tilde{\theta})$ and the resulting 256 field components from a D=10 non chiral (type IIA) or a D=10 chiral (type IIb) supergravity multiple.

Section 2.3: First - quantization

The propagator of a free open bosonic string is defined at the first qualified level by

$$K\ [\underline{X}_2(\sigma),\ t_2\ |\underline{X}_1\ (\sigma),\ t_1] = \langle\underline{X}_2(\sigma)|\ e^{-\hat{H}(t_2-t_1)}\ |\underline{X}_1(\sigma)\rangle \qquad (2.54)$$

where $|\underline{X}_1(\sigma)\rangle$, $|\underline{X}_2(\sigma)\rangle$ are the initial and final states of the string at times t_1 and t_2 respectively and

$$\hat{H} = \frac{1}{2}\int_0^{\pi} d\sigma\left[\pi\,\frac{\delta^2}{\delta\underline{X}(\sigma)^2} + \pi^{-1}\ \underline{X}'(\sigma)^2\right] \qquad (2.55)$$

is the string Hamiltonian (2.15) on replacing $\underline{P}(\sigma)$ by $-i\delta/\delta\underline{X}(\sigma)$. The propagator (2.54) may be evaluated by expanding into modes, equation (2.22), to give

$$K = \left[2\pi i\ T \right]^{\frac{-(D-2)}{2}} \exp\{i(\underline{x}_1 - \underline{x}_2)^2/2T\}$$

$$\prod_{n>0} \left[\frac{n}{2\pi i \sin(nT)} \right]^{(D-2)/2} f_n\ (\underline{X}_{n,2}\ |\underline{X}_{n,1}) \tag{2.56}$$

where $T = t_2 - t_1$, $X_n{}^i = i(\alpha_n{}^i - \alpha_{-n}{}^i)/\surd 2$ and

$$f_n(\underline{A}\ |\underline{B}) = \exp(in\ [(\underline{A}^2 + \underline{B}^2)\cos nT - 2\underline{A}.\underline{B}]/2\sin nT) \tag{2.57}$$

Note that K is a product of a free (non-relativistic) particle propagator and an infinite number of harmonic oscillator propagators. It follows that (2.54) may be rewritten in terms of a Feynman-Hibbs path-integral,

$$K = \int D\underline{X}(\sigma)\ \exp\left[i \int L d\sigma dt \right] \prod_{\sigma} \prod_{i=1}^{2} \delta(\underline{X}(\sigma,t_i) - \underline{X}_i(\sigma)) \tag{2.58}$$

where $L = 1/2\pi\ (\dot{\underline{X}}^2 - \underline{X}'^2)$ and the delta-functions enforce the boundary conditions $\underline{X}(\sigma,t_i) = \underline{X}_i(\sigma)$. The measure in equation (2.58) is given by

$$D\ \underline{X}\ (\sigma) = D\underline{x} \left[\prod_{n>1} D\underline{X}_n \right] \tag{2.59}$$

where $D\underline{x}$ and $D\underline{X}_n$ are the standard functional measure for a free particle and a harmonic oscillator.

One of the important features of strings is that first-quantised expressions for S-matrix elements, describing interacting strings, also exist (see Chapter 5). These are formally of the form

$$\int D\underline{X}(\sigma)\, e^{i\int_W L\, d\sigma dt} \tag{2.60}$$

(neglecting external states) where interactions are incorporated by allowing the worldsheet W to have a non-trivial topology. The free string propagator corresponds to W being a rectangle (open string) or a cylinder (closed string). More complex surfaces W are obtained by adding 'handles' representing internally propagating strings. Equation (2.60) must then be integrated over the parameters of W, which are given by the points of interaction of strings and the string lengths (section 3.4). In the case of closed strings, care must be taken over the constraint Λ=0, equation (2.34), expressing invariance under the choice of origin for σ. At the first-quantised level the operator $\hat{\Lambda}$, where

$$\hat{\Lambda} = \int_{-\pi}^{\pi} d\sigma \left[-i\underline{X}'(\sigma) . \frac{\delta}{\delta \underline{X}(\sigma)} \right] \tag{2.61}$$

is the generator of constant translations in σ,

$$e^{i\hat{\Lambda}\theta} \ |\underline{X}(\sigma)\rangle = |\underline{X}\ (\sigma+\theta)\rangle \tag{2.62}$$

Any state $|\underline{X}(\sigma)\rangle$ may be projected onto a physical state satisfying Λ=0 by defining

$$|\underline{X}(\sigma)\rangle_{phys} = (2\pi)^{-1} \int_{-\pi}^{\pi} d\theta\ e^{i\Lambda\theta}\, |\underline{X}(\sigma)\rangle \tag{2.63}$$

The range of θ reflects the fact that the eigen-values of $\hat{\Lambda}$ are integers To handle $\hat{\Lambda}$ properly in S-matrix elements requires turning to the second-quantised field theory of strings as developed by Kaku & Kikkawa [28] (see chapter 3). It is then found that the projection operator of equation (2.63) has to be inserted into each internal propogator of a scattering amplitude. (the projection operator is trivial on external legs since the physical external status satisfy $\Lambda=0$). This effectively twists the cylinder swept out by the propagating string by an amount θ, which is then integrated over. The 'twisted' propagator is of the form

$$\langle 2| e^{-i(t_2 - t_1)\hat{H}} e^{i\theta\hat{\Lambda}} |1\rangle \tag{2.64}$$

Note that the twists θ provide the extra variables, needed for a single covering of moduli space at the multiloop level (see chapters 3 and 5).

The functional equation (2.60) can be better defined by Euclideanising the time. The need to define a quantum theory in a Euclidean space-time is well-understood in quantum field theory. The reconstruction of the real-time theory from its Euclideanised version is also under control [51]. However, a similar analysis has not yet been performed in string theory. For the reconstruction theorem of [51] uses an explicitly covariant off-shell formalism; this is not available in the LC gauge approach. Thus a different approach to Euclideanisation may have to be developed, which will not be attempted here. Instead, we will proceed in the naive fashion used for all practitioners in the art of strings. Euclideanisation is achieved by the continuation

$$t = -i\tau \tag{2.65}$$

and the use of the corresponding world-sheet variable

$$\rho = \tau - i\,\sigma \tag{2.66}$$

In the process equation (2.60) becomes $\int D\underline{X}(\sigma)e^{-S_E}$ where S_E is the

Euclideanised action

$$S_E = \int_{W_E} d\sigma d\tau \left[\frac{1}{2\pi} \underline{X}.\Delta_o \underline{X} \right] \tag{2.67}$$

and Δ_o is the scalar Laplacian $\partial_\tau{}^2 + \partial_\sigma{}^2$, acting on scalar functions. W_E will later be shown to be identifiable with a closed (open) Riemann surface for closed (open) strings, with flat metric in the tree case. We may then obtain expressions for two-point functions, for example, as

$$\overline{X_1{}^{\ell}(P_1)X_2{}^{m}}(P_2) \equiv \int D\underline{X}(\sigma)X_1{}^{\ell}(P_1)X_2{}^{m}(P_1)e^{iS_E}$$

$$= -\frac{1}{2}\,\delta^{lm}\,G(P_1,P_2) \tag{2.68}$$

where G is the scalar Green's function on W_E.

A similar discussion may be given for the first-quantised superstring. In brief, S-matrix elements are written in the Euclideanised functional form

$$\int D\underline{X}\ \ D\underline{\lambda}\ D\underline{\theta}\ D\underline{\tilde{\lambda}}\ D\underline{\tilde{\theta}}\ \ e^{-S_E} \tag{2.69}$$

where

$$S_E = \int_{W_E} d\sigma\, d\tau \left[\frac{1}{2\pi} \underline{X}.\Delta_0\underline{X} - 2\partial_{\bar\rho}\theta\lambda + 2\partial_\rho\tilde\theta.\tilde\lambda \right] \tag{2.70}$$

However, the inclusion of interactions by requiring W_E to have handles is a delicate process, requiring the inclusion of suitable insertion factors to preserve the [10]-SUSY [29] (or the world sheet SUSY in the case of the NRS string). Moreover, there are added subtelties arising from higher order contact interactions [34)-37)], which we will discuss later. The insertion factors will lead to contractions between X'S or λ,θ at different points on W_E yielding the bosonic two-point function (2.68) and the corresponding fermionic two-point function

$$\bar\theta^A(P_1)\bar\lambda^B(P_2) = -\delta^{AB}\pi^{-1}\partial_{\rho_1} G(P_1,P_2) \tag{2.71}$$

(see chapter 5). We end this section by rewriting the real-time version of q^{-A} etc. in equation (2.46) as Euclideanised integrals in ρ:

$$q_1^{-A} = -\frac{2}{\pi}\oint d\rho\, [\surd 2\rho^i_{\bar A B}\partial_\rho X^i\theta^B + 2\pi\,\epsilon\,\partial_\rho X^L\lambda^A] \tag{2.72a}$$

$$q_1^{-\bar A} = -\frac{2}{\pi}\oint d\rho\, [2\partial_\rho X^R\theta^{\bar A} - \surd 2\pi\,\epsilon\,\partial_\rho X^i\rho^i_{AB}\lambda^B] \tag{2.72b}$$

$$q_2^{-A} = -\frac{2}{\pi}\oint d\rho\, [\surd 2\rho^i_{\bar A B}\partial_{\bar\rho} X^i\tilde\theta^B + 2\pi\,\epsilon\,\partial_{\bar\rho} X^L\tilde\lambda^A] \tag{2.72c}$$

$$q_2^{-A} = -\frac{2}{\pi}\oint d\rho\ [2\partial_{\bar{\rho}} X^R \tilde{\theta}^{\bar{A}} - \sqrt{2}\ \pi\ \epsilon\ \partial_{\bar{\rho}} X^i \rho^i_{AB} \tilde{\lambda}^B] \qquad (2.72d)$$

with first quantised Hamiltonian

$$h = -\frac{16i\epsilon}{\pi}\oint d\rho[(\partial_\rho X^i)^2 - \pi\theta\partial_\rho\lambda - \pi\tilde{\theta}\partial_{\bar{\rho}}\tilde{\lambda}] \qquad (2.73)$$

and $\epsilon = \pm 1$ for incoming (outgoing) strings. The integrals are taken along contours of constant time around the origin. The above rewriting has used that $V^\ell = 2i\pi^{-1}\partial_\rho X^\ell, \tilde{V}^\ell = 2i\pi^{-1}\partial_{\bar{\rho}} X^\ell$. The Euclideanised expression (2.72) will be used in our analysis of the insertion factors, section (5.2).

CHAPTER 3: FIELD THEORY FOR LIGHT-CONE SUPERSTRINGS.

Section 3.1. Free Bosonic String Field Theory

One of the features which justifies the string field theory approach is that it gives a framework which is expected to produce scattering amplitudes which are guaranteed to be unitary. The covariant approach to strings, as developed by Polyakov, has been to define amplitudes by means of summation over the two dimensional string world sheet and the variables on that world sheet such as the embedding variables $X^{\mu}(\sigma,\tau)$, etc. However the difficulty with this approach is that it is only a formal recipe for constructing a generating functional. In order to justify such a summation as a fundamental definition of string theory it would seem essential to formulate it in a canonical Hamiltonian manner. Such questions have been discussed carefully by Batalin, Fradkin and Vilkovisky and more recently by Henneaux [10], for a class of gauge field theories. There is no detailed construction of the canonical variables for covariant interacting string theory at first quantised level, to our knowledge.

There are two problems which arise here. The first is to construct a canonical formalism which incorporates the degree of freedom corresponding to the genus of the string world sheet. Each surface of genus g has a homology group with 2g generators and an associated function theory based on the Abelian differentials. It has not proved possible to define canonical variables and a Hamiltonian so that the resulting canonical theory is identical to that obtained by summing over the set of all Riemann surfaces of arbitrary genus. Even for a given genus such a formulation has not been obtained.

The second problem is that of defining the space of odd variables in supermoduli space, and of integration over them. Valiant attempts have been made recently to remedy this and a claim has appeared that this can be done. In any case there is a hidden assumption in these approaches about the symmetry of the theory in the construction of the amplitudes, which may be false. Thus the recent approach of Mandelstam [52] depends on the functional formalism giving covariant expressions, and this is used in [53]; that has not been justified in any publication.

It may be possible in the future to fill the gaps noted above. The purpose of this book is to indicate one way of doing so which avoids the basic problems raised by preventing their ever appearing. Thus unitarity is handled by working with a second quantised string field theory; symmetry (ten-dimensional super-Poincare in our case) is assured by constructing the field operator representation of the generators of the symmetry algebra. This latter will itself prove to be a non-trivial problem in its own right, as will be seen later, but it is one for which we can obtain a solution for the heterotic superstring. Thus we felt justified in using the construction of a second quantised field theory for superstrings in the light-cone gauge, since it is the only one presently available.

L.C. field theory for the bosonic string was originally constructed in the co-ordinate representation in [28], [54] by introducing the L.C. string field $\Phi(\underline{X},p^+,t)$ which satisfies the Schrödinger equation

$$i\frac{\partial}{\partial t}\Phi = H(\underline{X},\underline{P} = \frac{1}{i}\frac{\delta}{\delta \underline{X}}, p^+)\Phi \tag{3.1}$$

where H is given in (2.15). The corresponding Hamiltonian from which (3.1) arises is

$$H_2 = \int D\underline{X}\int_0^\infty dp^+\ \Phi^+\ H\Phi \tag{3.2}$$

where the suffix '2' on H_2 denotes that the expression is quadratic in the string field. A similar construction may be given of the Lorentz generators for the free string field theory, as

$$G_2 = \int D\underline{X} \int_0^\infty p+ \, dp^+ \, \Phi^+ \, g\Phi \tag{3.3}$$

where g is given by $j^{\mu\sigma}$ of (2.18) with X^- substituted using (2.14) (see (2.21)). The extra power of p^+ in the measure of (3.3) has arisen by replacement of H by p^+h, and the generalization of h as generator of translations in X^+ to all other generators of the Poincare algebra. The equal-time canonical commutation relation for Φ are given (in the open case) by

$$[\Phi(\underline{X}_1, p_1^+, t), \Phi(\underline{X}_1, p_2^+, t)]_- = (4p_2^+)^{-1} \; \delta(p_1^+ + p_2^+)$$

$$[\Delta^8(\underline{X}_1 - \underline{X}_2) - \Delta^8(\underline{X}_1 - \Omega\underline{X}_2) \tag{3.4}$$

where Φ satisfies the hermiticity property

$$\Phi\,(\underline{X}, p^+, t) = \Phi^+(\underline{X}, -p^+, t) \tag{3.5}$$

$\Omega\underline{X}(\sigma) = \underline{X}(2\pi\;|p^+| - \sigma)$, and $\Delta^8(\underline{x})$ is the string δ-function

$$\Delta^8(\underline{X}) = \prod_{\ell,\sigma} \delta(X^\ell(\sigma))$$

Then from (3.4) and the commutation relation of the first quantised generators g one may deduce that the set G_2 are a field representation of the same algebra. H_2 of (3.2) arises in this algebra from $g = H/p^+$ in (3.3.).

A final point to note is that the first and second quantised string propagators for free strings may be proven equal [28]. This has to be expected, but does not help set up the interacting theory by using first quantised theory. In other words we are going to have to depend on the second quantised theory when interaction is turned on. Thus we do not dwell on the identity between first and second quantised theory further here, but refer the reader to the original paper for the details.

Section 3.2. Free Superstring Field Theory

A similar construction [50)] can be given for L.C. superstring field theory. In the type I case the open string field $\Phi^{ab}(\underline{X}, \theta, \tilde{\theta}, p^+, t)$ can belong to the adjoint representations of SO(n) or USp(2n) gauge groups, with the generalisations of (3.4), (3.5) to

$$[\Phi^{ab}(Z_1), \Phi^{cd}(Z_2)]_- = (4p_2^+)^{-1} \delta(p_1^+ + p_2^+) \{\delta^{ad} \delta^{bc} \Delta^{16}(Z_1 - Z_2) - \delta^{ac} \delta^{bd} \Delta^{16}(Z_1 - \Omega Z_2)\} \tag{3.6}$$

$$\Phi^{ab}(Z) = - \Phi^{ba}(\Omega Z) \tag{3.7}$$

where $Z = (\underline{X}, \theta, \tilde{\theta})$, $\Omega Z(\sigma) = (\underline{X}(2\pi|p^+|-\sigma), \tilde{\theta}(2\pi|p^+|-\sigma), \theta(2\pi|p^+|-\sigma))$. For the closed string field Ψ there is the non-orientability constraint, for the type I case

$$\Psi(Z) = \Psi(\Omega Z) \tag{3.8}$$

and the condition of independence of the field on the origin of σ-parametrisation

$$\int d\sigma\, [Z'(\sigma)\, \delta/\delta Z(\sigma)]\, \Psi = 0 \tag{3.9}$$

The commutator bracket of two closed string fields contains an integration over the σ-origin in the Δ^{16}-function on the right hand side in order to be consistent with (3.9):

$$[\Psi(Z_1), \Psi(Z_2)]_- = (2\pi p_2^+)^{-1} \delta(p_1^+ + p_2^+) \int_0^{2\pi|p_2^+|} \frac{d\sigma_0}{2\pi|p_2^+|} \Delta^{16}(Z_1(\sigma)-Z_2(\sigma+\sigma_0)) \tag{3.10}$$

The total action in the L.C. gauge may be written down as

$$S = \int dt \int D^{16}Z \int_0^\infty p^+ dp^+ [i\Psi(Z,-p^+,t)\, \dot{\Psi}(Z,p^+,t) + i tr(\dot{\Phi}(Z,-p^+,t)\, \Phi(Z,p^+,t)] - \int dt\, H_2 \tag{3.11}$$

where

$$H_2 = \int D^{16}Z\, dp^+ [\Psi(Z,-p^+)\, H_c(Z,-i\frac{\delta}{\delta Z}, p^+)\, \Psi(Z,p^+) + tr(\Phi(Z,-p^+)\, H_o(Z,-i\frac{\delta}{\delta Z}, p^+) \times \Phi(Z,p^+))] \tag{3.12}$$

with H_c, H_o given by the expression (2.15) in the closed and open cases, respectively.

It is finally necessary to impose reality conditions on Φ and Ψ to be TCP self-conjugate, and so reduce the particle content (leading, for the open case to a single Yang-Mills multiplet) [50].

$$\hat{\Phi}^{ab}(\underline{X},\theta,\tilde{\theta}) = \Phi^{ab^*}(\underline{X},\theta/4p^+,\tilde{\theta}/4p^+) \tag{3.13}$$

$$\hat{\Psi}(\underline{X},\theta,\tilde{\theta}) = \Psi^*(\underline{X}, \theta/4p^+, \tilde{\theta}/4p^+) \tag{3.14}$$

where the symbol ^ denotes functional Fourier transform in the Grassmann variables:

$$\Psi(\underline{X},\lambda,\tilde{\lambda}) = \int D^4\theta\, D^4\tilde{\theta}\, e^{I(\lambda,\theta)}\, \Psi(\underline{X},\theta,\tilde{\theta}))$$

$$I(\lambda,\theta) = \int_0^{2\pi p^+} (\lambda^A(\sigma)\, \theta^{\bar{A}}(\sigma) + \tilde{\lambda}^A(\sigma)\, \tilde{\theta}^{\bar{A}}(\sigma))d\sigma$$

Finally we may extend the definition of $q^{\pm A}$; $q^{\pm \bar{A}}$ (the odd SUSY generators) and of the fermionic terms (2.42), (2.43) in the Lorentz generators to the free field level by the same definition (3.3), with $\underline{X}$ and Φ replaced by $\underline{Z}$ and Ψ.

Section 3.3. Interacting String and Superstrings Field Theories

Interaction can be introduced at the first quantised level by allowing strings to fuse or split at one point, so adding a handle to the world-sheet W in the closed case or a window in the open case. Since the σ-parametrisation length of a string has been taken to be $2\pi|p^+|$, conservation of p^+ requires that the total width of the set of all incoming or that of all outgoing L.C. diagrams is conserved. Therefore the fundamental interaction for strings appears as in figure 3, where there is continuity of string co-ordinates along the dashed line. It is possible to discuss such interactions purely at the first quantised level, as was done originally by Mandelstam for the bosonic string [52] and a little later for the NRS string [55]. However this approach seems somewhat limited if (a) as remarked at the beginning of the chapter, one wants to have unitarity of scattering amplitudes guaranteed, (b) one may also be interested in non-perturbative effects, (c) there are subtle questions of contact interactions, which may have to be regularised delicately, and may lead to higher order vertices, and (d) one would like to analyse the full symmetries of the theory. Field-theoretic techniques seem to be more appropriate in such a situation, and will be followed as far as possible here. In the previous section the free field theory of the superstring in the L.C. gauge was constructed. Here we will build the cubic interaction vertices, following the perturbative technique of [50], by means of satisfying the supersymmetry algebra

$$[Q^{-\bar{A}}, Q^{-\bar{B}}]_+ = [Q^{-A}, Q^{-B}]_+ = 0 \tag{3.15a}$$

$$[Q^{-\bar{A}}, Q^{-B}]_+ = 2\delta^{\bar{A}B}\, H \tag{3.15b}$$

To begin with, let us consider the bosonic string interaction. In order to reduce the possible ultra violet divergences to as low a level as possible we only let the strings interact at one point. For open strings this will be at their ends; for closed this will be at some point which becomes common to both of them. This results in the world sheet diagram of fig 3, and can be expressed simply in field theoretic terms as a cubic interaction term involving two annihilation operators and a creation operator and viceversa.

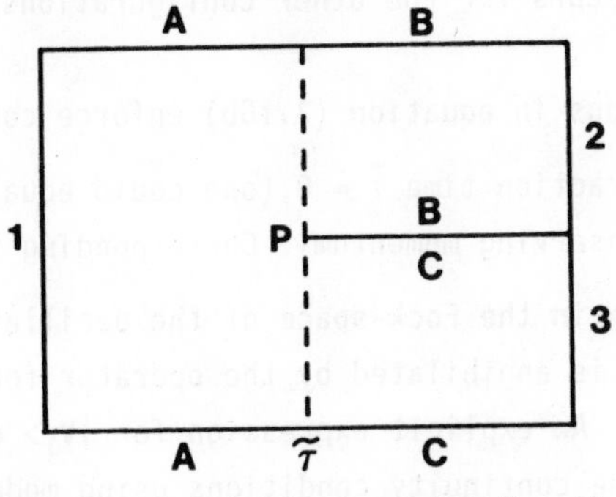

Fig: 3: The primitive interaction region for the splitting or fusion reaction of string 1 → string 2 and string 3; P is the interaction point and $\tilde{\tau}$ the associated interaction time.

Thus if X_1, X_2, X_3 are the three strings which are fusing together, the cubic interaction term H_3 to be added to H_2 of (3.12) is

$$H_3 = \int \prod_{r=1}^{3} dX_r \; dp_r^+ \; \Delta^{(3)}(X_1, X_2, X_3) \; \delta(\Sigma p_r^+)$$

$$\times \; \Phi(X_1, p_1^+) \; \Phi(X_2, p_2^+) \; \Phi(X_3, p_3^+) \qquad (3.16a)$$

where $\Delta^{(3)}$ is the δ-functional sewing the three strings together.

The integration over p_1^+, p_2^+, p_3^+ leads to various configurations according to which strings are incoming (when $p^+ > 0$) or outgoing (when $p^+ < 0$), and with $\Phi(X, -p^+) = \Phi^+(x,p^+)$ for $p^+ > 0$.
Thus for the configuration of figure 3 with $1 + 2 \to 3$,

$$\Delta^{(3)}(X_1, X_2, X_3) = \prod_{0 \le \sigma \le 2\pi p_3^+} \delta(X_1(\sigma) - X_3(\sigma)) \prod_{2\pi p_3^+ \le \sigma \le 2\pi p_1^+} \delta(X_1(\sigma) - X_2(\sigma - 2\pi p_3^+)) \qquad (3.16b)$$

A similar expression occurs for the other configurations.

The delta-functions in equation (3.16b) enforce coordinate continuity at the interaction time $\tilde{\tau} = 0$,[one could equally define H_3 in terms of conditions conserving momentum]. Corresponding to the action H_3 is a vertex vector $|V_3>$ in the Fock-space of the oscillators α_n^μ $\tilde{\alpha}_n^\mu$, of equation (2.29), which is annihilated by the operator form of the continuity conditions. An explicit expression for $|V_3>$ can be obtained by Fourier analysing the continuity conditions using mode expansions. The result is for closed strings

$$|V_3\rangle = \exp\left[\frac{1}{2}\sum_{r,s}\sum_{m,n=1}^{\infty}\underline{\alpha}^{(r)}_{-m}\cdot\bar{N}^{rs}_{mn}\,\underline{\alpha}^{(s)}_{-m}\right.$$

$$\left.+\sum_{r}\sum_{m=1}^{\infty}\bar{N}_m{}^{r}\,\alpha^{(r)}_{-m}\cdot\ \underline{P} + 2K\underline{P}^2\right]$$

$$\times\exp\left[\frac{1}{2}\sum_{r,s}\sum_{m,n=1}^{\infty}\underline{\tilde{\alpha}}^{(r)}_{-m}\,,\ \bar{N}^{rs}_{mn}\,\underline{\tilde{\alpha}}^{(s)}_{-m}\right.$$

$$\left.+\sum_{r}\sum_{m=1}^{\infty}\bar{N}_m{}^{r}\,\underline{\alpha}^{(r)}_{-m}\cdot\underline{P} + 2K\underline{P}^2\right]$$

$$\exp\left[\tau_o\sum_{r} 1/2p_r{}^{+}\right]\delta\left[\sum\underline{p}^{(r)}\right] \qquad (3.17)$$

where $P = 2p_1{}^{+}\underline{p}^{(2)} - 2p_2{}^{+}\underline{p}^{(1)}$ and the labels r,s denote three strings, r,s=1,2,3. The Neumann coefficients are,

$$\bar{N}^{rs}_{mn} = \frac{-4mn\ \ p_1{}^{+}p_2{}^{+}p_3{}^{+}\bar{N}_m{}^{r}\bar{N}_n{}^{s}}{np_r{}^{+} + mp_s{}^{+}}$$

$$\bar{N}_m{}^{r} = \bar{f}_{mn}{}^{r}\ /2p_r{}^{+}$$

with

$$\bar{f}_m^{(r)} = f_m\,(-p_{r+1}^{+}/p_r^{+})e^{m\tau_o/p_r^{+}}$$

$$f_m\,(T) = \left[\frac{1}{m!}\right]\Gamma(mT)/\Gamma(mT+1-m)$$

$$\tau_o = 2\sum_r p_r^{+}\,\log|2p_r^{+}|$$

also

$$K = -\,\tau_o/16p_1^{+}p_2^{+}p_3^{+}$$

Similarly, for open strings

$$|V_3> = \exp\left[\frac{1}{2}\sum_{r,s}\sum_{m,n=1}^{\infty}\underline{\alpha}_{-m}^{(r)}\,\bar{N}_{mn}^{rs}\cdot\underline{\alpha}_{-m}^{(s)}\right.$$

$$\left.+\sum_r\sum_{m=1}^{\infty}\bar{N}_m^{\,r}\,\underline{\alpha}_{-m}^{(r)}\cdot\underline{P} + 2K\underline{P}^2\right]$$

$$\exp\left[\tau_o\sum_r 1/4P_r^{+}\right]\delta\left[\sum\underline{p}^{(r)}\right] \tag{3.18}$$

Equations (3.17) and (3.18) will be needed for our discussion of left-right decomposition in section (4.3).

In order to create a superstring version of the cubic term H_3 of (3.16a) for the bosonic LC string field hamiltonian representing the interaction of fig 3, it is also necessary to consider cubic terms in the supersymmetry generators $Q^{-A,\bar{A}}$. From (3.15) these must satisfy the conditions

$$[Q_2^{-(\bar{A}} , Q_3^{-\bar{B}}]_+ = 0 = [Q_2^{-(A}, Q_3^{-B)}]_+ \tag{3.19a}$$

$$[Q_2^{-\bar{A}} , Q_3^{-B}]_+ + [Q_2^{-B} , Q_3^{-\bar{A}}]_+ = 2\delta^{A\bar{B}} H_3 \tag{3.19b}$$

This requirement, of a cubic contribution in the fields in Q^{-A}, $Q^{-\bar{A}}$ being related to a similar term in H, will also extend to the generators $J^{\ell-}$, since all of these non-L.C. sub-algebra generators are related by commutator brackets of the full [10]-SUSY algebra as displayed in Appendix IX. In other words any non-trivial interaction requires (and is in general required by) a non-linear realisation of the non L.C. sub-algebra of the [10]-SUSY algebra. This is not only true at the cubic level; we will see later that it is also valid at the quartic.

It is expected that the general form of any cubic generator is

$$G_3 = K \int \prod_{r=1}^{3} DZ_r \, dp_r^+ \, \Delta^{(3)}(Z_1, Z_2, Z_3) \, \delta(\Sigma p_r^+) \, \Psi_1 \, \Psi_2 \, \Psi_3 \times$$

$$\times V_3^{(g)}(Z) \tag{3.20}$$

where K is a coupling constant. In this expression $\Delta^{(3)}$ is the δ-functional sewing together the three superstrings. The integration

p_1^+, p_2^+, p_3^+ leads to various configurations according as to which of strings 1, 2 3 are incoming (with $p^+ > 0$) or outgoing (with $p^+ < 0$), as in the bosonic case. Thus for the configuration of figure 3,

$$\Delta^{(3)}(Z_1, Z_2, Z_3) = \prod_{0 \le \sigma \le 2\pi p_3^+} \delta(Z_1(\sigma) - Z_3(\sigma)) \prod_{2\pi p_3^+ \le \sigma \le 2\pi p_1^+} \tag{3.21}$$

$$\prod_{2\pi p_3^+ \le \sigma \le 2\pi p_3^+} \delta(Z_1(\sigma) - Z_2(\sigma - 2\pi p_3^+))$$

and a similar expression can be given for the other configurations. The insertion factor $V_3^{(g)}(Z)$ is to be constructed solely from the string co-ordinates $\underline{X}$, θ, $\tilde{\theta}$, at, or near, the interaction point. It is determined by the requirement that the algebra of the associated g's is satisfied to first order in K. In the case of the SUSY algebra involving H, Q^{-A}, $Q^{-\bar{A}}$ this means requiring the conditions (3.19).

Green and Schwarz [50] evaluated the SUSY insertion factors using normal mode techniques and found that, re-expressed in functional form, the insertion factors for right-movers are

$$V_3^{(Q^{-A})} = \frac{2}{3} \epsilon^{ABCD} Y^{\bar{B}} Y^{\bar{C}} Y^{\bar{D}} \tag{3.22a}$$

$$V_3^{(Q^{-\bar{A}})} = Y^{\bar{A}} \tag{3.22b}$$

$$V_3^{(H)} = \frac{Z^L}{\sqrt{2}} - Z^I \rho^I_{\bar{A}\bar{B}} Y^{\bar{A}} Y^{\bar{B}}$$

$$+ \frac{\sqrt{2}}{3} Z^R \epsilon^{ABCD} \bar{Y}^A \bar{Y}^B \bar{Y}^C \bar{Y}^D \tag{3.22c}$$

where

$$Z = \lim_{\sigma \to 2\pi p_3^+} e^{i\pi/4} (2\pi p_3^+ - \sigma)^{\frac{1}{2}} 2\partial_- X(\sigma) \tag{3.23a}$$

$$Y^A = \lim_{\sigma \to 2\pi p_3^+} e^{i\pi/4} (2\pi p_3^+ - \sigma)^{\frac{1}{2}} \bar{\theta}^A(\sigma) \tag{3.23b}$$

Similar expressions occur for the left-movers with Y^A replaced by $\tilde{\bar{Y}}^A$ and Z by $\tilde{Z}$ such that

$$\tilde{Z} = \lim_{\sigma \to 2\pi p_3^+} e^{-i\pi/4} (2\pi p_3^+ - \sigma)^{\frac{1}{2}} 2\partial_+ X(\sigma) \tag{3.23c}$$

For the type II superstring the total insertions are given by the product of left- and right-movers. Note the presence of the vanishing factor $\epsilon=(2\pi p_3{}^+ - \sigma)$ in equation (3.23). This factor is needed to cancel singularities in the operator $\partial_\pm X$, θ and $\tilde{\theta}$ arising from the singular nature of the interaction point, which has infinite curvature on the world-sheet.

An alternative derivation of the insertion factors may be obtained [33] using the functional techniques developed to construct multiloop amplitudes [32] (as discussed in chapters 5 and 7). These techniques make clearer the singular nature of the interaction points (section 5.2). Moreover, they appear powerful enough to deal with cases in which one has coalescences of several interaction points, as will arise in quartic or higher order contributions to (3.15) and similar commutators [36]. Such contributions are considerably more complex in the normal mode formalism. They will be discussed in more detail later in the book.

Section 3.4: Scattering Amplitudes

Open string scattering processes are represented by light-cone strip diagrams (LC diagrams) in which propagators and vertices are sewn together in parameter space. A particular diagram depicting a first order perturbative contribution to a four string scattering amplitude is shown in fig. 4. The internal loop is represented by a slit.

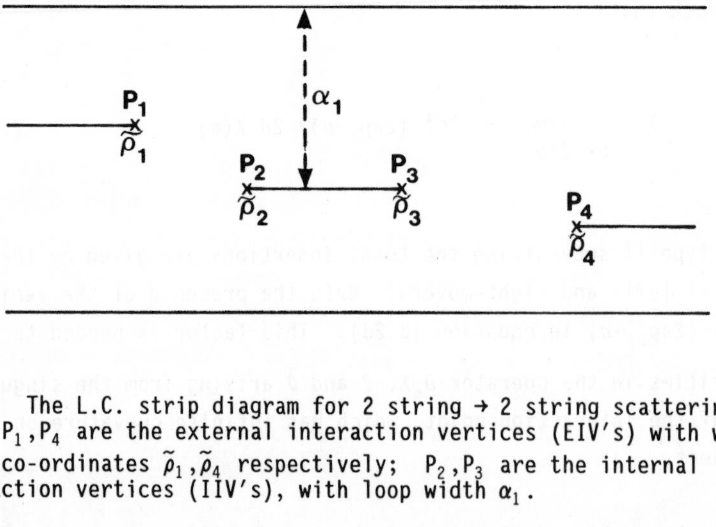

Fig 4: The L.C. strip diagram for 2 string → 2 string scattering at one loop; P_1, P_4 are the external interaction vertices (EIV's) with world sheet co-ordinates $\tilde{\rho}_1, \tilde{\rho}_4$ respectively; P_2, P_3 are the internal interaction vertices (IIV's), with loop width α_1.

The parametrisation of such diagrams (at the g-loop order) is in terms of N-2 external interaction times $\tilde{\tau}_r$, 2g-1 internal interaction times $\tilde{\tau}_\alpha$ (one of the $\tilde{\tau}_\alpha$ may be held fixed by overall translation invariance), and 2g loop parameters α_p. Here α_p is the width of one of the strings in the pth loop. Similarly, closed string LC digrams are obtained by replacing the rectangular strips by cylinders. Now, however, one must take into account the fact that when closed strings interact they have the freedom to join anywhere on their circumference. To obtain a vertex symmetric with respect to constant translations in σ it is necessary to integrate over all possible points of interaction. In a LC diagram an equivalent way to take into account this freedom is to allow each internal propagator to be twisted and to integrate each twist from 0 to 2π. This is the second-quantised argument for including the projection operator, equation (2.62), into each propagator. The effect of such twists is to double the number of parameters needed to specify a closed string LC diagram. Thus, Euclideanising the time along the lines of section 2.3 and denoting each point of the LC diagram by the complex variable $\rho = \tau - i\sigma$, we may express the parametrisation of a closed string diagram in terms of N-2 external complex interaction times $\tilde{\rho}_r$, 2g-1 internal complex interaction times $\tilde{\rho}_\alpha$ and 2g real loop parameters α_p, β_p where β_p is the relative twist in the pth loop. We shall consider the way the twist variable enter string scattering emplitudes in more detail in section 4.5.

Note that string LC diagrams can be considered in two ways. Firstly, as a string theoretic Feynman diagram whose contribution to a scattering amplitude is given by combining propagators and vertices as detailed in the diagram and summing over internal loop momenta. The complete scattering amplitude is then obtained by integrating over the LC diagram parameters defined above. Such an integration ensures that all relevant Feynman diagrams are included. Secondly, as a preimage in parameter space of a world-sheet in the target manifold swept out by interacting strings. This second picture is the appropriate one to use when considering the first-quantised formulation of interacting strings expressed in terms of path-integrals as in section 2.3. The field-theoretic and path-integral approaches are in fact equivalent. Thus the second-quantised field theory of strings allows us to construct unitary scattering amplitudes which may be reduced to first quantised expressions and the latter then evaluated by means of Riemann surface theory (see Chapter 5). Indeed a closed string LC diagram is conformally equivalent to a genus g Riemann surface with N punctures and the parameters $\tilde{\rho}_r$, $\tilde{\rho}_\alpha$, α_p, β_p correspond to the 3g-3+N complex moduli of the Riemann surface. Hence, various aspects of Riemann surfaces including homology cycles and inexact and exact one-forms can be used to analyse scattering amplitudes. These are sketched briefly in appendices.

CHAPTER 4: STRINGS ON TORI

Section 4.1: Quantum Mechanics on Multiply Connected Spaces

The configuration space of a string on a torus is multiply connected and care must be taken in quantising such a system. In this section we review a particular approach to quantum mechanics on multiply connected spaces due to Dowker [56)]. (For an alternative treatment, based on group theoretic quantisation, see Isham and Linden [57)]). It is the approach we shall follow in our treatment of strings and essentially involves lifting a multiply connected space M to its universal covering space $\tilde{M}$ and defining quantisation there. A prescription is then given on how to obtain quantisation in M when $\tilde{M} \to M$. In particular, the propagator in M is given by the superposition of propagators in $\tilde{M}$, summed over homotopically inequivalent paths in M. The coefficients in the sum give a one-dimensional representation of $\pi_1(M)$, the first homotopy or fundamental group of M.

The example of a particle on a circle is considered for which $M = S^1$, $\tilde{M} = \Re$ and $\pi_1(M) = Z$. The coefficients are now phases or θ-angles. It is shown that in momentum space the only essential difference between the propagator in $\Re$ and S^1 is that momenta are discretised in the latter case. This will be important in our analysis of strings on tori.

Let M denote a multiply connected, path-connected, locally path-connected, locally simply connected Riemannian space. Its universal covering space $\tilde{M}$ is defined by a map $\tilde{M} \to M = M/\Gamma$ where Γ is a properly discontinuous, discrete group of isometries, without fixed points. $\tilde{M}$ is simply connected and Γ is isomorphic to the first homotopy or fundamental group of M, $\pi_1(M)$. In other words, the following exists.

(i) A local homeomorphism p: $\tilde{M} \to M$ that each point $x \in M$ has an open neighbourhood U whose preimage $p^{-1}(U)$ can be represented by

$p^{-1}(U) = \bigcup_{j=1}^{N} V_j$ where the V_j are disjoint open sets of $\tilde{M}$ and $p|V_j \rightarrow U$ is a homeomorphism for all j=1,....,N. The number of sheets N may be finite or infinite.

(ii) A fibre corresponding to each point $x \in M$ defined by the discrete set of points $p^{-1}(x) = \{\tilde{x} \in \tilde{M} \mid p(\tilde{x}) = x\}$.

(iii) A discrete group of fibre preserving homeomorphisms such that for any two points $\tilde{x}, \tilde{y} \in \tilde{M}$ with $p(\tilde{x}) = p(\tilde{y})$ there exists $\gamma \in \Gamma$ for which $\tilde{x} = \tilde{y}.\gamma$. (We shall assume that Γ has a right action on M).

It is convenient to choose a fixed 'base' point $\tilde{x}_0$ in each fibre. The set B of all these base points is isomorphic to M and every point of $\tilde{M}$ may be written in the form $\tilde{x} = \tilde{x}_0.\gamma$. A lifting map may be defined by $s:M \rightarrow \tilde{M}$ such that $s(x) = \tilde{x}_0$.

Having defined the universal covering space $\tilde{M}$ we now consider quantum mechanics on the two spaces M and $\tilde{M}$. Since $\tilde{M}$ is simply connected a wavefunction on it, $\tilde{\psi}(\tilde{x})$ is single-valued. A multi-valued wavefunction on M, $\psi(x)$, may be defined by taking $\psi(x)$ to have the values $\tilde{\psi}(\tilde{x}_0\gamma)$ where $\tilde{x}_0 = s(x)$ and γ ranges over all Γ. In particular, choosing γ=1 gives a single-valued function on M, $\psi^{(1)}$, which satisfies $\psi^{(1)}(x) = \tilde{\psi}(\tilde{x}_0)$. The wavefunction $\psi^{(1)}$ may be considered as one of the N branches of the multi-valued wavefunction ψ defined on the N corresponding copies of B obtained from one another by a translation belonging to Γ.

For physical reasons we shall require that the branches of the multi-valued wavefunction on M be equivalent. This means that

$$\tilde{\psi}(\tilde{x}_0\gamma) = a(\gamma)\, \tilde{\psi}(\tilde{x}_0) \qquad (4.1)$$

where

$$|a(\gamma)| = 1 \tag{4.2}$$

It follows from (4.1) that

$$a(\gamma_1 \ \gamma_2) = a(\gamma_1)\ a(\gamma_2) \tag{4.3}$$

up to a constant phase. Hence $a(\gamma)$ is a representation of Γ.

We now introduce propagators. On $\tilde{M}$ we have

$$\tilde{\psi}(\tilde{x}'', t'') = \int_{\tilde{M}} \tilde{K}(\tilde{x}'', t'' \,|\tilde{x}', t')\ \tilde{\psi}(\tilde{x}', t')\ d\tilde{x}' \tag{4.4}$$

and the propagator $\tilde{K}$ has the path-integral form

$$\tilde{K}(\tilde{x}'', t'' \,|\tilde{x}', t') = \int D\tilde{x}\ e^{iS[\tilde{x}]} \tag{4.5}$$

where $S[\tilde{x}]$ is the action on $\tilde{M}$.

The integration over $\tilde{M}$ may be split up into an integration over B and a sum over Γ. Then rewriting $\tilde{x}'$ as $\tilde{x}_0{}'\ \gamma'$ and using (4.1),

$$\tilde{\psi}(\tilde{x}_0'') = \int_B \left\{ \sum_{\gamma'} \tilde{K}(\tilde{x}_0'' \,|\tilde{x}_0{}'\ \gamma')\ a(\gamma') \right\} \tilde{\psi}(\tilde{x}_0{}')\ d\tilde{x}_0{}' \tag{4.6}$$

where we have dropped the time variables. Using the isomorphism between

B and M and the correspondence between $\tilde{x}_o'$, $\tilde{x}_o''$ and x', x'' we reinterpret (4.6) as a propagation law on M,

$$\psi^{(1)}(x'') = \int_M K(x'' \mid x') \, \psi^{(1)}(x') \, dx' \tag{4.7}$$

with

$$K(x''|x') = \sum_{\gamma'} \tilde{K}(\tilde{x}_o'' \mid \tilde{x}_o' \, \gamma') \, a(\gamma') \tag{4.8}$$

Note that K propagates a single-valued wavefunction on M and that |K| is continuous on M.

For consistency, equation (4.1) must be true for all times and therefore be propagated by equation (4.6). This requires Γ to be a symmetry group of $\tilde{M}$ so that $\tilde{K}(\tilde{x}'' \gamma \mid \tilde{x}'\gamma) = \tilde{K}(\tilde{x}'' \mid \tilde{x}')$ or equivalently

$$\tilde{K}(\tilde{x}'' \, \gamma \mid \tilde{x}') = K(\tilde{x}'' \mid \tilde{x}' \, \gamma^{-1}) \tag{4.9}$$

Using equations (4.6) and (4.9),

$$\tilde{\psi}(\tilde{x}_o'' \, \gamma, t) = \int_B \left\{ \sum_{\gamma'} \tilde{K}(\tilde{x}_o'' \, \gamma \mid \tilde{x}_o' \, \gamma') \, a(\gamma') \right\} \tilde{\psi}(\tilde{x}_o') dx_o'$$

$$= \int_B \left\{ \sum_{\gamma'} \tilde{K}(\tilde{x}_o'' \mid \tilde{x}' \, \gamma'\gamma^{-1}) \, a(\gamma' \, \gamma^{-1}) \right\} a(\gamma) \tilde{\psi}(\tilde{x}_o') \, d\tilde{x}_o'$$

$$= a(\gamma) \, \tilde{\psi}(x_o'', t)$$

as required. (The sum over Γ is invariant under a constant translation by γ).

Since Γ is a group of isometries of M, it will be a symmetry group of Schrodinger's equation in the form

$$i\partial\, \tilde{\psi}(\tilde{x},t)/\partial t = -\, \tilde{\Delta}_2\, \tilde{\psi}(\tilde{x},t)/2 \qquad (4.10)$$

where $\tilde{\Delta}_2$ is the Laplacian on $\tilde{M}$. Furthermore, $\psi(x)$ will satisfy the same Schrodinger equation as $\tilde{\psi}(\tilde{x})$,

$$i\partial\, \psi(x,t)/\partial t = -\, \Delta_2\, \psi(x,t)/2 \qquad (4.11)$$

where $\tilde{\Delta}_2(\tilde{x})\,|_{\tilde{x}=x} = \Delta_2(x)$. The above derivation of the propagator on a multiply connected space M, equation (4.8), is based on the approach of Dowker [56]. However, one may link it with alternative treatments by exploiting the isomorphisms $\Gamma \equiv \pi_1(M)$.

Let $\alpha(x', x")$ be a fixed path in M between the two points x' and $x"$. Then any other path β between the same two points may be written as $\beta(x', x") = \alpha(x', x")\ \xi(x")$ where $\xi(x")$ is a loop based at $x"$. In particular, homotopy classes of paths may be labelled by elements [x] of the fundamental group $\pi_1(M)$. If each path between x' and $x"$ is lifted to a path in $\tilde{M}$ ending at $\tilde{x}_o"$, then each lifted path is unique and begins at one of the points on the fibre $p^{-1}(x')$. Moreover, there is a one-one correspondence between homotopy classes of paths between x' and x' and points on $p^{-1}(x')$. This provides an isomorphism between Γ and $\pi_1(M)$ as each point on $p^{-1}(x')$ may be written in the form $\tilde{x}_o'\ \gamma$, $\gamma \in \Gamma$.

Using the isomorphism $\Gamma \equiv \pi_1(M)$ and equation (4.5), each propagator $\tilde{K}(\tilde{x}_0"\ |\tilde{x}_0'\ \gamma')$ may be rewritten as a path-integral on M, $K^{[\gamma']}$, where

$$K^{[\gamma']}(x''|x') = \int_{x',[\gamma']}^{x''} Dx\, e^{iS[x]} \tag{4.12}$$

and the sum over paths is restricted to the homotopy class $[\gamma'] \in \pi_1(M)$ corresponding to $\gamma' \in \Gamma$. Hence equation (4.8) may be reformulated as a sum over homotopy classes of paths in M.

$$K(x'' \mid x' = \sum_{[\gamma']} K^{[\gamma']}(x'' \mid x')\, a([\gamma']) \tag{4.13}$$

where

$$a([\gamma]) = a(\gamma')$$

As an example we consider a non-relativistic particle of mass m constrained on a circle of radius r. The configuration space M is the circle S^1 with coordinate x, $0 \leq x \leq 2\pi r$ and the points x = 0 and x = $2\pi r$ identified. The universal covering space M is the real line $\mathfrak{R}$, the points $\tilde{x} = x + 2\pi m r$ of $\mathfrak{R}$ (integer m) all corresponding to the same point x of S^1. This is symbolically represented in fig. 5, where S^1 forms the base circle and $\mathfrak{R}$ the pre-image helix.

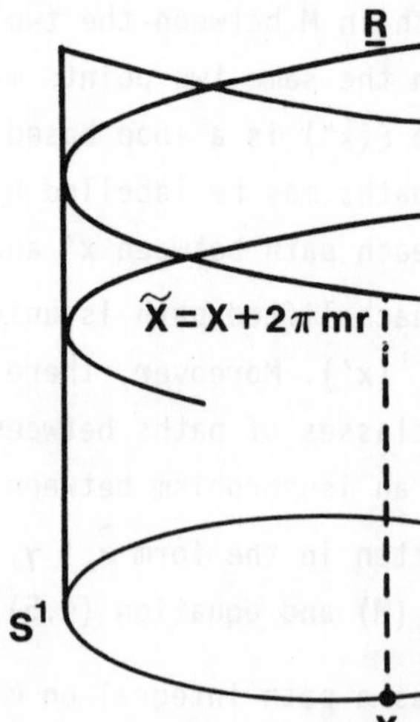

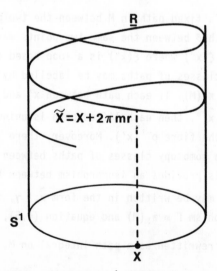

Fig. 5 The covering space $\mathfrak{R}$ of S^1 represented as an helix

Each fibre is isomorphic to the set of integers Z. Hence the discrete group Γ is isomorphic to the group Z and each element $\gamma_n \in \Gamma$ (integer n) has the action

$$(\gamma_n , \tilde{x}) \to \tilde{x}\gamma_n = \tilde{x} + 2\pi nr \qquad (4.14)$$

on $\tilde{M}$. Now consider closed paths in S^1. These divide into disjoint homotopy classes, labelled [n], for integer n. Paths in [n] go around the circle n times in an anticlockwise direction before returning to the starting point. The number n is the winding number of the paths. Given two closed paths taken in order, a third closed path may be formed by joining the end-point of the first to the start of the second. This induces a group structure on the [n], $[n] * [m] = [n+m]$, which is the group Z. Hence $\pi_1(S^1) = Z$.

Following equation (4.8) the propagator between two points x′ and x″ of S^1 is given by

$$K(x'',t''|x',t') = \sum_{n\epsilon Z} \tilde{K}(x'',t''|x' + 2\pi rn,t')\ a(n) \qquad (4.15)$$

where

$$\tilde{K}(x'',t''|x' + 2\pi rn,t') = \int_{x'+2\pi rn}^{x''} Dx\ \exp(i\int_{t'}^{t''} dt\ m\dot{x}^2/2) \qquad (4.16)$$

and

$$a(n)\ a(m) = a(n+m), \quad |a(n)| = 1 \qquad (4.17)$$

Equation (4.17) implies that $a(n) = e^{in\theta}$, θ arbitrary. The propagator in (4.16) is that of a free particle on $\Re$ and may be evaluated to give

$$\tilde{K}(x'',t''|x' + 2\pi rn,t')$$

$$= (m/2\pi iT)^{1/2} \exp\{im(X - 2\pi rn)^2/T\} \tag{4.18}$$

where $T = t'' - t'$, $X = x'' - x'$. Hence the full propagator on S^1 is of the form, for a given θ,

$$K^\theta(x'',t'' \mid x',t')$$

$$=(m/2\pi iT)^{1/2} \sum_n [\exp\{im(X - 2\pi rn)^2/_T\} \exp(in\theta) \tag{4.19}$$

Note that for each value of θ we have an inequivalent quantisation of the system and all values for θ are equally acceptable. In fact different values of θ correspond to different energy spectra. To see this recall the definition of the Jacobi theta function

$$\theta(z|\tau) = \sum_{-\infty}^{\infty} \exp(i\pi\tau n^2 + 2inz) \tag{4.20}$$

Then K^θ is expressible as

$$K^\theta(x'',t'' \mid x',t') = (m/2\pi iT)^{1/2} \exp(imX^2/2T)$$

$$\theta\ (\theta/2 - \pi rmX/T \mid 2mr^2\pi/T) \tag{4.21}$$

Using the Poisson summation formula

$$\theta(z|\tau) = (-i\tau)^{-1/2}\ \theta(z/\tau|-1/\tau) \exp(z^2/i\pi\tau) \tag{4.22}$$

equation (4.21) becomes

$$K^{\theta}(x'',t'' \mid x',t') = (2\pi r)^{-1} \sum_{n} \Big[\exp\{i(n + \theta/2\pi)\, X/r\}$$

$$\exp\{-i(n + \theta/2\pi)^2\, T/2mr^2\}\Big] \tag{4.23}$$

This is the Green's function for a system with energy eigenstates

$$\psi^{\theta}(x) = (2\pi r)^{-1/2} \exp\{i(n + \theta/2\pi)\, x/r\} \tag{4.24}$$

and energy eigenvalues

$$E_n = (n + \theta/2\pi)^2/2mr^2 \tag{4.25}$$

That is

$$K^{\theta}(x'',t'' \mid x',t') = \sum_{n} \psi_n(x'')\, \bar{\psi}_n(x') \exp\{-iE_n(t''-t')\} \tag{4.26}$$

The ψ_n have the periodicity condition

$$\psi_n(2\pi r) = \exp(i\theta)\, \psi_n(0) \tag{4.27}$$

(cf. equation (4.1)), and satisfy the Schrodinger equation on S^1

$$-\tfrac{1}{2}\ \partial^2\, \psi_n/\partial x^2 = E_n\, \psi_n \tag{4.28}$$

The eigenstates of equation (4.24) are also momentum eigenstates. Hence the momentum of a particle confined on a circle is discretized,

$$p = (n + \theta/2\pi)/r\ , \quad \text{integer } n. \tag{4.29}$$

For $\theta = 0$, p lies on the dual lattice Z^*.

Suppose we set $\theta = 0$ and Fourier transform equation (4.23). Then

$$\int_0^{2\pi r} dx'' \int_0^{2\pi r} dx' \, e^{ipx'/r} \, e^{-iqx''/r} \, K^{\theta}(x'',t''|x',t')$$

$$= 2\pi r \, \delta(p-q) \exp(-ip^2 T/2mr^2)$$

$$= K(q,t''|p,t') \qquad (4.30)$$

where $K(q|p)$ is the propagator in momentum space. The important point to note about equation (4.30) is that $K(q|p)$ has exactly the same form as for an unconstrained particle, the only difference being that p and q are discretized for a particle on a circle. Therefore, although $K(q|p)$ is a propagator defined in S^1 it has a simple functional form expressed in $\Re$,

$$K(q,t''|p,t') = \int Dx \exp\left[\int_{t'}^{t''} dt \, m\dot{x}^2/2\right] \exp\left[i\int_{t'}^{t''} dt \, Jx\right] \qquad (4.31)$$

where $J = -q\,\delta(t-t'') + p\,\delta(t-t')$ and the path-integral is performed over $\Re$ rather than S^1 i.e. x may be taken as a coordinate in $\Re$. We therefore have the following result:

The only difference between the momentum-space propagator of a particle in $\Re$ and in $S^1 \equiv \Re/Z$ is that for the latter case momenta are restricted to lie on the dual lattice Z^*(modulo θ-angles). This result generalises to a particle on a d-dimensional torus where Z is replaced by some d-dimensional lattice Γ and momenta are restricted to lie on the dual lattice Γ^* (see below).

Section 4.2: Bosonic String on a Torus

Consider a 26-dimensional closed bosonic string with d components compactified on a torus $T^d = \Re^d/\Gamma$ where Γ is an even, self-dual lattice generated by d independent basis vectors $\underline{e}_i$ $(i = 1,\ldots,d)$ of length $\sqrt{2}$. An even lattice is one whose lattice vectors have even length squared. Introducing the dual lattice Γ^* generated by the basis vectors $\underline{e}_i^*$ $(i=1,\ldots,d)$ which satisfy

$$\underline{e}_i^* \cdot \underline{e}_j = \delta_{ij} \tag{4.32}$$

then Γ is said to be self-dual if Γ is identical to Γ^* (the restriction to even, self-dual lattices will be explained later). In the LC gauge the 24-d transverse components may be quantised straightforwardly since the configuration space is topologically trivial (see section 2.3). On the other hand, the configuration space of the d compactified components is both multiply connected and disconnected. To show this note that such a configuration space is given by the space of loops $X^I : S^1 \to \frac{\Re^d}{\Gamma}$ $(I = 1,\ldots,d)$, which we denote by $L(\Re^d/\Gamma)$. This space may be decomposed as $L(\Re^d/\Gamma) = \Omega(\Re^d/\Gamma) \times \Re^d/\Gamma$ where $\Omega(\Re^d/\Gamma)$ is the space of loops with a fixed base point. The decomposition corresponds to resolving X^I into non-zero and zero modes respectively.

The disconnected components are given by

$$\pi_0(L(\Re^d/\Gamma)) \cong \pi_0\,(\Omega(\Re^d/\Gamma)) \times \pi_0(\Re^d/\Gamma)$$

$$\cong \pi_1(\Re^d/\Gamma) \times 1$$

$$\cong Z^d$$

Each disconnected component is therefore labelled by d integers n_i corresponding to the number of times the string wraps around the torus in the direction $\underline{e}_i (i = 1,\ldots,d)$. This is expressed in terms of points in $\Re^d$ as

$$X^I(\pi) = X^I(-\pi) + \pi \sum_{i=1}^{d} n_i \; e_i^I, \; I = 1,\ldots,d \tag{4.33}$$

The multiple connectedness of $L(\Re^d/\Gamma)$ is described in terms of the zero modes since

$$\pi_1(L(\Re^d/\Gamma)) \cong \pi_1(\Omega(\Re^d/\Gamma)) \times \pi_1(\Re^d/\Gamma)$$

$$\cong \pi_2(\Re^d/\Gamma) \times \pi_1(\Re^d/\Gamma)$$

$$\cong 1 \times \pi_1(\Re^d/\Gamma)$$

and the zero modes may be described in terms of a particle on a torus. Therefore, we may quantise the system by lifting the target manifold from $\Re^d/\Gamma$ to $\Re^d$ and using the techniques of section 4.1.

We begin by defining the mode expansion in $\Re^d$

$$X_w^I = \frac{1}{2} W^I \sigma + X^I(t,\sigma)$$

$$= \frac{1}{2} W^I \sigma + x^I + \frac{1}{2} p^I \; t + i \sum_{n \neq 0} \left\{ \tilde{\alpha}_n^I e^{-in(t+\sigma)} + \alpha_n^I e^{-in(t-\sigma)} \right\} / 2n \tag{4.34}$$

where the winding term W labels the particular disconnected sector satisfying equation (4.33) and hence

$$W^I = \sum_{i=1}^{d} n_i \, e_i^I \tag{4.35}$$

Since $X^I(\sigma + 2\pi) = X^I(\sigma) + \pi W^I$, we see that X^I is a multi-valued function on S^1, (a single-valued function is obtained on lifting S^1 to $\Re$ in the definition of X^I, equation (4.34). However, upon imposing the equivalence relation

$$x^I \equiv x^I + \pi \sum_{i=1}^{d} m_i \, e_i^I \tag{4.36}$$

so that $x^I \to [x^I] \in \Re^d/\Gamma$, X^I does project down to an element of $L(\Re^d/\Gamma)$. The Hamiltonian associated to the sector W is

$$H_w[X^I] = \frac{\pi}{2} \int d\sigma \, \{\underline{P}^2 + (\underline{X}_w'/\pi)^2\} \tag{4.37}$$

where $\underline{P} = \dot{\underline{X}}/\pi$ and $\underline{X} = (X^I)_{I=1,\ldots,d}$ etc.

To quantise X^I we shall consider quantum fluctuations about the classical background solution $W^I\sigma$,

$$\hat{X}_w^I(\sigma) = \frac{1}{2} W^I \sigma + \hat{X}^I(\sigma) \tag{4.38}$$

As $\hat{X}^I(\sigma)$ is a periodic function of σ, the set of eigenfunctions $\{e^{in\sigma}, n \in Z\}$ is a complete set and imposing

$$[\hat{X}^I(t,\sigma), \hat{P}^J(t,\sigma')] = i\delta(\sigma - \sigma')\,\delta^{IJ} \tag{4.39}$$

implies that

$$\left[\hat{x}^I, \hat{p}^J\right] = i\delta^{ij}, \quad \left[\hat{\alpha}^I_m, \hat{\alpha}^J_n\right] = \left[\hat{\tilde{\alpha}}^I_m, \hat{\tilde{\alpha}}^J_n\right] = m\delta_{m+n,0}\,\delta^{IJ} \tag{4.40}$$

The total Hilbert space h of the compactified components decomposes into a direct sum of Hilbert spaces, $h = \oplus h(W)$ where for each winding term W, h(W) has an energy spectrum treated by equation (4.37). At this stage the different sectors h(W) are completely disconnected from one another since W^I is considered as a purely classical object.

Having quantised in $\Re^d$ it is now necessary to project back down to $\Re^d/\Gamma$ using equation (4.36). The quantum mechanics of the non-zero modes is not affected by this and may be analysed as in the uncompactified case. The zero modes are then handled using section 4.1. In particular, the conjugate momenta p^I must lie on the dual lattice, so that

$$p^I = 2\sum_{i=1}^{d} m_i\, e_i^{*I} \tag{4.41}$$

(with all θ-angles set to zero). The discretisation of the momenta follows from the requirement that the eigenfunctions $e^{i\underline{p}.\underline{x}}$ are single-valued on $\Re^d/\Gamma$ (cf. equation (4.29)). Note that the spectrum of the string would be altered if non-trivial θ-angles were considered. For then equation (4.41) would become

$$p^I = 2 \sum_i (m_i + \theta_i/2\pi)\, e_i^{*I} \tag{4.42}$$

which changes the eigenvalues of H_W, equation (4.37). Our interest in compactified strings lies primarily in their application to the heterotic string, Section 4.6. Since non-zero θ-angles would give an unphysical spectrum for the heterotic string we shall assume throughout that all θ-angles are zero.

The propagator of the compactified components may be obtained by lifting to $\Re^d$, decomposing into zero and non-zero modes and then applying section 4.1 to the zero modes. Firstly, we write $X_w(\sigma) = \frac{1}{2} W\sigma + X(\sigma)$ (see equation (4.38)) with

$$X^I(\sigma) = x^I + \sum_{n>0} \left[X^I_{n\tau} \cos n\sigma - P^I_{n\sigma} \sin n\sigma \right] \frac{1}{n} \tag{4.43a}$$

$$P^I(\sigma) = \frac{1}{2\pi} \left[p^I + 2 \sum_{n>0} \left[P^I_{n\tau} \cos n\sigma + X^I_{n\sigma} \sin \sigma \right] \right] \tag{4.43b}$$

such that $\left[X^I_{n\tau}, P^J_{m\tau} \right] = im\delta_{mn}\delta^{IJ} = \left[X_{n\sigma}{}^I, P_{m\sigma}{}^J \right]$ and all other commutators vanish. The lifted propagator is then

$$\tilde{K}_w[\underline{X}_2(\sigma),\ t_2|\underline{X}_1(\sigma),\ t_1] \equiv < \underline{X}_2(\sigma)\ e^{-i\hat{H}_w(t_2-t_1)}|\underline{X}_1(\sigma)>$$

$$= e^{-\frac{i}{4}\underline{W}^2T} < \underline{X}_2(\sigma)|e^{-i\hat{H}T}|\underline{X}_1(\sigma)> \tag{4.44}$$

where $|\underline{X}_{1,2}(\sigma)> = |\underline{X}_{1,2}> \otimes \prod_{n>0} (\underline{X}_{n\tau,1,2}> \otimes \prod_{n>0} |\underline{P}_{n\sigma,1,2}>$ are the initial and final states of the string at the times t_1 and t_2 respectively, with $T = t_2 - t_1$, and $\hat{H}$ is the standard closed string Hamiltonian. Thus, $\tilde{K}_w$ is the usual closed string propagator multiplied by the winding term $\exp(-\frac{i}{4}\underline{W}^2T)$. In light of the discussion at the end of Section 4.1, we Fourier transform to momentum space,

$$\tilde{K}_w\ [\underline{P}_2(\sigma),\ t_2|\underline{P}_1(\sigma),\ t_1] = \int D\underline{X}_1(\sigma)\ D\underline{X}_2(\sigma)$$

$$\times \exp i\int d\sigma\ \underline{P}_2(\sigma).X_2(\sigma)\ \exp\left[-i\int d\sigma\ \underline{P}_1(\sigma).\underline{X}_1(\sigma)\right]$$

$$\tilde{K}_w[\underline{X}_2(\sigma),\ t_2|\underline{X}_1(\sigma),\ t_1] \tag{4.45}$$

Using standard expressions, the momentum space propagator has the mode expansion

$$K_w[\underline{P}_2(\sigma),\ t_2\,|\underline{P}_1(\sigma),\ t_1] = 2\pi\ \delta(p_2-p_1)\ e^{-\frac{i}{4}\underline{W}^2 T}\ e^{-\frac{i}{4}\underline{p}_1{}^2 T}$$

$$\times \prod_{n>0} \left[\frac{n}{2\pi i\ \sin(nT)}\right]^d f_n(\underline{X}_{n\sigma,2}|\underline{X}_{n\sigma,1})\ f_n(\underline{P}_{n\tau,2}|\underline{P}_{n\tau,1}) \qquad (4.46)$$

where $f_n(\underline{A}|\underline{B}) = \exp\,(in[(\underline{A}^2+\underline{B}^2)\cos nT - 2\underline{A}.\underline{B}]/2\sin(nT))$. Note that $\hat{K}_w$ in equation (4.46) can also be written in the path-integral form

$$\int D\underline{X}(\sigma)\ \exp\ i\int_\Sigma d\sigma\ dt\ \left[\underline{J}.\underline{X} - \frac{1}{2\pi}\,(\underline{X}'^2 - \dot{\underline{X}}^2)\right]\ \exp\left(-\frac{i}{4}\underline{W}^2 T\right) \qquad (4.47)$$

Here $\underline{J}(\sigma) = \underline{P}_1(\sigma)\ \delta(t-t_1) - \underline{P}_2(\sigma)\ \delta(t-t_2)$ is a source term and the domain of integrations Σ is the cylinder $S^1 \times [t_1, t_2]$.

Finally, the actual propagator in $\Re^d/\Gamma$ is obtained by imposing equation (4.36) on the zero-mode x^I in the mode-expansion (4.43a) and restricting the zero-mode momentum P^I to lie on the dual lattice Γ^*, equation (4.41). From the discussion of a particle on a circle, section 4.1, it follows that the momentum-space propagator for a bosonic string on a torus is given by equation (4.46) or (4.47), but with $\underline{P}$ restricted to Γ^*. In particular, the functional form for $K_w[\underline{P}_2(\sigma),t_2|\underline{P}_1(\sigma),t_1]$ equation (4.47) is given by a path-integral defined on the configuration space $\Re^d$.

Section 4.3: Left(L)-right(R) Decompositions

One of the main reasons for studying strings on tori is the possibility of separating out left- and right-moving compactified components. This enables construction of chiral string models such as the heterotic string (see section 4.6). The usual starting point for such an analysis is equation (4.34) which is rewritten as

$$X^I_W(t,\sigma) = X^I_R(t-\sigma) + X^I_L(t+\sigma) \tag{4.48}$$

where

$$X^I_L = x^I_L + \frac{1}{2} p^I_L(t+\sigma) + i \sum_{n\neq 0} \frac{1}{2n} \tilde{\alpha}^I_n e^{-in(t+\sigma)}$$

$$X^I_R = x^I_R + \frac{1}{2} p^I_R(t-\sigma) + i \sum_{n\neq 0} \frac{1}{2n} \alpha^I_n e^{-in(t-\sigma)} \tag{4.49}$$

$$p^I = p^I_L + p^I_R, \; W^I = p^I_L - p^I_R, \; x^I = x^I_L + x^I_R \tag{4.50}$$

At the classical level X^I_L and X^I_R are not independent variable as $p^I_L - p^I_R = W^I =$ constant. Furthermore x^I_L and x^I_R only have a meaning in the combination $x^I_L + x^I_R = x^I$.

However, once the string is quantised the possibility arises for strings to interact by splitting and joining (see Chapter 5). Then W^I is no longer constant and we may introduce the operator $\hat{W}^I$ such that $\langle\hat{W}^I\rangle = W^I$ in a given sector h(W). To describe the change of $\langle W^I\rangle$ for a string undergoing interactions a twist operator is introduced.

$$T(W) = \exp i \, \underline{W}.\underline{\hat{V}} \; , \; [\hat{V}^I, \hat{W}^J] = i\delta^{IJ} \tag{4.51}$$

If we introduce a vacuum state $|0\rangle \in h(0)$ so that $\hat{\underline{W}}|0\rangle = 0$, then $T(W)|0\rangle$ satisfies $\hat{\underline{W}}[T(W)|0\rangle]=\underline{W}T(W)|0\rangle$. Hence, $T(W)$ is an intertwining operator between the vacuum sector $h(0)$ and $h(W)$. Therefore, at the quantum level W^I is a dynamical degree of freedom and

$$\hat{p}^I_{L,R} = (\hat{p}^I \pm \hat{W}^I)/2 \tag{4.52}$$

become well defined, independent operators. The same holds true for $\hat{x}_L$ and $\hat{x}_R$ once we set

$$\hat{x}^I_{L,R} = (\hat{x}^I \pm \hat{V}^I)/2 \tag{4.53}$$

From these definitions,

$$\left[\, \hat{x}^I_L ,\ \hat{p}^J_L \,\right] = i\delta^{IJ}/2, \quad \left[\, \hat{x}^I_R ,\ \hat{p}^J_R \,\right] = i\delta^{IJ}/2 \tag{4.54}$$

The above analysis is not, however, complete as the operator $\hat{V}^I$ has been introduced in a rather ad-hoc manner. An attempt to remedy this was given in ref.[58] in which an explicit expression for $\hat{V}^I$ was suggested,

$$\hat{V}^I = -4 \int d\sigma \, \hat{P}^I(\sigma)\, \sigma \tag{4.55}$$

However, the momentum operator in equation (4.55) is defined with respect to $X^I_w(\sigma)$ rather than $X^I(\sigma) = X^I_W(\sigma) - \frac{1}{2} W^I\sigma$, such that

$$\left[\hat{X}^I_W(\sigma),\ \hat{P}^J(\sigma')\right] = i\delta^{IJ}\ \delta(\sigma-\sigma') \tag{4.56}$$

(cf. equation (4.34)). It is difficult to apply equation (4.55) and (4.56) as $X_W(\sigma)$ is not a periodic function of σ and therefore the usual mode analysis cannot be used. Equations (4.55) and (4.56) do, however, indicate that V^I arises from the unravelling of the mode expansions from $\Re^d/\Gamma$ to $\Re^d$. Therefore, we shall assume the existence of V^I and extend the quantisation of section 4.2 by imposing the CCR's

$$\left[\hat{X}^I(\sigma),\ \hat{P}^J(\sigma)\right] = i\delta^{IJ}\ \delta(\sigma-\sigma'),\ \left[\hat{V}^I,\ \hat{W}^J\right] = i\delta^{IJ} \tag{4.57}$$

Classically speaking $\underline{V}$ is interpreted as a point on the torus $\Re^d/\Gamma^*$. Quantisation proceeds by first lifting $\underline{V}$ from $\Re^d/\Gamma$ to $\Re^d$ and using equation (4.57). Then imposing the equivalence relation

$$V^I \equiv V^I + \pi\sum m_i\ e_i\ {}^{*I} \tag{4.58}$$

with W^I restricted to Γ (equation (4.35)), quantum mechanics on $\Re^d/\Gamma^*$ is defined following section 4.1. [Note that a more precise analysis of the extra coordinates $\underline{V}$ is given by Isham and Linden [57] using group theoretic quantisation directly in the space $\Re^d/\Gamma$].

Equations (4.36) and (4.58) may be rewritten in terms of x_L and x_R using equation (4.53) and the assumption that Γ is a self-dual lattice so that

$$\sum m_i \; e_i \; {}^{*I} = \sum n_i \; e_i{}^I \; , m_i \; , \; n_i \; \text{integers}$$

The result is

$$X^I_{L,R} \equiv \pi \sum m_i^{(L,R)} \; e_i^I + x^I_{L,R} \tag{4.59}$$

Similarly combining equations (4.41), (4.35) and (4.52), the momenta $p_{L,R}$ are restricted to lie on Γ^*,

$$p^I_{L,R} \equiv \sum n_i^{(L,R)} \; e_i^{*I} \tag{4.60}$$

Equations (4.59) and (4.60) are consistent with the commutation relations (4.54).

The above discussion has shown how separation into independent left and right movers $\underline{X}_L(\sigma)$ and $\underline{X}_R(\sigma)$ is only possible if the phase space $\{\underline{X}(\sigma), \underline{P}(\sigma)\}$ is extended to include $\underline{W}$ and $\underline{V}$. The CCR's are then given by equation (4.57) with $X^I(\sigma) = X_W{}^I(\sigma) - \frac{1}{2} W^I\sigma$. The extension is essentially a quantum effect relying on the possibility for strings to split and join. However, although $\{\underline{X}(\sigma), \underline{P}(\sigma), \underline{V}, \underline{W}\}$ is a set of canonical variables, the definitions of $\underline{X}_L$ and $\underline{X}_R$ are not canonical as they rely on the distinction betweeen $t+\sigma$ and $t-\sigma$. Moreover, the classical actions of $\underline{X}_L$ and $\underline{X}_R$ are identically zero. To proceed we need to identify the left-right split in a canonical fashion. To achieve this we return to the propagator of equation (4.46). As it stands it does not decompose into a product of left- and right-moving modes. Therefore, perform the Fourier transform

$$\tilde{K}_{p,w}\ [2|1] = \int \prod_{n>0} d\,\underline{X}_{n\sigma,1}\ d\underline{X}_{n\sigma,2}\ e^{i\sum_{n>0} \underline{X}_{n\sigma,1}\cdot\underline{P}_{n\sigma,1}}$$

$$e^{i\sum_{n>0} \underline{X}_{n\sigma,2}\cdot\underline{P}_{n\sigma,2}}\ \tilde{K}_w\ [\underline{P}_2(\sigma),t_2\,|\underline{P}_1(\sigma),t_1] \tag{4.61}$$

Defining a new set of non-zero modes

$$x^I_{nL} = (X^I_{n\tau} - X^I_{n\sigma})/\sqrt{2} = i(\tilde{\alpha}^I_n - \tilde{\alpha}^I_{-n})/\sqrt{2}$$

$$x^I_{nR} = (X^I_{n\tau} + X^I_{n\sigma})/\sqrt{2} = i(\alpha^I_n - \alpha^I_{-n})/\sqrt{2}$$

$$P^I_{nL} = (P^I_{n\tau} - P^I_{n\sigma})/\sqrt{2} = (\tilde{\alpha}^I_n + \tilde{\alpha}^I_{-n})/\sqrt{2}$$

$$P^I_{nR} = (P^I_{n\tau} + P^I_{n\sigma})/\sqrt{2} = (\alpha^I_n + \alpha^I_{-n})/\sqrt{2} \tag{4.62}$$

and using equation (4.50), $\tilde{K}_{p,w}$ may be rewritten as the product of a left-moving and a right-moving propagator,

$$\tilde{K}_{p,w}\left[\underline{P}_{n\tau,2},\ \underline{P}_{n\sigma,2}\ \ |\underline{P}_{n\tau,1},\ \underline{P}_{n\sigma,1}\right] = \tilde{K}_L\ \tilde{K}_R \tag{4.63}$$

where

$$\tilde{K}_L = \exp(-\frac{1}{2}iT\underline{p}_L^2) \prod_{n>0} \{(n/2\pi i \sin nT)^{d/2} f_n(\underline{P}_{nL,2}|\underline{P}_{nL,1})\} \tag{4.64}$$

and similarly for $\tilde{K}_R$. Furthermore, we recognise equation (4.64) as the momentum-space version of an open string propagator. Therefore decomposition into left and right movers corresponds to decomposition into two open strings. These open strings may be considered as independent of one another (but see section 4.4). This relies crucially on the presence of $\underline{V}$ and $\underline{W}$ which allows the separation of $\underline{x}_L$, $\underline{p}_L$ from $\underline{x}_R$, $\underline{p}_R$, (the uncompactified closed string propagator also decomposes into two open ones but they are no longer independent as $\underline{p}_L = \underline{p}_r = \frac{1}{2}\underline{p}$).

We conclude this section by identifying the string variables which describe the left and right open strings found above). We see from equation (4.64) that to achieve a left-right split it is necessary to work with the normal modes ($\underline{X}_{L,R}$, $\underline{P}_{L,R}$, $\underline{P}_{nL,R}$, $\underline{X}_{nL,R}$) defined by equations (4.52) (4.53) and (4.62) rather than ($\underline{x}$, $\underline{p}$, $\underline{W}$, $\underline{V}$, $\underline{X}_{n\sigma}$, $\underline{X}_{n\tau}$, $\underline{P}_{n\tau}$, $\underline{P}_{n\sigma}$). As the two sets of modes are related by a canonical transformation we apply a corresponding canonical transformation to the string variables $\underline{X}(\sigma)$, $\underline{P}(\sigma)$, $\underline{V}$ and $\underline{W}$, whose CCR's are given by equation (4.57). Note that although a mode analysis of the L,R decomposition is sufficient at the free level, it is not the most useful formalism at the interacting level. We would like to re-express K_L and K_R in terms of path-integrals involving left- and right-moving canonical string variables respectively. These path-integrals will then have a natural extension to the interacting case, based on a string field theory (see section 4.4).

Introduce the variables

$$\Phi^I_{L,R}(\sigma) = 2x^I_{L,R} + \sqrt{2}\sum \frac{1}{n} X^I_{nL,R} \cos n\sigma \tag{4.65}$$

with conjugate momenta

$$\Pi^I_{L,R}(\sigma) = \frac{1}{\pi}\left[p^I_{L,R} + \sqrt{2}\sum P^{\ I}_{nL,R} \cos n\sigma\right] \tag{4.66}$$

Since $\Phi_{L,R}(\sigma)$ and $\Pi_{L,R}(\sigma)$ are symmetric functions with respect to σ we are free to restrict the range of σ to be $0 \leq \sigma \leq \pi$. Then

$$\left[\ \Phi^I_{L,R}(\sigma),\ \Pi^J_{L,R}(\sigma')\right] = i\delta(\sigma-\sigma')\ \delta^{IJ} \tag{4.67}$$

and we may identify $\underline{\Phi}_{L,R}(\sigma)$, $\underline{\Pi}_{L,R}(\sigma)$ as conjugate variables of two d-dimensional open strings. The factor of two multiplying $\underline{X}_{L,R}$ in equation (4.65) arises from the CCR in equation (4.67) together with

$$[x^I_{L,R},\ p^J_{L,R}] = \frac{i}{2}\ \delta^{IJ}$$

The mode expansions (4.65) and (4.66) are defined on the target manifold $\underline{R}^d$. The equivalence relation (4.59) must then be imposed, so that $\underline{\Phi}_{L,R}$ $\rightarrow [\underline{\Phi}_{L,R}]$ describes open strings on the torus $\mathfrak{R}^d/\Gamma$. This leads to the zero mode momenta $\underline{p}_{L,R}$ satisfying equation (4.60). However, there are no winding terms in (4.65) and (4.66) as the end points of an open string on $\mathfrak{R}^d/\Gamma$ do not meet. This may be understood in terms of the symmetry of $[\underline{\Phi}_{L,R}](\sigma)$ under $\sigma \rightarrow -\sigma$. For going from $\sigma = -\pi$ to $\sigma = \pi$ corresponds to travelling from one end of the string to the other and then back again along the same path in $\mathfrak{R}^d/\Gamma$ (see fig. 6)

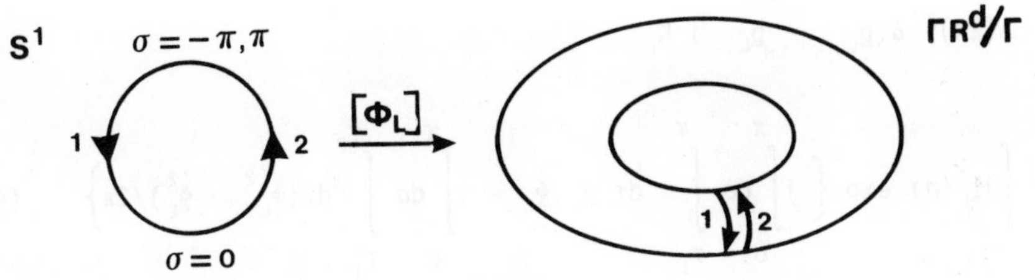

Fig. 6: The mapping $[\underline{\Phi}_L]$ from S^1 to $\Re^d/\Gamma$. The images of the semi-circles 1 and 2 are identified in $\Re^d/\Gamma$

The mappings $[\underline{\Phi}_{L,R}] : S^1 \to \Re^d/\Gamma$ are therefore immersions but become embeddings when the range of σ is restricted to $0 \le \sigma \le \pi$. Although it will usually be sufficient to take $0 \le \sigma \le \pi$ so that $\underline{\Phi}_L$ and $\underline{\Phi}_R$ are open string coordinates it should not be forgotten that they are originally defined over the full range $-\pi \le \sigma \le \pi$. This will be important when considering the symmetry, particular to the closed string, of invariance under choice of origin for σ parametrisation (see section 4.5).

The change of string variables given by $(\underline{X}(\sigma), (\underline{P}(\sigma), \underline{V}, \underline{W}) \to (\underline{\Pi}_{L,R}(\sigma), \underline{\Phi}_{L,R}(\sigma))$ is canonical. Under this transformation the Hamiltonian H_W of equation (4.37) is re-expressed as a sum of two open string Hamiltonians H_L and H_R,

$$H_w = \int_0^\pi d\sigma \left\{(\pi\, \underline{\Pi}_L(\sigma))^2 + \underline{\Phi}_L'(\sigma)^2 + (\pi\, \underline{\Pi}_R(\sigma))^2 + \underline{\Phi}_R'(\sigma)^2\right\}/2\pi$$

$$= H_L + H_R \tag{4.68}$$

In terms of ϕ_L and Π_L the propagator K_L of equation (4.64) has the string-functional form

$$(2\pi)^2\ \delta(\underline{p}_{L,1} - \underline{p}_{L,2})\ K_L$$

$$= \int D\underline{\Phi}_L(\sigma)\ \exp\left\{ i\int_0^\pi d\sigma \int_{\tau_1}^{\tau_2} d\tau\ \underline{J}_L\cdot\underline{\Phi}_L - i\int_0^\pi d\sigma \int_{\tau_1}^{\tau_2} d\tau(\underline{\Phi}_L'^2 - \dot{\underline{\Phi}}_L^2)/2\pi \right\} \qquad (4.69)$$

where $\underline{J}_L(\sigma) = \underline{\Pi}_{L,1}(\sigma)\ \delta(\tau-\tau_1) - \underline{\Pi}_{L,2}(\sigma)\ \delta(\tau-\tau_2)$ and $\underline{\Phi}_L(\sigma)$ is considered as a point in $\mathfrak{R}^d$. The propagator is nevertheless defined in $\mathfrak{R}^d/\Gamma$ provided $\underline{p}_{L,1}$ and $\underline{p}_{L,2}$ are restricted to lie on the dual lattice Γ^* (see end of section 4.1 and equation (4.60)). A similar expression holds for the right-moving propagator K_R. The action appearing in the product K_L K_R is

$$S[\Phi] = -\int_0^\pi d\sigma \int_{\tau_1}^{\tau_2} d\tau(\underline{\Phi}_L'^2 - \dot{\underline{\Phi}}_L^2 + \Phi_R'^2 - \dot{\underline{\Phi}}_R^2)/2\pi \qquad (4.70)$$

which is not equal to the light-cone gauge string action of equation (4.47),

$$S[x] = -\int_{-\pi}^\pi d\sigma \int_{\tau_1}^{\tau_2} d\tau(\underline{X}'^2 - \dot{\underline{X}}^2)/2\pi - \frac{1}{4}\underline{W}^2\ \ (t_2-t_1) \qquad (4.71)$$

This is because the canonical transformation

$$\int_{-\pi}^\pi \left[\underline{P}(\sigma)\dot{\underline{X}}_w(\sigma)\right]d\sigma + \frac{1}{2}\ \underline{W}.\dot{\underline{V}}-H_w = \int_0^\pi \left[\underline{\pi}_R(\sigma).\dot{\underline{\Phi}}_R(\sigma)+\underline{\pi}_L(\sigma).\dot{\underline{\Phi}}_L(\sigma)\right]d\sigma-H_R-H_L \quad (4.72)$$

mixes coordinate and momenta variables. Therefore, integrating out $\underline{\Pi}_{L,R}$ from a path-integral is not equivalent to integrating out $\underline{P}(\sigma)$ and $\underline{W}$.

Hence the resulting Lagrangians in the two cases are different. The action S[Φ] will be the appropriate one to use when we require explicit factorization into left and right-movers, as in the treatment of chiral bosons (see section 4.6).

Section 4.4 Field Theory on a Torus

The quantum theory of the free string on a torus is now extended to the interacting case using second quantised field theory. For convenience we shall drop the uncompactified components which can easily be reintroduced later to obtain the full light-cone scattering amplitudes. In terms of the string coordinates $x^I(\sigma)$ and V^I of equations (4.43) and (4.51), with zero modes x^I and V^I lifted to $\mathbf{R}^d$, Schrodingers equation takes the form

$$i \frac{\partial}{\partial t} \Psi = \hat{H} \left(\underline{X},\ \underline{P} = -i \frac{\partial}{\partial \underline{X}},\ \underline{W} = -i \frac{\partial}{\partial \underline{V}} \right) \Psi \qquad (4.73)$$

where $\hat{H}$ is the operator corresponding to the Hamiltonian (4.37) and $\Psi(\underline{X}(\sigma), \underline{V}, p^+, t)$ is the light-cone string field for the compactified components. Projection of $\underline{x}^I$ and $\underline{V}$ onto the $\mathbf{R}^d/\Gamma$ and $\mathbf{R}^d/\Gamma^*$ respectively is then defined by imposing equations (4.36) and (4.58) such that Φ satisfies the periodicity contraints

$$\Psi \left(x^I + \pi \sum_{i=1}^{d} m_i e_i{}^I \right) = \Psi \left(x^I \right)$$

$$\Psi \left(V^I + \pi \sum_{i=1}^{d} m_i e_i^{*I} \right) = \Psi \left(V^I \right)$$

To construct an interacting field theory we may proceed along similar lines to section 3.3, provided additional continuity conditions, associated with the extra degrees of freedom V^I, are included. For example, the cubic interaction term analogous to equation (3.16) is

$$H_3 = \int dV_r \int dX_r(\sigma) dp_r^+ \; \Delta^{(3)} \; (X_1, X_2, X_3) \delta(\Sigma p_i^+)$$

$$\delta(V_1 - V_3) \; \delta \; (V_1 - V_2)$$

$$\Psi \; (X_1, V_1, p_1{}^+) \; \Psi \; (X_2, V_2, p_2{}^+) \; \Psi \; (X_3, V_3, p_3{}^+) \qquad (4.74)$$

Similarly, the three point vertex for the compactifies components can be written in the form (3.17), with $\delta(\Sigma \underline{P}^{(r)})$ replaced by

$$\delta \left[\sum \underline{P}^{(r)} \right] \delta \left[\sum \underline{W}^{(r)} \right] = \delta \left[\sum \underline{P}_L{}^{(r)} \right] \delta \left[\sum \underline{P}_R{}^{(r)} \right]$$

and $\underline{P}_L{}^{(r)}$, $\underline{P}_R{}^{(r)}$ restricted to lie on the dual lattice Γ^* (see equation 4.60). Thus,

$$|V_3 \; > \; = \; |V > \otimes \; |\tilde{V} > \qquad (4.75)$$

where

$$|V> = \exp \left[\frac{1}{2} \sum_{r,s} \; \sum_{m,n=1} \underline{\alpha}_{-m}{}^{(r)} \; \bar{N}_{mn}{}^{rs} \; \underline{\alpha}_{-m}{}^{(s)} \right.$$

$$\left. + \sum_r \sum_{m=1}^{\infty} \bar{N}_m{}^r \; \underline{\alpha}_{-m}^{\;r} \; . \; \underline{P} + 2K \; \underline{P}^2 \right]$$

$$\exp \left(\tau_o \sum_r \frac{1}{4p_r{}^+} \right) \qquad (4.76)$$

and $|\tilde{V}\rangle$ is obtained from $|V\rangle$ by replacing $\underline{\alpha}_{-m}$ with $\underline{\tilde{\alpha}}_{-m}$ and $\underline{P}_R$ with $\underline{P}_L$.

The vertex (4.75) is essentially a product of left and right-moving parts coupled by the zero-mode momenta **P**. Such a vertex is not appropriate if complete separation into independent left and right moving modes is required, as in the construction of chiral strings. However, we have at our disposal an alternative set of string coordinates, Φ_L, and Φ_R, for which left-right separation is automatic, (section 4.3). Therefore, introduce the new string field $\Psi(\Phi_L(\sigma), \underline{\Phi}_R(\sigma), p^+, t)$ which satisfies the Schrodinger equation (4.73) rewritten as

$$\left[i \frac{\partial}{\partial t} - \hat{H}_L - \hat{H}_R \right] \Psi = 0 \tag{4.77a}$$

where

$$\hat{H}_{L,R} = \frac{\pi}{2} \int_o^\pi \left[\frac{\Phi_{L,R}'^{\,2}}{\pi^2} - \frac{\delta^2}{\delta \Phi_{L,R}} \right] d\sigma \tag{4.77b}$$

It follows that the field Ψ may be decomposed as $\Psi(\underline{\Phi}_L(\sigma), \underline{\Phi}_R(\sigma), p^+, t) = \Psi_L(\Phi_L(\sigma), p^+, t)\, \Psi_R(\underline{\Phi}_R(\sigma), p^+, t)$. Note at the free level, the string field theory expressed in terms of $\underline{\Phi}_{L,R}(\sigma)$ is equivalent to that expressed in terms of $\underline{X}(\sigma)$ and $\underline{V}$, eg. they yield the same propagator (4.63) (although the former leads more directly to explicit left-right decomposition). However, this equivalence breaks down at the interacting level since the cubic interaction terms expressing coordinate continuity differ in the two cases. In particular, the cubic Hamiltonian for $\Phi_{L,R}$ is

$$H_3{}' = \int \prod_r dp_r{}^+ \; d\Phi_{L,r}(\sigma) \; d\Phi_{R,r}(\sigma)$$

$$\Delta^{(3)}(\Phi_{L,1}, \Phi_{L,2}, \Phi_{L,3}) \, \Delta^{(3)}(\Phi_{R,1}, \Phi_{R,2}, \Phi_{R,3})$$

$$\delta\left[\sum p_r{}^+\right] \Pi_r \; \Psi_L \; [\Phi_{L,r}{}^{(\sigma)}, p_r{}^+] \; \Psi_R[\Phi_{R,r}{}^{(\sigma)}, p_r^+] \tag{4.78}$$

where $\Delta^{(3)}$ is defined in equation (3.16b). The associated string vertex is

$$|V_3{}'\rangle = |V_L\rangle \otimes |V_R\rangle \tag{4.79}$$

where

$$|V_R\rangle = \exp\left[\tau_o \sum \frac{1}{4p_r{}^+}\right] \delta\left[\sum \underline{P}_R{}^{(r)}\right]$$

$$\exp\left[\frac{1}{2} \sum_{r,s} \sum_{m,n=1}^{\infty} \underline{\alpha}_{-m}^{(r)} \; \bar{N}_{mn}^{rs} \; \underline{\alpha}_{-m}^{(s)}\right.$$

$$\left. + \sum_r \sum_{m=1}^{\infty} \bar{N}_m^r \; \underline{\alpha}_{-m}{}^{(r)} \underline{P}_R + 2K\underline{P}_R{}^2 \right] \tag{4.80}$$

and $\underline{P}_R = 2p_1^+ \underline{P}_R{}^{(2)} - 2p_2^+ \underline{P}_R{}^{(1)}$. Similarly, $|V_L\rangle$ corresponds to replacing α with $\tilde{\alpha}$ and $\underline{P}_R$ with $\underline{P}_L$. We see that equation (4.79) decomposes into a product of left and right - moving open string vertices (see equation 3.17) which are totally decoupled, in contrast to the vertex $|V_3\rangle$ of equation (4.75).

It is clear from the above analysis that in the case of compactified degrees of freedom, the cubic Hamiltonian is not uniquely determined by requirement such as locality and continuity. This ambiguity arises from the existence of an alternative set of string coordinates $\Phi_{L,R}(\sigma)$. Imposing the extra requirement of complete left-right decomposition then uniquely selects H_3' rather than H_3. For the latter leads to a mixing of left and right-moving modes on performing integrations over internal loop moments. Note that the interaction term H_3' is effectively the one used by Gross and Periwal [30], who modified the continuity conditions for the string directly at the mode level, rather than formulating the theory in terms of new open string coordinates. We shall only consider H_3' in the sequel.

Section 4.5: Scattering Amplitude on a Torus

Having obtained the propagator $\tilde{K}_L\ K_R$ of equation (4.63) and the vertex $|V_L>\otimes|V_R>$ of equation (4.79), we now turn to the construction of scattering amplitudes for the compactified components. Firstly, note that $K_{L,R}$ and $|V_{L,R}>$ may be interpreted as open string propagators and vertices derived using open string coordinates $\Phi_{L,R}(\sigma)$ with σ restricted to $o\leq\sigma\leq\pi p^+$. Hence, a scattering amplitute involving the left-movers, say, can be represented by an open-string light-cone diagram $\Sigma_\rho{}^+$ which may be obtained from the corresponding closed L.C. diagram Σ_ρ of figure 7 by opening the latter out and restricting σ to be positive [For convenience we take $-\pi p^+\leq\sigma\leq\pi p^+$ for closed strings]. The mirror image of $\Sigma_\rho{}^+$, denoted $\Sigma_\rho{}^-$ is then the region $\sigma\leq o$ (see figure 8)

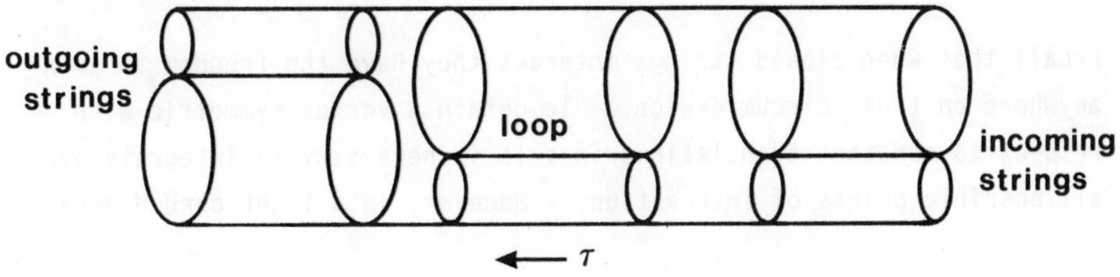

Fig. 7: One-loop closed string scattering diagram.

Both Σ_ρ and $\Sigma_\rho{}^\pm$ are mapped into the same open string world-sheet in R^d/Γ. This is a consequence of the symmetry of the immersions $\Phi_{L,R}(\sigma)$ under σ->-σ. (see section 4.3).

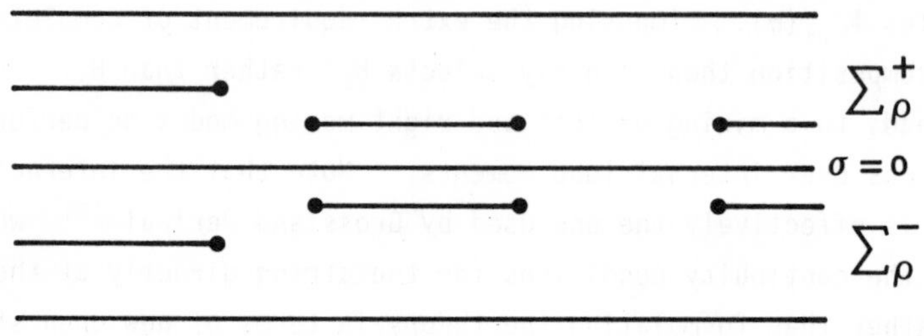

Fig. 8: One loop example of a closed LC diagram Σ_ρ, (fig 7), opened out to form the open LC diagram $\Sigma_\rho{}^+$ and its mirror image $\Sigma_\rho{}^-$

However, there would appear to be a difficulty with the above analysis related to the fact that Σ_ρ^+ is parametrised by $\tilde{\tau}$= Re$\tilde{\rho}$ and $\underline{\alpha}$ only, that is half the parameters of the closed LC diagram Σ_ρ. The missing parameters correspond to phase ambiguities in the identification of the boundaries of Σ_ρ^+ with those of Σ_ρ^- to reform Σ_ρ. These missing phases are recovered, and the difficulty resolved, by incorporating the twisted propagators derived in section 2.3. For, so far, our discussion has neglected the symmetry of closed string theory under constant translations in σ. Indeed, amplitudes constructed using $K_L K_R$ and $|V_L\rangle \otimes |V_R\rangle$ are not invariant under this symmetry.

Recall that when closed strings interact they have the freedom to join anywhere on their circumference. To obtain a vertex symmetric with respect to constant translations in σ it is necessary to integrate over all possible points of interaction. However, in a light-cone diagram an equivalent way to take into account this freedom is to allow each internal propagator to be twisted and to integrate each twist from 0 to 2π (see section 3.4). In terms of the compactified components, the propagator between two states $|1\rangle_L \otimes |1\rangle_R$ and $|2\rangle_L \otimes |2\rangle_R$ becomes [cf. equation (2.64)]

$$\langle 2|e^{-H_L(\tau_{12} + i\sigma_{12})}|1\rangle_L \ \langle 2|\ e^{-H_R(\tau_{12} - i\sigma_{12})}|1\rangle_R \quad (4.81)$$

with $K_L K_R$ corresponding to the untwisted case $\sigma_{12} = 0$. Amplitudes constructed from $|V_L\rangle \otimes |V_R\rangle$ and equation (4.81) are then invariant. It is important to note the difference in sign of σ_{12} for left and right-movers. This will play a crucial role in identification of right (left)-movers with holomorphic (anti-holomorphic) functions of the world-sheet moduli (see chapter 5).

To understand how the twists relate to the missing

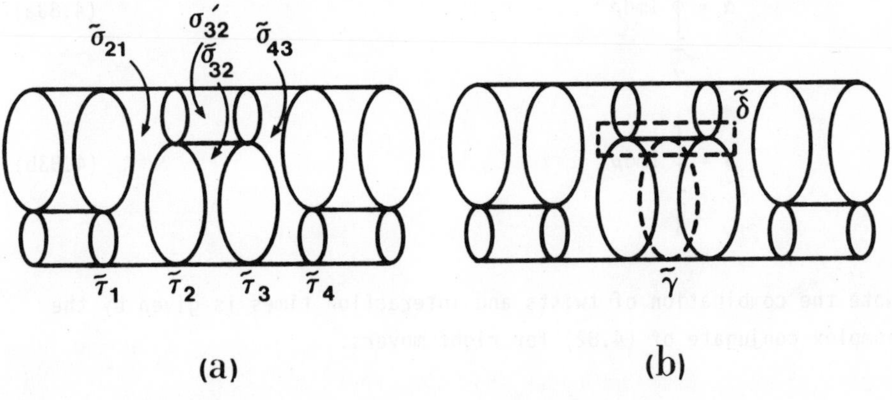

Fig. 9: (a) The internal twists and interaction times of a one-loop light-cone diagram. The twists $\tilde{\sigma}_{21}, \tilde{\sigma}_{32}, \tilde{\sigma}_{43}$ combine with the interaction times to form the complex interaction times $\tilde{\rho}_{21} = \tilde{\tau}_{21} - i\tilde{\sigma}_{21}$, $\tilde{\rho}_{32} = \tilde{\tau}_{32} - i\tilde{\sigma}_{32}$, $\tilde{\rho}_{43} = \tilde{\tau}_{43} - i\tilde{\sigma}_{43}$. The loop parameters α and β are given by the string length p^+_{32} of one the strings in the loop and the relative twists $\tilde{\sigma}'_{32} - \tilde{\sigma}_{32}$ respectively.
(b) The positions of canonical homology cycles $\{\tilde{\gamma}, \tilde{\delta}\}$ in the one-loop diagram.

parameters $\text{Im}\underline{\tilde{\rho}}$ and $\underline{\beta}$ of Σ_ρ consider the one-loop case, fig. 9a. From (4.81) the internal twists and interaction times for left-movers combine as

$$\tilde{\rho}_{21} = \tilde{\tau}_{21} - i\tilde{\sigma}_{21}, \quad \tilde{\rho}_{32} = \tilde{\tau}_{32} - i\tilde{\sigma}_{32}, \quad \tilde{\rho}_{43} = \tilde{\tau}_{43} - i\tilde{\sigma}_{43}$$

$$\tilde{\rho}'_{32} = \tilde{\tau}_{32} - i\tilde{\sigma}'_{32} = \tilde{\rho}_{32} - i\beta \tag{4.82}$$

where $\beta = \tilde{\sigma}'_{32} - \tilde{\sigma}_{32}$ is the relative twist between two strings in the loop. The points $\tilde{\rho}_{32}, \tilde{\rho}_{43}, \tilde{\rho}_{21}$ are the complex interaction times of Σ_ρ. Moreover the loop parameters are $\alpha = p^+_{32}$, where p^+_{32} is the length of one of the strings in the loop, and β. From fig 9 we can express α and β as integrals over the pair of homology cycles $\{\gamma, \delta\}$.

$$\alpha = \oint_{\tilde{\gamma}} \text{Im}d\rho \tag{4.83a}$$

$$\beta = \oint_{\tilde{\delta}} \text{Im}d\rho \tag{4.83b}$$

Note the combination of twists and interaction times is given by the complex conjugate of (4.82) for right movers.

The definition of the parameters of β and $\underline{\tilde{\rho}} = (\tilde{\rho}_{21}, \tilde{\rho}_{32}, \tilde{\rho}_{43})$ in equation (4.82) implies that the one-loop scattering amplitude may be constructed by combining the vertex $|V_L\rangle \otimes |V_R\rangle$ with the propagators $\langle j| e^{-H_L \bar{\tilde{\rho}}_{ij}} e^{-H_R \tilde{\rho}_{ij}} |i\rangle$ and inserting the twist operator $\exp i(H_L - H_R)\beta$ in the loop. This leads to the expression

$$A^{(1)} = \int d^2\tilde{\rho}\ d\alpha\ d\beta\ f_L^{(1)}(\bar{\tilde{\rho}},\alpha,\beta)\ f_R^{(1)}(\tilde{\rho},\alpha,\beta) \tag{4.84}$$

where $f_R^{(1)}$ is an analytic function of the complex interaction points $\tilde{\underline{\rho}}$. Similarly $f_L^{(1)}$ is anti-analytic. Note that is we had taken $\tilde{\tau}_{32}'$ as the loop twist β, then $\tilde{\rho}_{32}' = \mathrm{Re}\ \tilde{\rho}_{32} - i\beta$ rather than $\tilde{\rho}_{32}' = \tilde{\rho}_{32} - i\beta$ and the statement concerning (anti) analyticity would no longer hold.

The discussion at one-loop has a straight-forward generalization at the multi-loop level. There are now g pairs of canonical homology cycles $\{\tilde{\gamma}_p, \tilde{\delta}_p\}$ and the 2g loop parameters are given by

$$\alpha_p = \oint_{\tilde{\gamma}_p} \mathrm{Im} d\rho \tag{4.85a}$$

$$\beta_p = \oint_{\tilde{\delta}_p} \mathrm{Im} d\rho \tag{4.85b}$$

This is illustrated in fig 10 for two loops

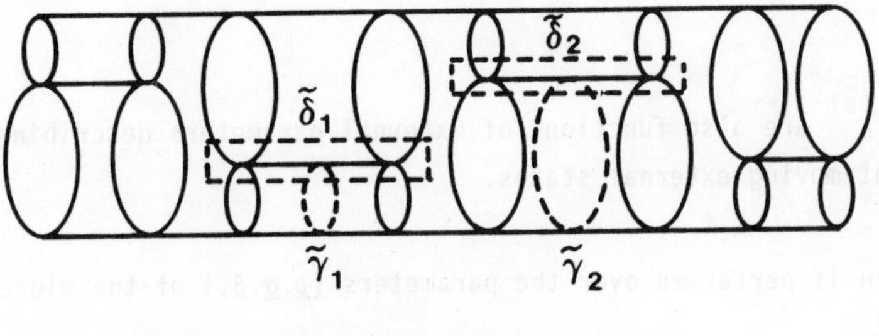

Fig. 10: Homology cycles at two loops.

The multiloop scattering amplitude is

$$A^{(g)} = \lambda^{N-2+2g} \prod_{r=1}^{N-2} \int d^2\tilde{\rho}_r \prod_{\alpha=1}^{2g-1} \int d^2\tilde{\rho}_\alpha \prod_{p=1}^{g} \int d\alpha_p \, d\beta_p$$

$$f_L^{(g)}(\tilde{\underline{\rho}},\underline{\alpha},\underline{\beta}) \; f_R^{(g)}(\bar{\tilde{\underline{\rho}}},\underline{\alpha},\underline{\beta}) \qquad (4.86)$$

where

(a) $f_L^{(g)}$ $f_R^{(g)}$ is obtained by combining propagators of the form $<j|\exp[-H_L \tilde{\rho}_{ij}] \exp[-H_R \bar{\tilde{\rho}}_{ij}]|i>$ with the vertices $|V_L> \otimes |V_R>$ and inserting the twist operator $\exp\{-i(H_L - H_R)\beta_p\}$ in each loop p; left and right-movers remain decoupled at the multi-loop level.

(b) $f_R^{(g)}$ and $f_L^{(g)}$ are respectively analytic and anti-analytic functions of the complex interaction times $\tilde{\underline{\rho}}$; this follows from the structure of(4.81) and the definition of $\underline{\beta}$.

(c) the functions $f_L^{(g)}(\tilde{\underline{T}},\underline{\alpha}) = f_L^{(g)}(\tilde{\underline{\rho}},\underline{\alpha},\underline{\beta})|_{\underline{\beta}=0,\, \mathrm{Im}\tilde{\underline{\rho}}=0}$ and $f_R^{(g)}(\tilde{\underline{T}},\underline{\alpha}) = f_R^{(g)}(\tilde{\underline{\rho}},\underline{\alpha},\underline{\beta})|_{\underline{\beta}=0,\, \mathrm{Im}\tilde{\underline{\rho}}=0}$ are both open string amplitudes defines on Σ_ρ^+, the top half of the closed LC diagram at g loops, Σ_ρ.

(d) $f_L^{(g)}$ and $f_R^{(g)}$ are also functions of external parameters describing left- and right-moving external states.

(e) integration is performed over the parameters $(\tilde{\underline{\rho}},\underline{\alpha},\underline{\beta},)$ of the closed L.C. diagram Σ_ρ

(f) a coupling constant λ has been included.

The structure of equation (4.86) will play a crucial role in our analysis of the heterotic string and chiral bosons to which we now turn.

Section 4.6: The Heterotic String.

The heterotic string consists of a 10-dimensional right-moving type II superstring combined with a 26-dimensional left-moving bosonic string (see Chapter 2). The physical degrees of freedom of the right-moving sector consists of 8 transverse coordinates $X^i(t - \sigma)$, $(i = 1,\ldots,8)$ and 8 Majorana-Weyl fermions $\theta^a(t - \sigma)$, $(a = 1,\ldots,8)$. The physical degrees of freedom of the left-moving sector consists of 8 transverse coordinates $X^i(t + \sigma)$, $(i = 1,\ldots,8)$ and 16 internal compactified components $X^I(t + \sigma)$, $(I = 1,\ldots,16)$. The transverse coordinates $X^i(t + \sigma)$ and $X^i(t - \sigma)$ combined with the LC coordinates $X^\pm$ to describe the embedding of the string in 10-dimensional flat space-time. The configuration space of the 16 compactified components is the 16-dimensional torus R^{16}/Γ. As in section 4.3, we shall assume that Γ is an even, self-dual lattice. The reason for such a restriction is the requirement that multi-loop amplitudes are modular invariant, as will shortly be discussed in Chapter 5. In 16 dimensions there are only two even, self-dual lattices, the root lattice of $E_8 \times E_8$ and $SO(32)/Z_2$. The heterotic string construction enables these two groups to be incorporated as gauge groups due to the presence of the associated gauge mesons in the string spectrum (see below).

The LC gauge action for the heterotic string is

$$S = - \int d\tau \int d\sigma \, \{\partial_\alpha X^i \partial^\alpha X^i + \partial_{\tilde{\alpha}} X^I \partial^{\tilde{\alpha}} X^I - 4\pi\theta^a \partial_+ \theta^a\}/2\pi \qquad (4.87)$$

where θ^a are the eight fermionic string fields introduced in section 2.2. Although it is a simple matter to define chiral fermions, (by dropping the components), an extra condition must be imposed to enforce chirality on the compactified bosons. This takes the form

$$\Lambda^I = (\partial_\tau - \partial_\sigma) \, \tilde{X}^I = 0, \qquad I = 1,\ldots,16 \tag{4.88}$$

and has to be taken into account when constructing multi-loop amplitudes. As in the case of the superstring it is useful to replaced θ^a by the conjugate variables $\theta^{\bar{A}}$, λ^A ($A,\bar{A}=1,..,4$)as discussed in section 2.2, so that fermionic part of the action becomes $2i \int d\tau \int d\sigma \, \lambda^A \partial_+ \theta^{\bar{A}}$. The complete action is invariant under one global SUSY generated by $q^{\pm A}$, $q^{\pm \bar{A}}$ of equation (2.45), (the components $\tilde{X}^I$ are invariant under the SUSY transformation).

The quantisation of X^i, $\theta^{\bar{A}}$ and λ^A proceeds as in Chapter 2. The compactified components have the mode expansions, after imposing (4.88),

$$\tilde{X}^I(\tau + \sigma) = \tilde{x}^I + \frac{1}{2}\tilde{p}^I(\tau + \sigma) + i \sum_{n\neq 0} \tilde{\alpha}_n \, e^{-in(\tau + \sigma)}/2n$$
$$\tag{4.89}$$

with

$$[\tilde{\alpha}_n^I, \, \tilde{\alpha}_m^J] = n\delta(n+m) \, \delta^{IJ} \tag{4.90}$$

$$[\tilde{x}^I, \tilde{p}^J] = i\delta^{IJ}/2 \tag{4.91}$$

The factor of 1/2 in (4.91) corresponds to the fact that Λ^I of (4.88) is a second-class constraint. As in equation (4.60), $\tilde{p}^I$ is restricted to lie on the dual lattice Γ^*,

$$\tilde{p}^I = \Sigma \, n_i e^*{}_i{}^I \tag{4.92}$$

The spectrum of the heterotic string is determined by the mass-shell condition

$$\frac{1}{2} m^2 = 2(N + \tilde{N}) + \sum_{I=1}^{16} (p^I)^2 - 2 \tag{4.93}$$

where (for string length 2π),

$$N = \sum_{n>o} \alpha^i_{-n} \alpha^i_n + \sum_{m>o} m(Q^A_{-m} Q^A_m + \bar{Q}^A_{-m} \bar{Q}^A_m) \tag{4.94}$$

$$\tilde{N} = \sum_{n>o} \tilde{\alpha}^i_{-n} \tilde{\alpha}^i_n + \sum_{n>o} \tilde{\alpha}^I_{-n} \tilde{\alpha}^I_n \tag{4.95}$$

As usual, invariance under choice of origin for σ enforces an extra condition, which for the heterotic string takes the form

$$N - \tilde{N} - \frac{1}{2} \sum_{I=1}^{16} (\tilde{p}^I)^2 + 1 = 0 \tag{4.96}$$

The normal ordering terms in (4.93) and (4.96) arise from the zero-point energies of the α_n's for the left-moving bosonic string. Equation (4.96) is crucial for excluding tachyons from the theory.

The ground-state is massless and consists of the direct product of $|i>_R$ or $|a>_R$ with $\tilde{\alpha}^j_{-1}|0>_L$, $\tilde{\alpha}^I_{-1}|0>_L$ or $|p^I>_L$ where $|i,a>_R$ are the 8 fermionic and 8 bosonic states of the right-moving superstring ground-state. The states $|i,a>_R \times \tilde{\alpha}^j_{-1}|0>_L$ form the irreducible N = 1, D = 10 supergravity multiplet. The states $|i>_R \times \tilde{\alpha}^I_{-1}|0>_L$ correspond to 16 massless vector mesons which are the expected Kaluza-Klein gauge bosons associated with the isometry group $[U(1)]^{16}$ of R^{16}/Γ. However, these bosons are supplemented by 480 additional vector mesons arising from the states $|i\rangle_R \times |p^I, (p^I)^2 = 2\rangle_L$. Together these fill out the adjoint representation of the gauge group G (where $G = E_8 \times E_8$ or $SO(32)/Z_2$). Combined with the fermionic states $|a\rangle_R \times \tilde{\alpha}_{-1}^I|0\rangle_L$ and $|a\rangle_R \times |p^I, (p^I)^2 = 2\rangle_L$ they form an irreducible N = 1, D = 10 super-Yang-Mills multiplet of G.

Implicit in the above construction is the requirement that the lattice Γ be integral so that the zero mode momenta and winding numbers of the string on a torus may be combined to form the left-moving momenta of equation (4.60). Furthermore, Γ must be even so that the additional massless gauge mesons satisfying $(p^I)^2 = 2$ are generated. Gross et al [21] show that the states of the heterotic string form representations of G. Following the work of Frenkel and Kac [59], they construct operators $E(K^I)$ which act on the left-moving states and represent the generators of G which translate states on the weight lattice by a root vector K^I, $(K^I)^2 = 1$. The Cartan subalgebra is generated by p^I. $E(K^I)$ has the explicit form

$$E(K^I) = \oint_{c_1} \frac{dz}{2\pi i} : \exp\{2iK^I X^I(z)\} : C(K^I) \tag{4.97}$$

where $z = e^{i(\tau+\sigma)}$ and the contour c_1 is the unit circle. The additional term $C(K^I)$ is an operator 1-cocycle which is required for the closure of the Lie algebra of G; in the Chevalley basis this takes the form

$$[E(K), E(L)] = \begin{cases} \epsilon(K,L)E(K+L) & (K+L)^2 = 1 \\ K^I p^I & K + L = 0 \\ 0 & \text{otherwise} \end{cases} \tag{4.98}$$

where ϵ is a 2-cocycle satisfying

$$\epsilon(K,p)\epsilon(K + p, q) = \epsilon(p,q)\epsilon(K, p + q) \tag{4.99}$$

with $\epsilon = \pm 1$. C(K) obeys the operator cocycle

$$C(K)C(p) = \epsilon(K,p)C(K + p) \tag{4.100}$$

(for more details see ref.[59]).

We end this section by briefly mentioning an alternative formulation of the heterotic string in which the 16 compactified bosons are fermionised; that is, they are replaced by 32 left-moving Neveu-Schwarz-Ramond (NSR) fermions [20] $\psi^I(\tau + \sigma)$, I = 1,...,32. The heterotic LC gauge action then becomes

$$S = \int d\tau \int d\sigma\{ -\partial_\alpha X^i \partial^\alpha X^i/2\pi + 2i\lambda^A \partial_+ \theta^A + \psi^I \partial_- \psi^I \} \tag{4.101}$$

The fermions ψ^I have conformal weight 1/2 and are hence world-sheet spinors. They can obey either periodic or anti-periodic boundary conditions leading to two distinct Fock spaces. In the periodic (R) sector the ψ^I have the mode expansions

$$\psi^I(\tau + \sigma) = \sum_{n \in Z} \psi_n \, e^{-in(\tau + \sigma)} \tag{4.102}$$

where

$$[\psi^I_n, \psi^J_{-m}]_+ = \delta^{IJ} \delta_{nm} \tag{4.103}$$

In the antiperiodic (NS) sector, the mode expansion is

$$\psi^I(\tau + \sigma) = \sum_{r \in Z + \frac{1}{2}} \psi^I \; e^{-ir(\tau + \sigma)} \tag{4.104}$$

where

$$[\psi^I_r, \psi^J_{-s}]_+ = \delta^{IJ} \; \delta_{rs} \tag{4.105}$$

The mass-shell condition in the NS sector is

$$\frac{1}{2} m^2 = 2(N + \tilde{N}) - 2, \qquad N = \tilde{N} - 1 \tag{4.106}$$

$$\tilde{N} = \sum_{n>0} \tilde{\alpha}^i_{-n} \; \tilde{\alpha}^1_n + \sum_{\frac{1}{2}}^{\infty} r \; \psi^I_{-r} \psi^I_r \tag{4.107}$$

whereas in the R sector it is

$$\frac{1}{2} m^2 = 2(N + \tilde{N}) + 2, \qquad \tilde{N} = N + 1 \tag{4.108}$$

$$\tilde{N} = \sum_{n>0} \tilde{\alpha}^i_{-n} \; \tilde{\alpha}^i_n + \sum_{m>0} m \; \psi^I_{-m} \psi^I_m \tag{4.109}$$

The normal-ordering terms are different in the two cases and we see that the R sector does not contribute to the massless ground-state. The ground-sate is constructed from the direct product of $|i>_R$ or $|a>_R$ with $\psi^I_{-1/2} \; \psi^J_{-1/2} \; |0>_L$ or $\tilde{\alpha}^j_{-1}|0>$ and gives the same multiplets as in the bosonised version. In particular, the states $\psi^I_{-1/2} \; \psi^J_{-1/2} \; |0>_L$ are Lorentz singlets and transform in the adjoint representation of $SO(32)/Z_2$. Finally note that $E_8 \times E_8$ is generated by taking mixed boundary conditions for the NSR fermions [20].

One of the main problems in the construction of multi-loop amplitudes for the heterotic string has been the presence of the second-class constraints (4.88) enforcing chirality on the compactified bosons X^I. We now present a direct way of handling these constraints and of obtaining the chiral contribution to the heterotic string amplitudes based on the construction developed in sections 4.2 and 4.3.

Recall that working with the open string coordinates $\Phi_{L,R}$ made possible the separation of left and right-moving compactifies modes at the multi-loop level. The contribution of d compactified components to a scattering amplitude could then be written as

$$f_L(\bar{\underset{\sim}{\rho}}, \underline{\alpha}, \underline{\beta}) \quad f_R(\underset{\sim}{\rho}, \underline{\alpha}, \underline{\beta}) \tag{4.110}$$

We therefore expect that the contribution of d left-moving chiral bosons to a scattering amplitude would be

$$f_L(\bar{\underset{\sim}{\rho}}, \underline{\alpha}, \underline{\beta}) \tag{4.111}$$

This turns out to be the case as we now demonstrate.

To link up with section 4.2 note that $\tilde{X}^I$ in equation (4.88) corresponds to X_W^I in equation (4.34) whereas $\tilde{X}^I$ in (4.89) corresponds to X_L^I in (4.49). The constraint Λ^I may be re-expressed in the canonical form

$$\Lambda^I = 2\pi P^I(\sigma) - \tilde{X}^{I\prime}(\sigma) - W^I \tag{4.112}$$

In terms of $\Phi_{L,R}$ and $\Pi_{L,R}$ of equations (4.65) and (4.66) this becomes

$$\Pi_R^I(\sigma) = 0 \qquad \Phi_R^{I\prime}(\sigma) = 0 \tag{4.113}$$

It would appear from (4.113) that the right-moving zero modes x_R^I are unconstrained. However note that the constraints $\Lambda^I = 0$ in (4.112) set all right-moving modes to zero except for x_R^I and therefore should be supplemented by the extra constraints

$$x_R^I = 0 \tag{4.114}$$

to obtain a truly chiral theory. Thus, combining (4.113) and (4.114) we conclude that under the canonical change of coordinates, the constraints enforcing chirality take the form

$$\Pi_R^I(\sigma) = 0 \qquad \Phi_R^I(\sigma) = 0 \tag{4.115}$$

These second-class constraint decouple from the left-moving coordinates and consequently the left-moving Dirac bracket is equal to the original Poisson bracket which is given by $\{\Phi_L^I(\sigma),\ \Pi_L^J(\sigma)\} = \delta(\sigma - \sigma')\ \delta^{IJ}$. The Dirac bracket of the right-movers is zero; this allows (4.115) to be imposed consistently at the quantum level. Hence using the string coordinates Φ_L^I, Φ_R^I and conjugate momenta Π_L^I, Π_R^I the chiral constraints (4.88) may be solved by setting $\Phi_R^I = \Pi_R^I = 0$.

The associated field theory consists of just the left-moving field $\psi_L(\Phi_L)$ defined in section 4.4. Therefore scattering amplitudes in the compactified sector of the heterotic string are constructed from the left-moving propagator K_L and vertex $|V_L>$ defined in equation (4.63) and (4.79). The propogator is twisted by the operator constraint corresponding to (4.96)

$$\hat{\Lambda} = \hat{H}_L - \text{superstring terms} \tag{4.116}$$

where $\hat{H}_L$, defined in (4.77), is the Hamiltonian of the chiral bosons. Including these twists as described in section 2.3, the resulting propagator is given by the left-moving part of equation (4.81). Finally, following through the arguments of section 4.4, we see that the contribution of the left-moving chiral bosons to the scattering amplitude is just $f_L(\tilde{\underline{p}}, \underline{\alpha}, \underline{\beta})$.

CHAPTER 5: CONSTRUCTING MULTI-LOOP AMPLITUDES

Section 5.1: Reduction to Functional Form.

In the preceding two chapters second-quantised superstring field theories have been constructed which satisfy various symmetries. In Chapter 3 the symmetry group was that of the ten-dimensional super-Poincare group, whilst in Chapter 4 this was reduced to that associated with certain compactified dimensions, so leading to the heterotic string. It is the purpose of this chapter to use these various second quantised string field theories to evaluate superstring amplitudes by reducing them, by perturbation theory, to first quantised expressions. These latter are more directly evaluated by means of Riemann surface theory. (A brief summary of some definitions and results concerning Riemann surface theory is presented in appendices I-VII). One may question the convergence of the resulting perturbation expansion. Indeed for the bosonic string the perturbation series is not even Borel summable [60]. However it is well known from point quantum field theories that perturbation theory is highly useful. In particular it enables analysis of possible infinities of a field theory to be carried out explicitly; that will be performed in the superstring context in the following section. How to go beyond perturbation theory for superstrings is presently an open question; our constructions in this book lead to a formulation which could be a basis for an attempt to answer that question.

The reduction of perturbation amplitudes from second to first quantised form was achieved for the bosonic string in 1974 [28]. Various aspects of this reduction were briefly noted in Chapters 3 and 4; they will be summarised and extended here. The reduction was accomplished in two stages. Firstly, a particular contribution at g^{th} order to the 2nd quantised perturbation expansion may be seen to correspond to a light-cone strip diagram with g slits and N external strings. This L.C.

diagram can be decomposed into either rectangular strips (single string propagators) or into primitive interaction regions, in which two strings fuse to make one, or one string splits to make two others. This has already been illustrated in figure 3; one loop L.C. strip diagram for 4 external strings is given in figure 4. The contributions from each of these decomposed parts of the L.C. strip diagram may then be separately re-expressed as a first quantised expression. This may be done individually for the rectangles (free strings) and the primitive interaction regions.

Consider first the rectangular strip S for which on the lines $\tau = \tau_i$ (i = 1, 2) the values of the transverse modes $\underline{X}(\sigma,\tau)$ are given as $\underline{X}(\sigma,\tau) = \underline{X}_i(\sigma)$ (i = 1, 2) and p^+ is fixed as $p^+ = p_i^+$. It may be shown by explicit calculation, as noted in Section 3.1, that for S, the first and second quantised propagators are identical. Thus

$$< 0 \mid \Phi(\underline{X}(\sigma , \tau_2)) \, \Phi^+(\underline{X}(\sigma , \tau_1)) \mid 0> =$$

$$\int_S D\underline{X} \exp \left[+A(S)\right] \prod_{\sigma,i=1}^{2} \delta(\underline{X}(\sigma,\tau_i) - \underline{X}_i(\sigma)) \tag{5.1}$$

where the integration on the right-hand side of (5.1) is over all transverse modes $\underline{X}$ for (σ,τ) on the strip S, the δ-functionals restricting the boundary values of $\underline{X}(\sigma,\tau)$ at $\tau = \tau_i$ to be the required values $\underline{X}_i(\sigma)$. A(S) is the world sheet action

$$A(S) = \frac{1}{2\pi} \int_S \underline{X} \cdot \Delta\underline{X} \, d\sigma \, d\tau \tag{5.2}$$

where Δ is the Laplacian on S. For the primitive interaction regions of figure 3 it may similarly be shown using the notation of that figure, that

$$\lim_{\alpha\to 0} \int_I D\underline{X} \exp[A(I)] \prod_{\sigma, i=1}^{3} \delta(\underline{X}(\sigma_i, \tau_i) - \underline{X}_i(\sigma_i))$$

$$= \prod_{\sigma} \delta(\underline{X}_3(\sigma) - \underline{X}_1(\sigma))\, \delta(\underline{X}_3(\sigma) - \underline{X}_2(\sigma)) \tag{5.3}$$

The expression on the left hand side of (5.3) is that of the first quantised theory, that on the right hand side being that of the second quantised theory. The mode expansion version of (5.3) has already been presented in section 3.3 for the bosonic string; we refer the reader back to that for an explicit expression of (5.3). We will not however, use mode expansions much (except for external states) in this discussion.

In the second stage the various first-quantised factors of a multi-loop amplitude are re-assembled into a form which allows a global first quantised description to be given on the Riemann surface corresponding to the L.C. strip diagram. It will be shown in this section how the Riemann surface formulation of the bosonic string amplitude, developed briefly in Section 3.4, can be obtained in more detail and adapted to the superstring case. This is achieved as follows.

The bosonic superstring second quantised propagator may be reduced to that of the first quantised one, and the bosonic δ-function at the vertex also reduces to the first quantised expression, using the method of the preceding paragraph. Both of these results may be extended rapidly to the superstring variables, θ, $\tilde{\theta}$. This is immediate for the interaction vertex (on inclusion of the insertion factors H_3^V at each intersection point), by continuity. For the propagator, the first quantised expression for propagation from the boundary value $\theta_1(\sigma)$ at τ_1 to $\theta_2(\sigma)$ at τ_2 ($\tilde{\theta}$ is dropped explicitly) is

$$\int D\theta\, D\lambda \exp\left[-2i\int_{\tau_1}^{\tau_2} d\tau \int d\sigma\, \lambda \partial_{\bar{\rho}} \theta \right]$$

$$\prod_{i=1}^{2} \prod_{\sigma} \delta(\theta(\sigma,\tau_i) - \theta_i(\sigma)) \tag{5.4}$$

The variable λ acts as a Lagrange multiplier (this is discussed in more detail in the next section) and leads to a factor $\delta(\partial_{\bar{\rho}}\theta)$ at each σ,τ. This may be solved for θ as

$$\theta = \sum_{n\in \underline{Z}} A_n\, e^{n(\tau+i\sigma)} \tag{5.5}$$

where only integers are allowed in the r.h.s. of (5.5) if θ is to satisfy the periodic boundary conditions $\theta(\sigma,\tau) = \theta(\sigma+2\pi,\tau)$. The expression (5.4) reduces, by change of variable to $\{A_n\}$, to

$$\int \prod_n dA_n\, \delta(A_n e^{n\tau_1} - \theta_{n1})\, \delta(A_n e^{n\tau_2} - \theta_{n2}) \tag{5.6}$$

where θ_{ni} are the Fourier coefficients of $\theta_i(\sigma)$. Finally (5.6) may be evaluated to have the value

$$\prod_n (\theta_{n1} e^{n(\tau_2-\tau_1)} - \theta_{n2}) = \sum_{\underline{m}} F_{\underline{m}}(\theta_1)\, e^{|m|(\tau_2-\tau_1)}\, F^{+}_{\underline{m}}(\theta_2) \tag{5.7}$$

The functions $F_{\underline{m}}$ in (5.7) are the complete set of occupied states given in terms of products of the set of Fourier coefficients $\theta_{m\underline{1}}$ whilst $F^{+}_{\underline{m}}$ is the adjoint set. The r.h.s. of (5.7) is exactly the expression which would arise in the second quantised field expression $\langle 0 \,|T(\Phi(-, \theta_1, \tau_1)\, \Phi^{+}(-, \theta_2, \tau_2))\, |0\rangle$, following the methods of [28].

These results may be used to reduce any second quantised perturbation diagram to a first quantised expression by using the decomposition of a general L.C. strip diagram into rectangles and primitive interaction vertices. It is necessary to rewrite the resulting factors in the exponential in a global fashion, since otherwise modes may be missed which have global but no local character. These will be solitonic in form, and arise from the non-simply connected character of the L.C. strip diagram.

To see how these solitonic modes arise, a global form of the superstring action may be given, after Euclideanisation, by rewriting the bosonic part of (2.70) in terms of one-forms $d\underline{X}$, ${}^{*}d\underline{X}$, with (appendix IV)

$$d\underline{X} = \partial_{\rho}\underline{X}\, d\rho + \partial_{\bar{\rho}}\underline{X}\, d\bar{\rho}, \quad {}^{*}d\underline{X} = -i[\partial_{\rho}\underline{X}\, d\rho - \partial_{\bar{\rho}}\underline{X}\, d\bar{\rho}]$$

The fermionic part of (2.70) will already be globally expressed if, as we choose, θ and λ are (1,0) and (0,0) forms in $d\rho$, $d\bar{\rho}$ (so $\theta d\rho$ and λ are invariant under analytic changes of co-ordinates) and $\tilde{\theta}$, $\tilde{\lambda}$ are (0,1) and (0,0) forms. The global form of (2.70) becomes

$$S_E = \iint_{W_E} \left[\frac{1}{\pi}\, d\underline{X} \wedge {}^{*}d\underline{X} + i\, d\rho \wedge d\bar{\rho}(\partial_{\rho}\tilde{\theta}.\tilde{\lambda} - \partial_{\bar{\rho}}\theta.\lambda)\right] \qquad (5.8)$$

The globally defined one-form $d\underline{X}$ may be expressed as

$$d\underline{X} = d\underline{X}_e + \sum_{i=1}^{g} (z_i w_i + \text{herm-conj.}) \qquad (5.9)$$

where the suffix e denotes exact, the w_i are g independent analytic (1st Abelian) differentials and the z_i are g complex co-ordinates for the solitonic modes. The w_i may also be written globally on W_E as $w_i = u_i{}'dz$ with u_i the corresponding set of (multivalued) first Abelian integrals. The bosonic part of the action (5.8) may then be written as

$$\frac{1}{2\pi} \iint_{W_E} d\sigma \, d\tau \; \underline{X}_e \; \Delta \; \underline{X}_e \; - \frac{2}{\pi} \; \underline{z}^+ \; (\mathrm{Im}\Pi)\underline{z} \qquad (5.10)$$

In order to obtain the second term in (5.10), with Π the period matrix on W_E, the usual normalization

$$\oint_{\gamma_j} w_k = \delta_{jk} \qquad (5.11a)$$

has been used, so that the period matrix is given by

$$\oint_{\delta_j} w_k = \Pi_{jk} \qquad (5.11b)$$

(see appendix IV), together with the Riemann bilinear relation

$$\int_{W_E} \theta \wedge {*\theta} = \sum_{j=1}^{g} \left[\oint_{\gamma_j} \theta \oint_{\delta_j} \bar{\theta} - \oint_{\delta_j} \theta \oint_{\gamma_j} \bar{\theta} \right] \qquad (5.12)$$

for any holomorphic one-form θ. Note that for uncompactified strings the bosonic components $\underline{X}$ are single-valued everywhere implying that there is no contribution from the solitonic modes z_i (which may be dropped). However, a non-zero contribution will occur for compactified strings as discussed in section 5.7.

Section 5.2: Superstring Feynman Rules

We are now in a position to deduce rules for writing down perturbation diagram contributions to the S-matrix at any multiloop order for superstring theories. This will allow us to complete and expand the discussions of Section 3.4. This may be done using the results of the preceding section to give the first quantised expression

$$\left[\prod_{r=1}^{N}(\pi p_r^+)\right]\int\prod_{P} d^2\tilde{\rho}\int\prod_{i=1}^{g} d\,\alpha_i\;d\beta_i\int D\underline{X}\,D\lambda\,D\theta\,D\tilde{\lambda}\,D\tilde{\theta}\,\times$$

$$\exp\left\{\iint_{W_E}\left[\frac{1}{\pi}\,d\underline{X}\wedge{}^*d\underline{X}+i\,d\rho\wedge d\bar{\rho}(\partial_{\rho}\tilde{\theta}\tilde{\lambda}-\partial_{\bar{\rho}}\theta.\lambda)+\underline{J}.\underline{X}+Q\theta+\tilde{Q}\tilde{\theta}-\sum_t p_t^-\tau_t\right\}$$

$$\times\prod_{P} V_3^H(\surd\epsilon\;\partial X,\surd\epsilon\theta)\;V_3^H(\surd\epsilon\;\bar{\partial}X,\surd\epsilon\tilde{\theta}) \qquad (5.13)$$

The various terms and factors in (5.13) are as follows:

(i) N denotes the number of external strings, with sources $\underline{J}$, Q and $\tilde{Q}$ denoting the Fourier transforms of the δ-functions enforcing boundary conditions on $\underline{X}$, θ and $\tilde{\theta}$ at initial and final times τ_i, τ_f respectively, as τ_i, $\tau_f \to \pm\infty$. Thus $\underline{J}$, Q and $\tilde{Q}$ will have supports at τ_i, τ_f; for the zero modes, for example,

$$\underline{J}(\sigma,\tau) = \sum_r \underline{p}^{(r)}\,\delta(\tau-\tau^i) + \sum_s \underline{p}^{(s)}\,\delta(\tau-\tau_f) \qquad (5.14)$$

In (5.14) the set r(s) denotes the incoming (outgoing) momenta of the zero modes. The term $p_t^- \tau_t$ (t = i or f) is a factor to remove the time-dependence of the external states (see Chapter 11 GSW).

(ii) Wave functions depending on $\underline{J}$, Q, $\tilde{Q}$ have still to be folded in to the expression (5.13). This may be done for the zero modes in terms of superfield corresponding to the full supergravity (or Super-Yang-Mills) multiplets, as discussed in [50]. There is an asymmetry between λ and θ (or $\tilde{\lambda}$ and $\tilde{\theta}$) which allows us to regard only θ, $\tilde{\theta}$ as propagating modes and exclude λ, $\tilde{\lambda}$ from external sources or wave-functions. This is essential to obtain 10-dimensional supersymmetry; it will be shown below why it is correct to neglect propagation of λ and $\tilde{\lambda}$. The left-moving external wave functions each have a normalisation factor $(\pi p_r^+)^{-1}$ associated with the expansion (2.53).

(iii) The first factor $\prod_{r=1}^{N} (\pi p_r^+)$ arises from corresponding factors $(\pi p_r^+)^{1/2}$ for each external closed string, which takes into account integration around the circumference of the string, and a Lorentz flux factor $(\prod_{r=1}^{N} \pi p_r^+)^{\frac{1}{2}}$ (see Chapter 11 of GSW).

(iv) Integration is to be performed over all positions $\tilde{\rho}$ of the interaction vertices P, as expressed by the measure $\prod_P d^2\tilde{\rho}$. The integration over Re$\tilde{\rho}$ = $\tilde{\tau}$ corresponds to the usual integration over interaction times in perturbation theory. That over Im$\tilde{\rho}$ = $\tilde{\sigma}$ is also necessary due to the need to sum over all possible positions of the interaction points (without double counting).

(v) Integration is also to be performed over the widths and twists denoted by the variables α_i and β_i, for the g internal loops or handles.

The quantities α_i and β_i are given as the changes of $\rho(P)$ at the point P on taking P around the loops γ_i or δ_i ($| \leq i \leq g$), which form a basis of the homology group $H_1(W_E)$ of W_E, aand satisfy equation (4.85).

A modular transformation $T = \begin{pmatrix} A & B \\ C & D \end{pmatrix}$ changes the set of α_i and β_i to

form another homology basis of the surface (see section 5.8). It therefore corresponds to an element of the symplectic group $Sp(2g,\underline{z})$, where A, B, C, D, are integer

valued g×g matrices with detT=1 and $T^T T = \begin{pmatrix} 0 & 1 \\ -1 & 0 \end{pmatrix}$. It then follows

from (4.85) that under the change of basis,

$$\begin{pmatrix} \underline{\alpha} \\ \underline{\beta} \end{pmatrix} \to \begin{pmatrix} A & B \\ C & D \end{pmatrix} \begin{pmatrix} \underline{\alpha} \\ \underline{\beta} \end{pmatrix} \tag{5.15}$$

so that $\prod_{i=1}^{g} d\alpha_i d\beta_i \to (\det T) \prod_{i=1}^{g} d\alpha_i d\beta_i$, and the measure on the α's and β's is invariant.This indicates that (5.13) is independent of any choice of homology basis (since the measure $\prod_P d\tilde{\rho}$ will be); such independence is necessary if (5.13) is to be obtained from a second quantised field theory which has no specification of choice of such a basis $\{\underline{\gamma}, \underline{\delta}\}$. The integration over $\tilde{\rho}$'s, the α's and the β's corresponds to that over the moduli of a Riemann surface, there being 3g-1+N variables after translation invariance has been used to remove one of the $\tilde{\rho}$'s. The integration domain of $\tilde{\rho}$'s, α's and β's over the L.C. strip diagram corrresponding to a fundamental domain in moduli space, has been discussed in detail in [53] and [61].

(vi) Finally the vertex factors V_3^H are as given in (3.22c); they have been written explicitly in terms of the variables $\underline{X}$, θ and $\tilde{\theta}$, so need to contain the regularising factors $\sqrt{\epsilon}$, as discussed in association with equations (3.23).

It is now necessary to perform the integration over $\underline{X}$, λ, θ, $\tilde{\lambda}$, $\tilde{\theta}$. The $\underline{X}$-integration may be performed simply, since $\underline{X}$ enters the exponential in (5.13) as a Gaussian, and there is no contribution from the inexact part dX in equation (5.9). On performing the Gaussian integration over $\underline{X}$, it is necessary to include contractions between the insertion factors in (5.13) involving $\partial_\rho \underline{X}$ or $\partial_{\bar{\rho}} \underline{X}$, which are of the form

$$\overbrace{X_e^\ell(\rho_1) X_e^m}(\rho_2) = -\frac{1}{2}\delta^{\ell m}\, G(\rho_1,\rho_2), \quad (1\le\ell,\ m\le d-2) \tag{5.16}$$

where G is the inverse of Δ on W_E, $\Delta G = -2\pi\delta^2$, and may be identified as the real part of the third Abelian integral with purely imaginary normalization, $\Omega_{\rho_2\ \rho_0}(\rho_1)$, ρ_o an arbitrary point on W_E.
That is, $G(\rho_1,\rho_2) = \mathrm{Re}\ \underline{G}(\rho_1,\rho_2)$ where $\underline{G}(\rho_1,\rho_2) = \Omega_{\rho_2\rho_0}(\rho_1)$ and

$$\mathrm{Re}\oint_{\gamma_i} d\Omega_{\rho_2\rho_0} = \mathrm{Re}\oint_{\delta_i} d\Omega_{\rho_2\rho_0} = 0$$

$$\overbrace{\partial_{\rho_1} X^\ell\, \partial_{\rho_2} X^m} = -\frac{1}{2}\delta^{\ell m}\, \partial_{\rho_1}\partial_{\rho_2} G(\rho_1,\rho_2) \tag{5.17a}$$

$$\partial_{\rho_1} \overline{X^\ell \partial_{\bar\rho_2} X^m} = 0 \tag{5.17b}$$

It is to be remarked that the $\delta^2(\rho_1 - \rho_2)$ contribution to the left hand side of (5.17b) is not included in the right hand side; this corresponds to preventing ρ_1 and ρ_2 coinciding by extension of small discs around them.

The λ integration may be performed immediately, to give the functional δ-function $\delta(\partial_{\bar\rho}\theta)$. This simplicity indicates an important feature of the SU(4) × U(1) version of the superstring, in that in the SO(8) version [49] the fermionic variables do not seem to be decomposable into a Lagrange multiplier and its constrained associate variable. The remaining θ integration is over the zero modes of $\partial_{\bar\rho}$. The complete solution of this problem was shown in the second reference in [32] to be

$$\theta(\rho) = \sum_{r=1}^{N} \partial_\rho \, \underline{G}(\rho,\rho_r) \, \pi p_r^+ \, \theta_r + \sum_{i=1}^{g} \theta_i \, u_i'(\rho) \tag{5.18}$$

The first term on the r.h.s. of (5.18) has boundary value θ_r as $\rho \to \rho_r$; θ_r is the zero-mode value of the external string variable $\theta_r(\sigma)$ and ρ_r is the source. (the non-zero mode boundary values will be discussed in section 5.9). The second term in (5.18) incorporates the fermionic solitonic modes (with the correct conformal weight of (1,0). A more complete analysis is given in Appendix XIV as to the uniqueness of this solution. We note that if we had integrated on θ (neglecting θ dependence of the vertex insertion factors) the resulting constraint $\partial_\rho\lambda$ = 0 cannot have a similar solution to (5.18) if it has conformal weight (0,0), and it can be seen that no suitable solution to the constraint can exist. This justifies the earlier remark that λ is to be regarded as a non-propagating mode.

There remains to perform the integration over the free variables θ_r, θ_i in (5.18). This may be justified by carefullly following through a change of variables (see Appendix XV and [62]) but more rapidly it can be seen that the measure

$$\left[\Pi p_r^+\right]^{-1} \prod_{r=1}^{N} d^4\theta_r \;.\; \delta^4 \left[\sum p_r^+ \,\theta_r\right] \;.\; \prod_{i=1} d^4\theta_i \;\times\; (\theta \leftrightarrow \tilde{\theta})$$

$$(\det \partial_\rho)^4 \;(\det \partial_{\bar{\rho}})^4 \qquad (5.19)$$

is invariant under the modular transformation (5.15) since invariance of the last term in (5.18) at $\rho = \tilde{\rho}$ implies the transformation properties

$$\prod_{i=1}^{g} d^4\theta_i \rightarrow [\det(C\Pi + D)]^{-4} \prod_{i=1}^{g} d^4\theta_i \qquad (5.20a)$$

and we use also that

$$\det \partial_{\bar{\rho}} \rightarrow \det\,(C\Pi + D).\det \partial_{\bar{\rho}} \qquad (5.20b)$$

For a discussion of chiral determinants see appendix XVI and [63]. Modular invariance is discussed in detail in section 5.8. Finally the δ-functions in (5.19) correspond to the condition that the sum of the residues of the right hand side of (5.18) is zero. Further aspects of this measure are discussed in Appendix XV.

The above results may be collected together to formulate the superstring multiloop diagram rules. These will only be expressed here for massless external states; the massive case will be considered in detail in section 5.9. The amplitude rules are more conveniently formulated in the z-plane obtained by a suitable mapping called the Mandelstam mapping, which is now discussed. The Mandelstam mapping F is the mapping from W to a Riemann surface Σ with uniformisation (local complex co-ordinates) z:

$$\rho = F(z) = \sum \pi p_r^+ \, \underline{G}\,(z,\, z_r) \tag{5.21}$$

Under this we take θ, λ, $\underline{X}$ to have conformal weights 1, 0, 0 respectively, so that

$$\theta(\rho) = [F'(z)]^{-1}\; \theta(z) \tag{5.22}$$

It is clear that at an interaction point P

$$F'(\tilde{z}) = 0, \quad \tilde{\rho} = F(\tilde{z}) \tag{5.23}$$

The variable $\theta(\tilde{\rho})$ is therefore poorly defined if $\theta(z)$ is (as we take) well-defined.

Since for $z \sim \tilde{z}$,

$$\rho - \tilde{\rho} \sim \frac{1}{2}\,(z-\tilde{z})^2\;F''(\tilde{z}) \tag{5.24a}$$

then

$$[F'(z)]^{-1} \sim [2(\tilde{\rho}-\rho)\;F''(\tilde{z})]^{-\frac{1}{2}} \tag{5.24b}$$

Thus if $\rho-\tilde{\rho} = i\epsilon$, then

$$\bar{Y}^{A}(P) = \lim_{\epsilon\to 0} \sqrt{\epsilon}.\ \eta\ \bar{\theta}^{A}(\rho) = [2F''(\tilde{z})]^{-\frac{1}{2}}\ \bar{\theta}^{A}(\tilde{z}) \tag{5.25}$$

is well defined (as discussed in detail, using modes, in [50]), where $\eta = i^{\frac{1}{2}}$. It is the finite quantity $\bar{Y}^A$ which appears as the primary one in the cubic insertion factor construction in [50]. In a similar manner the well-defined quantity related to $\partial_\rho \underline{X}$ will be

$$\underline{Z}(P) = \lim_{\epsilon\to 0} 2i\ \sqrt{\epsilon}\ \eta\ \partial_\rho \underline{X}(\rho) = 2i\ [2F''(\tilde{z})]^{-\frac{1}{2}}\ \partial_z \underline{X}(\tilde{z}) \tag{5.26}$$

It is to be noted that Y and $\underline{Z}$ are defined on the string, at a given time τ, and not necessarily only in a perturbation theoretic form in which time development is also needed (as the third term in (5.25) or (5.26) would indicate, due to the presence of the complex variable $\tilde{z}$). Similar expressions hold for $\tilde{Y}$, $\tilde{\underline{Z}}$:

$$\tilde{\bar{Y}}^{A}(p) = \lim_{\epsilon\to 0} \sqrt{\epsilon}\ \eta^*\ \tilde{\bar{\theta}}^{A}(\rho) = [2F''(z)]^{-\frac{1}{2}}\ \tilde{\bar{\theta}}^{A}(\tilde{z}) \tag{5.27}$$

$$\tilde{\underline{Z}}(p) = \lim_{\epsilon\to 0} 2i\sqrt{\epsilon}\ \eta^*\ \partial_{\bar{\rho}}\ \underline{X}(\rho) = 2i[2\bar{F}''(\tilde{z})]^{-\frac{1}{2}}\ \partial_{\bar{z}}\ \tilde{\underline{X}}(\tilde{z}) \tag{5.28}$$

Various other useful expressions are collected in Appendix VIII, using expansion of co-ordinates around interaction points; these will be found helpful later. The above has given a deriviation of the vanishing factors in (3.23) which is independent of modes, as had been promised in section 3.3.

The above expressions may now be used to deduce the following factors in a multi-loop amplitude:

(a) the insertion factors

$$\left\langle \prod_P ([2F''(\tilde{z})]^{-1/2}\, V_3^H \left[\partial_{\tilde{z}} \underline{X} + \sum_{r=1}^{N} \frac{1}{2} \partial_{\tilde{z}} G(\tilde{z},z_r)\, \underline{p}_r\,,\ [2F''(\tilde{z})]^{-1/2}\ \theta(\tilde{z})\right]\right. \tag{5.29a}$$

$$\left. ([2\bar{F}''(\tilde{\bar{z}})]^{-1/2}\, V_3^H \left[\partial_{\bar{\tilde{z}}} \underline{X} + \sum_{r=1}^{N} \frac{1}{2} \partial_{\bar{\tilde{z}}} G(\tilde{\bar{z}},z_r)\, \underline{p}_r\,,\ [2\bar{F}''(\tilde{\bar{z}})]^{-1/2}\ \tilde{\theta}(\tilde{\bar{z}})\right]\right\rangle$$

where the signs < > denote contractions, as given by (5.17) evaluated in the z-plane, the contractions being in all possible ways between pairs of the variables $\partial_z \underline{X}$, $\partial_{\bar{z}} \underline{X}$, and $\theta(\tilde{z})$ is given by (5.18) with $\tilde{\rho}$, ρ_r, etc. replaced by $\tilde{z}$, z_r, etc.

(b) the Veneziano-Virasoro-Shapiro (VVS) factor

$$\exp[-\frac{1}{2}\sum_{r<s} p_r . p_s\ G(z_r,z_s)] \tag{5.29b}$$

arising from the external bosonic sources (where the Lorentz-invariant inner products $p_r\ p_s$ arise from similar products involving only transverse momenta by inclusion of wave function factors, as made explicit by Mandelstam [52]. Note that there is no contribution from the singular terms $G(z_r,z_r)$ since $p_r^{\,2} = 0$ for external superstrings in the ground-state.

(c) integration over the internal Grassmann-valued solitonic modes, with modular invariant measure

$$\prod_{i=1}^{g} d^4\theta_i\ d^4\tilde{\theta}_i\ (\det \mathrm{Im}\Pi)^{-4} \tag{5.29c}$$

(d) integration over the external Grassmann-valued variables θ_r, $\tilde{\theta}_r$, after multiplication by external superfields $\Phi^{(r)}(\zeta^{(r)}, u^{(r)}, \theta_r, \tilde{\theta}_r, p_r)$ describing external massless modes with helicities or polarizations $u^{(r)}$, $\zeta^{(r)}$, with measure

$$\left[\prod_{r=1} p_r^+\right]^{-2} \prod_{r=1}^{N} d^4\theta_r \; d^4\tilde{\theta}_r \;\; \delta^4 \left[\sum p_r^+ \theta\right] \delta^4 \left[\sum p_r^+ \; \tilde{\theta}_r\right] \qquad (5.29d)$$

(e) integration over the interaction points and internal string widths and twists, with measure

$$\prod_{P-1} \; d^2\tilde{\rho} \; \prod_{i=1}^{g} d\,\alpha_i \; d\beta_i \qquad (5.29e)$$

where P-1 denotes removal of one of the interaction positions to be integrated over, on use of translation invariance on the L.C. strip diagram.

(f) a determinental factor

$$\Delta_1 \; = [(\det \Delta/\det \partial_\rho \; \det \partial_{\bar{\rho}} \;). \det \operatorname{Im} \Pi^{-1}]^{-4} \qquad (5.29f)$$

arising from the bosonic and fermionic kinetic terms in the first quantised action. (This determinental factor satisfies $\Delta_1 = 1$ as shown in appendix XVI.

The net effect of the above superstring Feynman rules is to lead to ordinary (point field theory) propagators being drawn between interaction positions on W_E. Those of bosonic character are of two sorts, either as given by (5.17), or arising as the expression

$$\sum_{r=1}^{N} \partial_{\tilde{z}} \; G(\tilde{z}, \; z_r) \; \underline{p}_r$$ (or its hermitian conjugate) from the Gaussian

integration over $\underline{X}$. It is also possible to recognise fermion propagators as in the first term on the r.h.s. of (5.18). These propagators lead to various lines ruled on the world sheet in the manner shown in figure 11.

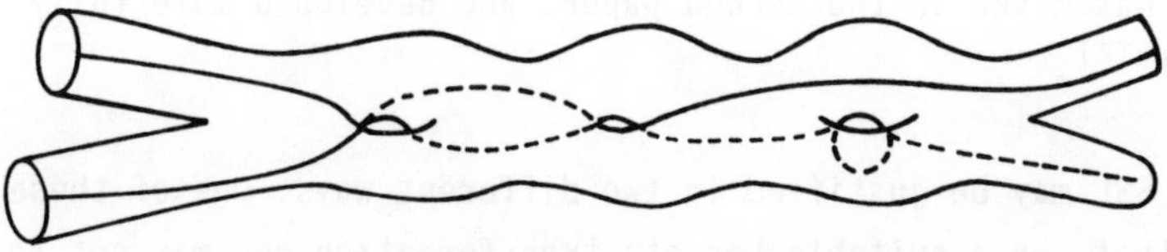

Fig. 11 A multi-loop contribution to the 4-string scattering amplitude, with ruled lines on the surface, the continuous lines denoting "fermionic" and the dashed ones "bosonic". They are chosen to obey the rules (a)-(b) of this section.

It is clear that these lines may cause difficulties when the interaction points coincide, due to possible divergences. These questions will be considered in more detail in Chapter 6.

Section 5.3: The Short-String Limit (SSL).

The above superstring Feynman rules are compact, but involve as integration variables the interaction points $\tilde{\rho}$ and the internal loop width and twist variables α_i and β_i. Both of these variables depend on the external variables - the positions of the external strings - in a complicated manner. Expressions for tree and 1-loop superstring amplitudes [48] do not have any dependence on the $\tilde{\rho}$ (or $\tilde{z}$) variables. Moreover the above expressions are not explicitly Lorentz-covariant. Thus equation (5.23) for the interaction positions only involves the p^+-components; on the other hand the d-2 transverse components alone enter the insertion factors (5.29a). In this section the use of external strings which have $p_r^+ \sim 0$ for (N-2) of them will be shown to lead to simpler expressions for the multi-loop amplitudes, in which (N-2) of the interaction positions are eliminated in favour of the external source variables. Furthermore the various factors involving p^+-components will disappear. This approach was developed for the bosonic string in [52] at tree level and in [64] for the 1-loop level; the multi-loop superstring extension was given in the second paper, and developed more fully in the fourth, of [32].

The SSL may be justified in two different ways. One of these uses the fact that, by a suitable Lorentz transformation one may set to zero p_r^+ for a set of vectors p_r; this is seen to be possible for $r \leq d$, where d is the dimension of the embedding space-time. The amplitudes can then be evaluated in this Lorentz frame, and transformed to any other on appeal to the Lorentz invariance of the total field theory. More than d external states (or ½d if vector bosons are considered with $\zeta_r^+=0$ as well) cannot lead to general forms of matrix elements. In the superstring case it is necessary to appeal to the full sP_{10} supersymmetry of the embedding space-time, since a supersymmetry transformation is also required to take the short-string limit (as is discussed in more detail in Appendix XVII).

The second approach is more general, being valid for any number of external states. All non-covariant inner products are made covariant in this case by combining non-covariant terms together in the amplitude in an explicit manner, as done by Mandelstam at tree level for NSR [55] or by ourselves (unpublished) for the tree level superstring amplitudes. Independence of volume elements multiplying the covariants on the p_r^+ then allows these latter to be evaluated in any configuration. It is to be expected that this approach is valid also at multi-loop level; it has not been proven, however, and we rely on the first approach for justification. That is one reason why we must also close the sP_{10} algebra which we discuss in Chapter 8.

The equation determining the positions of the interaction points in the z-plane is (5.23). Written out more completely this takes the form:

$$\sum_{r=1}^{N} p_r^+ \, \underline{G}'(\tilde{z},z_r) = 0 \tag{5.30}$$

If the source point z_o involved in the definition of $\underline{G}(z,z^1)$ as $\Omega_{z^1 z_o}(z)$, the third Abelian integral with purely imaginary normalization, is taken as z_N the summation in (5.30) is only up to N-1. In the short string limit

$$p_r^+ \sim 0 \qquad 2 \le r \le N\text{-}1 \tag{5.31}$$

a set of (N-2) values $\tilde{z}$ satisfying (5.30) will approach to within $O(p_r^+)$ of one or other of the z_r's so that (5.30) may be satisfied. The corresponding $\tilde{z}$ will be denoted $\tilde{z}_r$ and called an external interaction point. It will be one of the initial or final interactions in which incoming or outgoing strings fuse or split. The value of $\tilde{z}_r$ may be obtained from (5.30) and (5.31) by using that as $z \sim z^1$

$$\underline{G}'(z,z^1) \sim -(z-z^1)^{-1} + O(1) \tag{5.32}$$

The result is that

$$\tilde{z}_r = z_r + p_r^+ \, [p_1^+ \, \underline{G}'(z_r,z_1)]^{-1} + O(p_r^{+2}) \tag{5.33}$$

It is possible to determine the higher order terms in (5.33) if so desired [32], although they will not needed here.

The leading order terms in $O(p_r^{+-1})$ in the various terms in (5.29a) may now be determined, using (5.33). Thus

$$F''(\tilde{z}_r) = (a_r^2 p_r^+)^{-1} + O(1) \tag{5.34a}$$

$$G'(\tilde{z}_r, z_r) = -(a_r p_r^+)^{-1} + O(1) \tag{5.34b}$$

with

$$a_r = [p_1^+ \, \underline{G}'(z_r, z_1)]^{-1} \tag{5.34c}$$

Combining (5.33) and (5.34) it is possible to replace the external interaction positions $\tilde{z}_2, \ldots, \tilde{z}_{N-1}$ by the same positions $z_2, \ldots, z_{N-1}$ as integration variables. This still leaves the 2g 'internal' interaction vertices, which are denoted generically by $\tilde{z}_\alpha$. Thus the total set of variables will be the (N-3) complex variables $z_2, \ldots, z_{N-2}$ (fixing one of the sources, say z_{N-1}, by translation invariance), the 2g complex variables $\tilde{z}_\alpha$ and the 2g variables α_i, β_i. This gives the correct number of moduli, 3g-3+N, for integration over conformally inequivalent Riemann surfaces.

It is still necessary to discuss the Jacobian of the transformation to the new variables which, when the $\tilde{\rho}$'s, α_i's and β_i's are rescaled by p_1^+, with $\tilde{\rho}_\alpha = p_1^+ \hat{\rho}_\alpha$, $\alpha_1 = p_1^+ \hat{\alpha}_i$ $\beta_i = p_1^+ \hat{\beta}_i$, is

$$(p_1^+)^{6g-2+2(n-2)} \prod_{r=2}^{N-1} |\underline{G}'(z_r, z_1)|^2 \qquad (5.35a)$$

From each external interaction vertex there is a factor $(p_1^+)^{-2}(p_r^+)$ by (5.34a), counting only the quadratic factors in $\tilde{\theta}(z)$ in each v_3^H (this is justified in the next section) and from each internal interaction vertex the factor $(p_1^+)^{-3}$. This leads , with (5.35a) to a net factor

$$\left[\prod_{r=1}^{N} p_r^+ \right] \prod_{r=2}^{N-1} |\underline{G}'(z_r, z_1)|^2 \qquad (5.35b)$$

On incorporating the first factor (Πp_r^+) of equation (5.13) together with the wave-function normalisation, the net non-lorentz-covariant overall factors involving p^+-components in the superstring Feynman rules (5.35b) cancel to within the residual factor

$$\prod_{r=1}^{N-1} |\underline{G}'(z_r, z_1)|^2 \qquad (5.36)$$

There is still the remaining involvement of the transverse components of the p_r's in the insertion factors (5.29a). These contributions become simplified in the short-string limit, when only their leading order terms, in the vanishing p_r^+'s, need be considered. Such contributions were analysed in some detail in the fourth paper of [32)] for a general multi-loop diagram, and a set of reduced superstring Feynman rules deduced for them. This method will be explained in detail for tree amplitudes in the next secion.

Section 5.4: Tree amplitudes

The amplitudes are obviously the first multi-amplitudes to calculate explicitly, and the results of such calculations using short string limit were made originally by Green and Schwartz.The amplitudes are not, however, the simplest to evaluate in this manner as will be seen from the easier 1 loop discussion of the next section. We will not give a complete description of the calculation here, but refer the reader to [38] for further details and earlier references; only the most crucial steps will be outlined here.

In the case of tree amplitudes there are only two external interaction vertices. For the case of four external strings N=4, the bosonic contribution to the external superfield Φ_r describing the L modes will be taken as $\zeta_r^{i_r} \cdot \theta_r^{2i_r}$, with polarisation vector ζ_r^μ with $\zeta_r^7 = \zeta_r^8 = 0$ $(1 \le r \le 4)$; the R modes may betaken similarly, and type II modes obtained by taking products of L and R modes. The full SO(1,9)-covariant amplitude vectors or tensors may be reconstructed from the SO(6)-covariant part directly. This explains why only quadratic factors in $\tilde{\theta}(\rho)$ were considered in association with (5.35b). Only the L modes will be considered in detail here. The S-matrix elements in this case using the superstring Feynman rules of section 5.2 and the short-string limit of section 5.3 are (to within the Veneziano-Virasoro-Shapiro factor)

$$M_{tree} = \int \prod_{r=1}^{3} d^4\theta_r \cdot \zeta_r^{i_r} \theta_r^{2i_r} \cdot \zeta_4^{i_4} \left[p_1^+\theta_1 + p_2^+\theta_2 + p_3^+\theta_3 \right]^{2i_4} \times$$

$$\left\langle V_3^H \left[\partial_{z_2} \underline{X} + (a_2 p_2^+)^{-1} \underline{p}_2 + \sum_{r=1,3} \underline{G}'(z_2, z_r)\underline{p}_r , \right.\right.$$

$$\left. - p_1^+ \, \underline{G}'(z_2, z_1) + p_3^+\theta_{31} \, \underline{G}'(z_2, z_3) \right] \times$$

$$\times \; V_3^H\Big[\, \partial_{z_3}\underline{X} + (a_3 p_3^+)^{-1}\underline{p}_3 + \sum_{r=1,2} \underline{G}'(z_3,z_r)\underline{p}_r ,$$

$$- p_1^+ \,\underline{G}'(z_3,z_1) + p_2^+\theta_{21}\,\underline{G}'(z_1,z_3)\Big] \Big\rangle \qquad (5.37)$$

where only the O(1) terms in p_2^+, p_3^+ are to be kept in (5.37), and the Veneziano-Virasoro-Shapiro factor has also to be included, as well as integration over the z_r's. The only non-zero contribution in (5.37) may be evaluated directly (useful formulae on combinations of powers of θ are given in Appendix X), and after some algebra and Berezhin integration over θ_1, θ_2, θ_3 one obtains

$$-8 \prod_{r=1}^{4} \zeta_r^{i_r}\, p_2^i\, p_3^j\, G'_{21}\, G'_{31}\, t^{i i_1 j i_3 i_4 i_2} - 64\, G''_{23}\, (\zeta_1 . \zeta_4)(\zeta_2 . \zeta_3)$$

$$+16\, G'_{23}\, G'_{32}\, (\zeta_1 \zeta_4)\, \zeta_2^{i_2}\, \zeta_3^{i_3}\, p_2^i\, p_3^j\, t^{i i_2 j i_3} + 16 G'_{31} \times$$

$$\times\, (\zeta_2 . p_3\, G'_{23} + p_1\, G'_{21})\, p_3^j\, \zeta_1^{i_1}\, \zeta_3^{i_3}\, \zeta_4^{i_4}\, t^{i_1 i_4 i_3 j}$$

$$-8\, G'_{32}\, G'_{21}\, p_2^i\, p_3^j \prod_{r=1}^{4} \zeta_r^{i_r} .\, t^{j i_2 i i_1 i_4 i_3} + 16\, G'_{21} \times$$

$$\times\, (\zeta_3 . p_2\, G'_{32} + p_1\, G'_{31})\, p_2^j\, \zeta_1^{i_1}\, \zeta_2^{i_2}\, \zeta_4^{i_4}\, t^{i_1 i_4 i_2 j}$$

$$-8\, G'_{23}\, G'_{31}\, p_3^i\, p_2^j \prod_{r=1}^{4} \zeta_r^{i_r}\, t^{j i_3 i i_1 i_4 i_2} - 64(\zeta_1\zeta_4) \times$$

$$\times\, (\zeta_2 \cdot p_1\, G'_{21} + p_3\, G'_{23})(\zeta_3 \cdot p_1\, G'_{31} + p_2\, G'_{32}) \qquad (5.38)$$

where

$$t^{i_1 i_2 i_3 i_4} = tr(\rho^{i_1}\, \rho^{i_2}\, \rho^{i_3}\, \rho^{i_4}),\ t^{i_1 \ldots i_6} = t^r(\rho^{i_1 \ldots\ 6})$$

The total expression (5.38) may be reduced to

$$M_{tree}(N=4) = 64\Big\{-G''_{23}(\zeta_1\zeta_4)(\zeta_2\zeta_3) + G'_{23}\, G'_{32}(\zeta_1\zeta_4)[p_2, \zeta_2, p_3, \zeta_3]$$

$$- (\zeta_1\zeta_4)(\zeta_2 \cdot p_1\, G'_{21} + p_3\, G'_{23})(\zeta_3 \cdot p_1\, G'_{31} + p_2\, G'_{32})$$

$$+ G'_{31}(\zeta_2 \cdot p_3\, G'_{23} + p_1\, G'_{21})\ [\zeta_1\zeta_4\zeta_3 p_3]$$

$$+ G'_{21}(\zeta_3 \cdot p_2\, G'_{32} + p_1\, G'_{31})\ [\zeta_1\zeta_4\zeta_2 p_2] - \frac{1}{2}\, G'_{21}\, G'_{31} \times$$

$$\times\,[\zeta_4\zeta_2 p_2\zeta_1 p_3\zeta_3] - \frac{1}{2}G'_{21}G'_{32}\,[\zeta_4\zeta_1 p_2\zeta_2 p_3\zeta_3]$$

$$-\frac{1}{2}G'_{31}G'_{23}\,[\zeta_4\zeta_1 p_3\zeta_3 p_2\zeta_2]\Big\} \tag{5.39}$$

in terms of

$$[a\ b\ c\ d] = a^i\ b^j\ c^k\ d^l\ t^{ijkl},$$
$$[a\ b\ c\ d\ e\ f] = a^i\ b^j\ c^k\ d^l\ e^m\ f^n\ t^{ijklmn}$$

The coefficients of $(\zeta_1\ \zeta_2)(\zeta_3\cdot\zeta_4)$, $(\zeta_1\zeta_3)(\zeta_2\zeta_4)$, $(\zeta_1\zeta_4)(\zeta_2\zeta_3)$, for example may be calculated from (5.39) to be respectively

$$Y_{12/34} = -16t(G'_{21}G'_{31} + G'_{21}G'_{32} - G'_{31}G'_{23}) \tag{5.40a}$$

$$Y_{13/24} = -16t(G'_{21}G'_{31} - G'_{21}G'_{32} + G'_{31}G'_{23}) \tag{5.40b}$$

$$Y_{14/23} = -16t(-G'_{21}G'_{31} + G'_{21}G'_{32} + G'_{31}G'_{23}) \tag{5.40c}$$

$$-64\,G'_{23} + 32t\,G'_{23}\,G'_{32}$$

where the Mandelstam variables are taken to be

$$s = -(p_1+p_2)^2\ ,\ t = -(p_2+p_3)^2,\ u = -(p_1+p_3)^2$$

There is only one independent source variable, say $z_3 = x$, and we may set $z_1 = 0$, $z_2 = 1$ ($z_4 = \infty$ has already been taken implicitly), so the VVS factor (with $G(z,z^1) = \ln |z-z^1|$) becomes

$$x^{-\frac{u}{2}} (1-x)^{-\frac{t}{2}}$$

Using partial integration,

$$\int_1^\infty dx\, x^{-\frac{u}{2}} (x-1)^{-\frac{t}{2}} x^{-a}(x-1^{-b} = \left(\frac{u}{2} \text{ or } -\frac{s}{2}\right) \times \frac{\Gamma\left(-\frac{s}{2}\right)\Gamma\left(-\frac{t}{2}\right)}{\Gamma\left(1-\frac{s}{2}-\frac{t}{2}\right)} \tag{5.41a}$$

where the first or second choice in the bracket is for (a,b) = (0,1) or (1,1) respectively. Also

$$\int_1^\infty dx\, x^{-\frac{u}{2}} (x-1)^{-2-\frac{t}{2}} = \frac{1}{4} su(1 + \tfrac{1}{2} t)^{-1} \frac{\Gamma\left(-\frac{s}{2}\right)\Gamma\left(-\frac{t}{2}\right)}{\Gamma\left(1-\frac{s}{2}-\frac{t}{2}\right)} m \tag{5.41b}$$

Then the various terms (5.40), on integration over $x \in [1,\infty]$, the s-t diagram [54], gives the contribution to $M_{tree}(N=4)$ of value

$$-16[ut(\zeta_1\zeta_2)(\zeta_3\zeta_4) + st(\zeta_1\zeta_3)(\zeta_2\zeta_4) + su(\zeta_1\zeta_4)(\zeta_2\zeta_3)] \times$$

$$\times \Gamma(-\frac{s}{2}) \Gamma(-\frac{t}{2})/\Gamma(1 - \frac{s}{2} - \frac{t}{2}) \qquad (5.42)$$

This agrees exactly with the value (4.23) for the tree amplitude in [63]. The other terms in (5.39) reduce similarly.

The above evaluation for the open superstring tree amplitude may be repeated for the closed by taking products of the above factors for the L and R modes (since there is complete factorisation at tree level, with $\overbrace{\partial_{z_1} X \partial_{\bar{z}_2} X} = 0$ from (5.17). Also external spinors may be included in the short-string limit construction (though with care on the spinor normalizations as $p_r^+ \sim 0$).

Section 5.5: One-loop amplitudes.

It is convenient to extend the short-string limit which was used in the previous sub-section to all of the N external strings. This is possible for one- or higher-loop amplitudes by use of a mapping first introduced in [64] for the bosonic string. In the one-loop case one uses a modification of the Mandelstam map (5.21) as

$$\hat{\rho} = F_1(z) = \sum_{r=1}^{N} \pi p_r^+ G(z,z_r) - \gamma \ln z \qquad (5.43)$$

where the parameter γ wil be discussed shortly. It is now possible to let all of the external string widths

$$p_r^+ \sim 0 \quad (1 \le r \le N) \qquad (5.44)$$

In this short string limit (5.44) all of the N interaction points $\tilde{z}_r$ behave as

$$\tilde{z}_r \sim z_r + a_r\, \alpha_r + O(\alpha_r^2) \tag{5.45a}$$

$$a_r = \gamma^{-1}\, z_r \tag{5.45b}$$

and the measure $d^2\tilde{\rho}_r = (|\gamma|^2/|z_r|^2)\, d^2 z_r$. The remaining loop length and twist variables are given by

$$\begin{bmatrix} \hat{\alpha} \\ \hat{\beta} \end{bmatrix} = \text{Im} \oint_{\begin{bmatrix} \gamma \\ \delta \end{bmatrix}} \left(\frac{1}{2\pi i}\right) F_1'(z)\, dz = \begin{bmatrix} \text{Im}\, \gamma \\ \text{Im}(\tau\gamma) \end{bmatrix} \tag{5.46}$$

As in the case with γ=0, the measure for $\hat{\alpha}$ and $\hat{\beta}$ is taken to be $d\hat{\alpha}\, d\,\hat{\beta}$. Therefore the new variables are (N-1) of $(z_1, \ldots, z_N)$ and τ. The conformally equivalent L.C. strip diagram for the mapping (5.43) is shown in figure 12(a), and the SSL of this in fig. 12(b), specialized to the case N=4. The corresponding region in the z-plane is the annulus of fig 13.

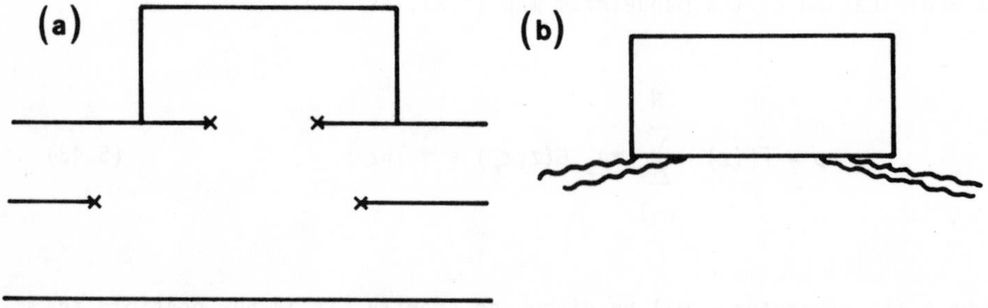

fig. 12 (a) The L.C.D. for the Arfaei mapping (5.43)
(b) The L.C.D. of (a) in the short-string limit (5.44)

The points z and $w^n z$ are identified, for any integer n, where $w = e^{2\pi i\tau}$. Moreover, $|w| = r_1/r_2$, the ratio of the internal and external radii of the annulus. The value of arg w gives the amount of angular twist to allow the identification of point a on the inner circle with b on the outer circle, as shown in fig 13.

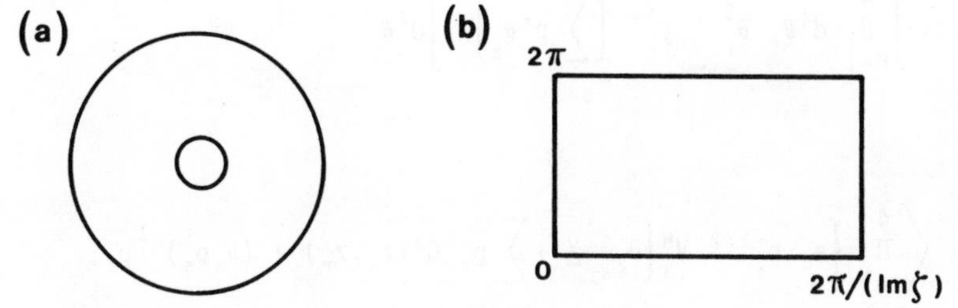

fig. 13 (a) The one-loop annulus in the z-plane
(b) The L.C.D. in the $\hat{\rho}$-plane arising from the Arfaei map of (a)

In this latter case, and following the method of the previous section,

$$F''(\tilde{z}_r) \sim -(a_r^2\, p_r^+)^{-1} \tag{5.47a}$$

$$\sum_{r=1}^{N} \underline{p}_r\, G'(\tilde{z}_s, z_r) = (a_s p_s^+)^{-1}\, \underline{p}_s + \sum_{r\neq s} \underline{p}_r\, G'(z_s, z_r) + O(1) \tag{5.47b}$$

$$\sum_{r=1}^{N} p_r^+ \theta_r G'(\tilde{z}_s, z_r) = a_s^{-1} \theta_s + \sum_{r\neq s} p_r^+ \theta_r G'(z_s, z_r) + O(p_r^+) \qquad (5.47c)$$

Keeping only the quadratic terms $\zeta^i \theta^{2i}$ in the external superfields as in the previous section, the S-matrix element for the L-going massless modes is

$$M_4 = \int \prod_{r=1}^{4} d^4\theta_r \, \theta_r^{2ir} \zeta_r^{ir} \left[\sum_{s=1}^{4} p_s^+ \theta_s\right]^4 \int d^4\theta$$

$$\left\langle \prod_{s=1}^{4} \left\{ a_s p_s^{+1/2} V_3^H \left[\partial_{z_s} \underline{X} + \sum_{r\neq s} \underline{p}_r G'(z_s, z_r) + (a_s p_s^+)^{-1} \underline{p}_s , \right.\right.\right.$$

$$\left.\left.\left. a_s p_s^{+1/2} \left[a_s^{-1} \theta_s + \sum_{r\neq s} p_r^+ \theta_r G'(z_s, z_r) + \theta z_s^{-1} \right]\right\}\right\rangle \qquad (5.48)$$

The only non-zero contribution, when $p_r^7 = p_s^8 = 0$, will be from the terms quadratic in the Grassman variables in the insertion factors, the net factor from them being $\prod_{s=1}^{4} p_s^+$. Combined with the factors $(F'')^{-\frac{1}{2}}$ at each interaction point, there is a net power of $\prod_{s=1}^{4} (p_s^+)^{3/2}$. This must be combined with a factor $(\Pi p_r^+)^{\frac{1}{2}}$ from the first term in equation (5.13) together with a factor $(\Pi p_r^+)^{-1}$ arising from the normalisation of external states. , so as to give a net vanishing factor (Πp_r^+).

Therefore the only non-zero term in (5.48) arises from the terms in the bosonic parts of V_3^H which are singular as $(p_r^+)^{-1}$ and from the non-vanishing terms in the Grassman part. The net result of this is, to within bosonic factors,

$$\int \prod_{r=1}^{4} d^4\theta_r \, \theta_r^{2i_r} \zeta_r^{i_r} \prod_{s=1}^{4} \theta_s^{\bar{E}_s} \epsilon_{\bar{E}_1 \bar{E}_2 \bar{E}_3 \bar{E}_4} \prod_{t=1}^{4} p_t^{j_t} (\theta_t \gamma + \theta)^{2j_s} \, d^4\theta \tag{5.49}$$

Evaluation of the Grassman integration leads to

$$\gamma^4 \prod_{r=1}^{4} \zeta_r^{i_r} p_r^{j_r} (\rho^{i_r} \rho^{j_r})_{\bar{D}_r}{}^{\bar{E}_r} \epsilon_{\bar{E}_1 \ldots \bar{E}_4} \epsilon^{\bar{D}_1 \ldots \bar{D}_4} \tag{5.50}$$

The symmetric terms in i_r, j_r in the product $(\rho^{i_r} \rho^{j_r})$ give zero since $(\underline{\zeta}_r \cdot \underline{p}_r) = 0$, whilst $\rho^{[i_r} \rho^{j_r]} = \gamma^{i_r j_r}$, and the expression (5.50) is seen to be equal to the scalar product of $\prod_{r=1}^{4} \zeta_r^{i_r} p_r^{j_r}$ with the tensor $t^{i_1 j_1 i_2 j_2 i_3 j_3 i_4 j_4}$. This latter tensor may be recognized as the zero-mode trace of $\prod_{r=1}^{4} (\bar{S}_o \gamma^{i_r j_r} S_o)$ as discussed in [63], since the trace on the S_o, $\bar{S}_o$ variables leads to the product of the two ϵ-factors in (5.50). Thus the expression (5.49) is identical to the kinematic factor K(1,2,3,4) of that reference. Including the remaining V-V-S factor and variables for integration and taking account of both L- and R-modes, leads to the expression (setting $z_4 = 1$)

$$K_{L+R}(1,2,3,4) \int \prod_{r=1}^{3} \frac{d^2 z_r}{|z_r|^2} \exp\Big[-\frac{1}{2} \sum_{r<s} p_r p_s \, G(z_r, z_s)\Big] \cdot \frac{d^2\tau}{(\mathrm{Im}\tau)^5} \tag{5.51}$$

and K_{L+R} is the kinematic factor for both L and R modes. This agrees exactly with the 1-loop closed superstring amplitude in [65]. There was still freedom in the choice of γ; the value

$$\gamma = \mu(\mathrm{Im}\ \tau)^{-1} \tag{5.52}$$

was used in the derivation of (5.51). It is important to obtain the independence of (5.43) on μ. With that choice, (5.51) may be seen to be identical to equation (6.28) of [65]. Similar expressions may be developed for other one-loop scattering amplitudes for other massless external states, both for N=4 and other values of N.

The reason for the choice (5.52) may be obtained by considering the form of mapping $z \to \rho$ which takes the annulus of figure 13(a) into the rectangle of (b) in the ρ-plane. The mapping is precisely that given by (5.52). It is to be expected that integration over the modulus γ in the SSL is best achieved by integrating over the conformally inequivalent rectangles (b), where not only the length $2\pi(\mathrm{Im}\tau)^{-1}$ but the twist $(\mathrm{Re}\tau/\mathrm{Im}\tau)$ is to be integrated over. In the next section we will give a general argument at higher loops, which reduces to (5.52) when g = 1. A further discussion is given in Appendix X.

It is to be noted that for higher N than 4 the 1 loop amplitudes avoid vanishing factors $(\zeta_r . p_r)$ by picking up terms of next to leading order in the various terms inside < > in (5.48). After some algebra this may be seen to produce a sum of terms of which the simplest has the extra kinematic factors

$$K(S,\ldots,N) = \Pi\left[\partial_{z_s}\underline{X} + \sum_{r\neq s} \underline{p}_r G'(z_s,z_r)\right]^{i_r} \zeta_r^{i_r} \times \tag{5.53}$$

$$\times \text{ similar R-moving factors}$$

which must be inserted in the integrand of (5.52) and summation taken over all possible ways of dividing the N external strings into 4 and (N-4) in order to arrive at a non-zero expression The details are left to the reader.

Section 5.6: Multi-loop amplitudes.

The short-string limit of the previous section may also be applied to obtain covariant expressions for massless state multi-loop amplitudes for superstrings. For a g-loop amplitude the mapping (5.43) may be extended to

$$\rho = F_g(z) = \sum_{r=1}^{N} \pi p_r^+ \; G(z,z_r) - \sum_{i=1}^{g} \gamma_i \; u_i(z) \tag{5.54}$$

In the SSL (5.44), the coalescence of N interaction points with speed of approach given by (5.45) still occurs, though now (5.45b) becomes

$$a_r = \left[\sum_{i=1}^{g} \gamma_i \; u_i'(z_r)\right]^{-1}$$

and the measure becomes $d^2\tilde{\rho}_r = |\Sigma\, \gamma_i \; u_i'(z_r)|^2 \; d^2 z_r$. Moreover (5.46) is replaced by

$$\begin{bmatrix} \underline{\alpha} \\ \underline{\beta} \end{bmatrix} = \mathrm{Im} \oint_{\begin{bmatrix} \underline{\gamma} \\ \underline{\delta} \end{bmatrix}} F_1'(z)\, dz = \begin{bmatrix} \mathrm{Im}\, \underline{\gamma} \\ \mathrm{Im}(\Pi \underline{\gamma}) \end{bmatrix} \tag{5.55}$$

The new variables are (N-1) of $(z_1 \ldots z_N)$ (with $z_N = 1$, say), the g variables $\underline{\gamma}$ and the (2g-z) zeros $\tilde{z}_1, \ldots, \tilde{z}_{2g}$, of the 1st Abelian differential

$$\phi_g'(z) = \sum_{i=1}^{g} \gamma_i\, u_i'(z) = 0 \tag{5.56}$$

with the Jacobian

$$\frac{\partial(\underline{\alpha}, \underline{\beta})}{\partial(\mathrm{Re}\underline{\gamma}, \mathrm{Im}\underline{\gamma})} = \det \mathrm{Im}\Pi \tag{5.57}$$

The multi-loop extension of (5.48) involves inclusion of further factors inside the expectation value < > equal, in the SSL, to

$$\prod_{i=1}^{2g-2} \phi_g''(\tilde{z}_i)^{-1/2}\, V\Big[\partial_{\tilde{z}_i} \underline{X} + \Sigma \underline{p}_r\, G'(\tilde{z}_i, z_r),$$

$$\phi_g''(\tilde{z}_i)^{-1/2} \Big[\sum_{j=1}^{2g-2} \theta_j\, u_j'(\tilde{z}_i)\Big]\Big] \tag{5.58}$$

It is to be noted that the integration over the internal θ_j's can be done explicitly, as discussed in detail in the fourth paper in [32]. In particular a definition was given of the tensor $T_{\underline{j}}(\underline{\tilde{z}})$ as

$$T_{\underline{j}}(\tilde{\underline{z}}) = \int \prod_{i=1}^{g-1} d^4\theta_i \prod_{k=1}^{2g-2} \left[\Sigma\, \theta_i\, u_i'(\tilde{z}_k)\right]^{2j_k} \qquad (5.59)$$

where $\underline{j} = (j_1,\ldots,j_{2j-2})$, $\underline{\tilde{z}} = (\tilde{z}_1,\ldots,\tilde{z}_{2g-2})$. Then the Grassman integral of (5.58) can be written as

$$\prod_{i=1}^{2g-2} \left[\phi_g''(\tilde{z}_i)\right]^{-3/2} \left(\partial_{\tilde{z}_i} \underline{X} + \sum \underline{p}_r\, G'(\tilde{z}_i, z_r)\right)^{j_i} . \; T_j(\tilde{z}) \qquad (5.60)$$

The expectation value of (5.60) must still be taken on the $\partial_{\tilde{z}} \underline{X}$'s, using (5.17). However, it is necessary to extend the definition of $t_{\underline{j}}$ in order to take account of the presence of $\theta_i\, u_i'(z_t)$ in the last factor in (5.49), replacing (θ/z_S) there. This extension will be done shortly.

It is next necessary to choose the variables $\underline{\gamma}$ in (5.54) in a suitable manner. One way to achieve that is to notice that the moduli α_i, β_i are close to those first introduced by Ahlfors as an analytic co-ordinatisation of moduli space [66] (see also [67]), in terms of a first Abelian differential θ with 2g-2 zeros $z_1,\ldots,z_{2g-2}$ and moduli

$$\tau_i = \int_{\delta_i} \theta \qquad (i = 1,\ldots.,g) \qquad (5.61a)$$

$$\tau_{i+g} = \int_{z_1}^{z_i} \theta \qquad (i = 2,\ldots,2g-2) \qquad (5.61b)$$

where the quantities $\int_{\gamma_i} \theta$ as taken as fixed. The string uses the moduli (5.61b) but modifies (5.61a) to

$$\mathrm{Re}\, \tau'_i = \mathrm{Im} \int_{\gamma_i} \theta \tag{5.61c}$$

$$\mathrm{Im}\, \tau'_i = \mathrm{Im} \int_{\delta_i} \theta \tag{5.61d}$$

with the 2g real quantities

$$\mathrm{Re} \int_{\gamma_i, \delta_i} \theta \tag{5.61e}$$

being fixed for conformally inequivalent surfaces.

One way of keeping the quantities (5.61e) independent of the conformal inequivalence class is to fix them to be numerical constants. A simple choice is

$$\gamma_i = i \sum_j (\mathrm{Im}\Pi)^{-1}_{ij} \tag{5.62}$$

(which reduces for g = 1 to (5.52)). For then

$$\int_{\gamma_i} d\phi_g = i \sum_j (\mathrm{Im}\pi)^{-1}_{ij} \tag{5.62a}$$

$$\int_{\delta_i} d\phi_g = i \sum_{j,k} (\mathrm{Im}\Pi)^{-1}_{ij} \Pi_{jk} \tag{5.62b}$$

so that the quantities (5.61c) are 0 and 1 for γ_i, δ_j respectively. They are thus fixed, independent of the moduli. The quantities (5.61c), (5.61d) are thus expected to be good Teichmüller co-ordinates in the surface, along with the interaction points $\tilde{\rho}$, the zeros of $d\phi_g$, following a suitable adaptation of the Ahlfors co-ordinates (5.61a,b) to the non-analytic case along the lines of that discussed, for example, in prescription II′ in [67]. It is to be noted that the interaction points ρ will not depend on the moduli of the surfaces being integrated over. This arises from the net change of variables occurring in (5.54); the modular invariance of the total amplitude cannot, of course, be destroyed, in the map (5.54).

The resulting expression for the g-loop closed superstring amplitude may now be written down, for the scattering of N=4 massless modes, as

$$\sum_j K^j(1,2,3,4) \int \prod_{r=1}^{4} d^2 z_r \; |\phi_g'(z_r)|^{-2} \; \exp[-\tfrac{1}{2} \sum_{r<s} p_r p_s \; G(z_r, z_s)] \times$$

$$\prod_{r=1}^{2g-3} d^2\tilde{z}_r \; |\partial\phi_g(\tilde{z}_r)| \; \prod_{i=1}^{g} d^2\hat{\alpha}_i \times$$

$$\times \left\langle \prod_{r=1}^{2g-2} \left[\partial_{\tilde{z}_r} \underline{X} + \sum_s \underline{p}_s \, G'(\tilde{z}_r, z_s)\right]^{j_r} \times\right.$$

$$\times \left.\left[\bar{\partial}_{\tilde{z}_r} \underline{X} + \sum_s \underline{p}_s \, \bar{G}'(\tilde{z}_r, z_s)\right]^{j_r}\right\rangle \times \Pi|\phi_g''(\tilde{z}_i)|^{-3} \tag{5.63}$$

where $\partial\phi_g(\tilde{z}_r)$ denotes the total derivative of ϕ_g w.r.t. the variable $\tilde{z}_r$. The kinematic factor $K^{\underline{j}}(1,2,3,4)$ is still a product of L and R factors, the R-factors being defined as

$$\left[\prod_{r=1}^{4} p_r^{i_r} \zeta_r^{j_r} (\gamma^{i_r j_r})^{\bar{E}_r}{}_{\bar{D}_r}\right] \epsilon_{\bar{E}_1\bar{E}_2\bar{E}_3\bar{E}_\epsilon} T^{\underline{j}\bar{D}_1\bar{D}_2\bar{D}_2\bar{D}_2} \tag{5.63a}$$

where the tensor $T^{\underline{j}\bar{D}_1,\ldots,D_4}$ ($\underline{j}$ denoting the internal vector indices j_i in (5.60)) is defined by integrating out the zero Grassmann modes as

$$T^{\underline{j}\bar{D}_1\ldots\bar{D}_4} = \int \prod_{k=1}^{g} d^4\theta_k \prod_{r=1}^{2g-2} \left[\sum_k \theta_k u_k'(\tilde{z})\right]^{2j_r} . \prod_{s=1}^{4} \left[\sum_i \theta_i u_i'\right]^{\bar{D}_s} \tag{5.63b}$$

and generalizes that for $g = 1$, when $T^{\underline{J}\bar{D}_1\ldots\bar{D}_4} = \epsilon^{\bar{D}_1\ldots\bar{D}_4} . (\Pi z_r)^{-1}$. It also generalizes the tensor $T_{\underline{i}}$ of (5.59), as was promised.

The generalization of the above expressions (5.63) is clearly possible for higher N. It leads to a sum of terms similar in form to (5.53) although now including further factors $(\partial_{\tilde{z}_j}\underline{X}+\Sigma \underline{p}_r G'(z_j,z_r))^{\tilde{j}}$ at the (2g-2) internal insertion factors. The simplest term extending (5.45) to g > 1 contains simply a product of these further factors, the $\tilde{j}$'s to be saturated when multiplied by the tensor (5.63).

Finally it is possible to see that there is zero contribution, at all loop orders, for N=2 or 3 in the SSL. For example for N=2 the non-cancelling factors of p_r^+ occur, at every loop order (for $g \geq 0$) as

$$\left[\sum_{r=1}^{2} p_r^+ \theta_r\right]^4 / \prod_{r=1}^{2} p_r^+ \tag{5.64}$$

which vanishes in the SSL. A similar vanishing factor arises if the products and summation in (5.64) are for $1 \leq r \leq 3$. Only for N=4 does (5.64) give a finite value. This proves the oft-quoted and very important non-renormalisation theorems, and resolves the problems raised by Weinberg [68].

This approach may be related to that already used at 1-loop [65], based on operator methods. The latter utilises that for N=2,3 there are not enough zero mode factors θ_i to give a non-zero amplitude. That fact can be related to the somewhat different feature of (5.64) by noting that the contributions (5.64) in the SSL (5.44) has a satisfactory number of zero modes θ_i but is not lorentz invariant. Any lorentz-invariant contribution would be expected to require too few zero modes θ_i, as described in [65].

Section 5.7: Heterotic Amplitudes.

A heterotic string field theory was constructed in Chapter 4. This will now be used to obtain first quantised expressions for multi-loop amplitudes similar to those obtained so far in this section for types I and II superstrings. Since reduction from second to first quantised form of the R-moving 10-dimensional uncompactified superstring components have already been discussed earlier in this section let us consider initially the contribution of the compactified L-moving bosonic components to a multi-loop amplitude, as in [31)].

There will only be a difference between uncompactified and compactified bosonic strings at a global level on the L.C. strip diagram W_E (for a given perturbation diagram) since locally there is no difference between the two types of strings. Thus the decomposition to first quantised form will proceed for the compactified string as for the uncompactified one. It will be necessary to include the solitonic modes of (5.9) with care, however, since they describe the details of the compactification.

The first and second quantised approach to bosonic strings on tori was developed in Chapter 4. The lattice Γ may be specialized here to the root lattice of $E_8 \times E_8$ or Spin $32/Z_2$, though such detail will not be used explicitly except for the even self-dual character of Γ. Recall from sections 4.5 and 4.6 that the contribution of D=16 left-moving chiral bosons on a torus to a light-cone scattering amplitude may be formulated in terms of D open string coordinates ϕ_L^I, = 1,...,D. Moreover, this contribution is of the form $f_L\ (\tilde{\rho},\underline{\alpha},\underline{\beta})$ which may be obtained from a corresponding open string term $f_L(\tilde{\underline{\tau}},\underline{\alpha})$ by (anti) analytic continuation in the interaction points $\tilde{\underline{p}}$ and inclusion of the loop twists $\underline{\beta}$. Proceeding as in the uncompactified case, section 5.2, we can write $f_L(\tilde{\underline{\tau}},\underline{\alpha})$ in the first-quantised functional form

$$f_L(\tilde{\underline{\tau}},\underline{\alpha}) = \int D\underline{\phi}_L \ \exp\left\{\iint_{\Sigma^+_\rho} \left[\frac{1}{\pi} \ d\underline{\phi}_L \wedge {}^*d\underline{\phi}_L \ + \underline{J}_L \cdot \underline{\phi}_L \right]\right\} \tag{5.65a}$$

where Σ^+_ρ is the open string L.C. diagram corresponding to the half-region of the closed string L.C. diagram defined by $0 \leq \tau \leq \pi$, and $\underline{J}_L$ is a source term analogous to (5.14). Using the fact that $\underline{\phi}_L(-\pi) = \underline{\phi}_L(\pi)$ we have

$$\iint_{\Sigma^+_\rho} d\phi_L \ \wedge {}^*d\phi_L \ - \frac{1}{2} \iint_{\Sigma^R_\rho} d\phi_L \wedge {}^*d\phi_L$$

where Σ^R_ρ is a closed L.C. diagram formed by identifying the boundaries of Σ^+_ρ and its mirror Σ^-_ρ without introducing any twists (see section 4.5). In other words, Σ^R_ρ is the double of the open string diagram Σ^+_ρ. Following equation (5.9) we may globally express $d\phi_L$ on Σ^R_ρ as

$$d\phi^I_L = d\phi^I_{ex} + \left(\sum_{\ell=1}^{g} z^I_\ell w_\ell + \text{h.c.}\right) \tag{5.66}$$

with the periodicity conditions

$$\int_{\gamma_\ell} d\phi_L = 0 \tag{5.67a}$$

$$\int_{\delta_\ell} d\phi^I_L = 2\pi \sum_i n^{(\ell)}_i e^I_i \tag{5.67b}$$

giving

$$z^I_\ell = - i\pi\, n^{(m)}_i \,(\mathrm{Im}\Pi)^{-1}_{m\ell}\, e^I_i = -\bar{z}^I_\ell \tag{5.67c}$$

The change in ϕ_L around each γ-cycle is zero as it corresponds to going from one end of an open string to the other and back again. Equation (5.67b) follows from the quantization condition (4.59).

If we use the expression (5.10) together with equation (5.12) and include an external bosonic source with momenta $\tilde{p}_r$ on the rth string (as in (5.13), the resulting amplitude contribution from the compactified L-movers is

$$f_L(\tilde{\underline{\tau}},\underline{\alpha}) = \sum_{\underline{n}} (\det{}'\Delta_p)^{-\frac{D}{4}} \exp[-\frac{1}{2} \sum_{r,s} \tilde{p}_r\tilde{p}_s\, G(z_r,z_s)] \tag{5.68}$$

$$\exp\left[-\pi\underline{n}^T_i(\mathrm{Im}\Pi)^{-1}\,\underline{n}_j\,\Gamma_{ij} + 2i\pi\,\tilde{p}_{ir}\,\underline{n}^T_i\,(\mathrm{Im}\pi)^{-1}\,\mathrm{Im}\underline{u}(z_r)\right]$$

where $\Gamma_{ij} = \Sigma\, e^I_i e^I_j$ is the Cartan matrix of the root lattice $\tilde{p}^I_r = \Sigma\,\tilde{p}_{ir} e^{*I}_i$, and for simplicity any object with an index $\iota(1 \le \iota \le g)$ has been written as a vector. (We are using the Einstein summation convention for the indices i,j).

The Greens function G may be expressed as $G = R_e\, \underline{G}$ and $\underline{G}$ decomposed as

$$\underline{G}(z,z') = \underline{H}(z,z') + 2\pi i\, \underline{u}^T(z)(\mathrm{Im}\pi)^{-1}\,\mathrm{Im}\underline{u}(z') \tag{5.69a}$$

where $\underline{H} = H_{z'a}(z)$ is the third Abelian integral with complex normalization

$$\int_{\underline{\gamma}} dH_{z'a}(z) = 0, \quad \int_{\underline{\delta}} dH_{z'a}(z) = 0 = 2\pi i[\underline{u}(a) - \underline{u}(z')] \tag{5.69b}$$

We may replace G by the real part of (5.69a), and use a Poisson re-summation formula to obtain

$$f_L(\tilde{\underline{\tau}},\underline{\alpha}) = \sum_n \exp[\, -\pi \underline{n}_i^T (\mathrm{Im}\pi)\, \underline{n}_j\, \Gamma_{ij}^{-1} - 2\pi \sum_r \tilde{p}_{j,r}\, \Gamma_{ij}^{-1}\, \underline{n}_i^T\, \mathrm{Im}\underline{u}(z_r)] \;\; x$$

$$x\; (\det \mathrm{Im}\Pi)^{D/2}\; (\det{}'\Delta_p)^{-D/2}\; \exp[-\tfrac{1}{2} \sum_{r,s} \tilde{p}_r \tilde{p}_s\; \mathrm{Re}\; \underline{H}(z_r,z_s)] \tag{5.70}$$

It is important to note that (5.70) is defined on the double Σ_p^R of the open string L.C. diagram Σ_ρ^+. In paricular, Σ_ρ^R is itself parametrised by 2g-3+N interaction times $\tilde{\underline{\tau}}$ and g string widths $\underline{\alpha}$. To obtain the full L.C. amplitude contribution $f_{L}(\tilde{\underline{\tilde{\rho}}},\underline{\alpha},\underline{\beta})$, it is necessary to incorporate the imaginary part of the interaction points and the loop twists $\underline{\beta}$, so recovering the full parametrisation of the corresponding Riemann surface description of the amplitude. We shall show how this can be achieved by mapping the L.C. diagram Σ_ρ to a corresponding Riemann surface represented by a region Σ_z in the z-plane and performing an analytic continuation from real to complex moduli of Σ_z

Recall that a g-loop L.C. diagram Σ_ρ has g pairs of canonical homology cycles $\{\tilde{\gamma}_p,\tilde{\delta}_p\}$ where $\tilde{\gamma}_p$ is taken to be one of the strings in the pth loop and $\tilde{\delta}_p$ to be the cycle surrounding the same loop, (see fig 10). Now break the L.C. diagram along the $\tilde{\gamma}$-cycles and map it conformally onto the region of the z-plane, Σ_z, exterior to g pairs of circles. The cycles $\{\tilde{\gamma}_p,\tilde{\delta}_p\}$ are mapped to $\{\gamma_p,\delta_p\}$ such that the γ-cycles surround the circles on the δ-cycles join pairs of circles. This is illustrated in fig 14 for two loops.

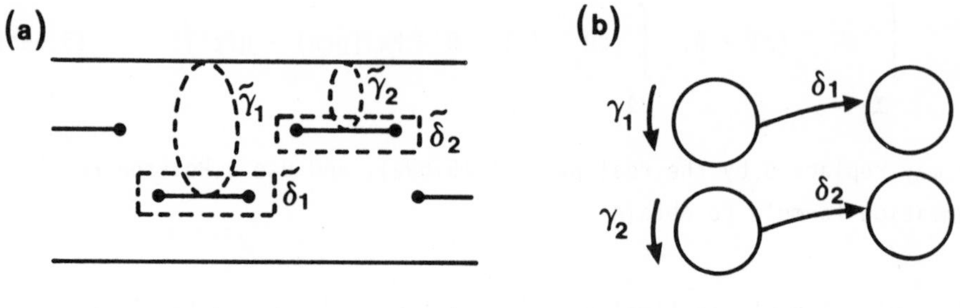

Fig 14: The Schottky uniformisation of a two-loop closed string diagram. The L.C. diagram fig 14a is conformally equivalent to the region in the z-plane exterior to two pairs of circles, fig 14b. The positions of the canonical homology cycles have been indicated

With each pair of circles we associate a complex projective transformation T_p that maps the circumferences onto one another. Points which transform into one another under T_p are identified. T_p has the form,

$$\{T_p(z) - z_{1p}\}/\{T_p(z) - z_{2p}\} = \omega_p(z - z_{1p})/(z - z_{2p})$$

Here z_{1p} z_{2p} are the fixed points of the tranformation T_p and ω_p is its multiplier with $|\omega_p| < 1$. All three parameters are complex. The tranformation T_p and their reciprocals generate an infinite, discrete group of projective transformations denoted by Λ. The group Λ is a Schottky group. Note that each of the fixed points z_{1p} and z_{2p} lies within one of the circles associated with T_p and that T_p maps points outside the circle containing z_{1p} into the circle containing z_{2p}. Hence the region exterior to the 2g circles, Σ_z, is a fundamental region of Λ. A g-loop Riemann surface is recovered if the boundary circles within each pair are identified. The fact that any Riemann surface can be so described is a consequence of a particular version of the uniformisation theorem of Poincare, Klein and Koebe [69] which states that the only simply connected Riemann surfaces are the sphere $\hat{C}$, the complex C, and the upper-half plane U. Any other compact Riemann surface is given by C/Λ where Λ is a Schottky group (see appendix III).

The generators T_p may be subjected to an overall projective transformation A i.e. $T_p' = A^{-1} T_p A$. As each projective tranformation has three complex parameters, a Riemann surface with g handles is thus characterised by 3g-3 complex moduli. There will also be N punctures corresponding to the N external strings giving a total of 3g-3+N moduli as expected. These moduli are identified as z_{1p}, z_{2p}, ω_p $(1 \leq p \leq g-1)$ and the sources z_r $(1 \leq r \leq N)$. For more information about projective transformations see [70].

The conformal transformation described above is achieved by the Mandelstam map of equation (5.21). If equation (5.21) is restricted to real moduli then we obtain a map from Σ_ρ^R to Σ_z^R or Σ_ρ^+ to Σ_z^+. The open Riemann surface Σ_z^+ consists of the region in the upper-half plane exterior to 2g circles whose diameters lie on the real axis. The double Σ_z^R is constructed by identifying Σ_z^+ with its mirror image Σ_z^- in the lower-half plane along the boundary diameters. Just as in the one-loop case, introducing twists into Σ_ρ^R to form Σ_ρ corresponds to complexifying the moduli of Σ_z^R so that $\Sigma_z^R = \Sigma_z \mid$ real moduli. In the case of one-loop, Σ_ρ maps to an annulus Σ_Z with inner and outer circles identified with a twist, as in section 5.5. the complex moduli Σ_z are $\{z_r, w\}$ where $|w|$ is the ratio of circle radii and ArgW is the twist. Setting the twist to zero gives Σ_z^R. (see fig 13).

Associated with the 2g homology cycles are g linearly independent first Abelian integrals u_p satisfying (5.11). Explicit expressions for u_p and the period matrix Π have been obtained by Mandelstam. Let V_i be a general element of the Schottky goup Λ. Then

$$u_p(z) = \frac{1}{2\pi i}\sum_i^{(p)} \ln \frac{z - V_i z_{1p}}{z - V_i z_{2p}} \tag{5.71a}$$

The summation $\overset{(p)}{\Sigma}$ is restricted to those elements of V_i which cannot be expanded as a product of T's with $T_p^{\pm 1}$ as the rightmost factor. Similarly,

$$\Pi_{pq} = \frac{1}{2\pi i}\sum_i \ln \frac{(z_{1q} - V_i z_{1p})(z_{2q} - V_i z_{2p})}{(z_{1q} - V_i z_{2p})(z_{2q} - V_i z_{1p})} \tag{5.71b}$$

where products of T's with $T_p^{\pm 1}$ rightmost or $T_q^{\pm 1}$ leftmost are now excluded. When $p = q$, $V_i = 1$ is also excluded,

$$\Pi_{pp} = \frac{1}{2\pi i}\left[\sum_{i\neq 1}^{(p,q)} \ln \frac{(z_{1p} - V_i z_{1p})\,(z_{2p} - V_i z_{2p})}{(z_{1p} - V_i z_{2p})\,(z_{2p} - V_i z_{1p})} + \ln\omega\right] \tag{5.71c}$$

Similarly the third Abelian integral H has the form [71]

$$H(z,z') = \sum_i \ln\left[\frac{(z - V_i z')(z_o - V_i z_N)}{(z - V_i z_N)(z_o - V_i z')}\right] \tag{5.71d}$$

where z_o and z_N are fixed points. Equation (5.71) implies that u_p, Π_{pq} and H are holomorphic functions of the moduli on Σ_z. Moreover, if the moduli are restricted to be real then for any real points x,y,

$$\mathrm{Im}\,\Pi = -i\Pi,\ \mathrm{Im}\,u(x) = -iu(x),\ \mathrm{Re}H(x,y) = H(x,y) \tag{5.72}$$

In the case of one-loop, equations (5.71) simplify considerably with

$$u(z) = \frac{1}{2\pi i}\ln z,\ \Pi = \frac{1}{2\pi i}\ln w \tag{5.73a}$$

$$H(z,z') = \ln(1-z'/z) - \frac{1}{2}\ln z'/z + \sum_{n=1}^{\infty}\left[\ln(1-w^n\frac{z'}{z}) + \ln(1-w^n\frac{z}{z'}) + 2\ln(1-w^n)\right] \tag{5.73b}$$

Note that the definitions of H, Π_{pq} and u_p in (5.71) are particular examples of a Poincaré series [72] and care must be taken over their convergence properties. We shall assume that convergency can be obtained by appropriate analytic continuation. Alternatively, the amplitudes could be re-expressed in terms of a different set of moduli for which these problems do not arise [66].

The determinant of the Laplacian on Σ_z takes the general form

$$\det{}'\Delta_z = \det \operatorname{Im} \Pi \, |Q(\underline{w},\underline{z}_1,\underline{z}_2)|^2 \tag{5.74}$$

where Q is a holomorphic functions of the moduli. In the case of one-loop Q takes the familiar form

$$Q(w) = \prod_{m=1}^{\infty} (1-w^m)^2 \; w^{1/12} = \eta^2(w) \tag{5.75}$$

where η is the Dedekind function. Finally, the simple relationship between Σ_ρ^+ (Σ_z^+) and the double Σ_ρ^R (Σ_z^R) implies that [73], on Σ_ρ^R

$$\det{}' \Delta_\rho = [|A|^2 \det \operatorname{Im} \Pi \det{}'\Delta_z]^{\frac{1}{2}}|_{\text{real moduli}} \tag{5.76}$$

real moduli. Here A is the closed string conformal anomaly [72] (see also Appendix XVI)

$$A = \left(\prod_p F''\right)^{1/24} \left(\prod_{r=1}^{N} \pi P_r^+\right)^{1/12} \left(\prod_{r=1}^{N} \epsilon_r\right)^{1/12} \tag{5.77}$$

which results from the fact that the metric on Σ_p, given by

$$d\rho \, d\bar{\rho} = |F'|^2 \, dz \, d\bar{z}, \; F' = \frac{\partial F}{\partial z} \, ,$$

has isolated singularities at the sources z_r and zeros at the interaction points P. The factor $(\prod_r \epsilon_r)^{1/12}$, where ϵ_r is the radius of a small disc surrounding the source point z_r, is cancelled by the singular part of $\exp(-\frac{1}{2}\sum_r \tilde{p}_r^2 G(z_r,z_r))$ in equation (5.68) and will be dropped in the following.

Combining equations (5.71)-(5.77) with equation (5.70) and using the fact that Γ is self-dual gives

$$f_L(\tilde{\underline{\tau}},\underline{\alpha}) = \left[\left(\prod_p |F''|\right)^{-D/48} g(\underline{w},\underline{z}_1,\underline{z}_2,z_r)\right]\Bigg|_{\text{real moduli}}$$

$$\times\left(\prod_{r=1}^{N} \pi p_r^+\right)^{-D/24} \tag{5.78}$$

where

$$g(\underline{w},\underline{z}_1,\underline{z}_2,z_r) = Q(\underline{w},\underline{z}_1,\underline{z}_2)^{-D/2}$$

$$\times \exp\left[-\frac{1}{2}\sum_{r,s} H(z_r,z_s)\tilde{P}_r\tilde{P}_s\right]$$

$$\times\, \theta(\underline{V}|\Gamma\otimes\Pi) \tag{5.79a}$$

and $\theta(\underline{V}|\Gamma\otimes\Pi)$ is a generalised theta function [74] associated with the quadratic form Γ,

$$\theta(\underline{V}|\Gamma\otimes\Pi) = \sum_{\underline{n}} \exp(i\pi \sum_{i,j} \underline{n}_i^t\, \Pi \underline{n}_j\, \Gamma_{ij} + 2\pi i \sum_i \underline{n}_i^t\, \underline{V}_i) \tag{5.79b}$$

$$\underline{V}_i = \sum_r \tilde{P}_{ir}\, \underline{u}(z_r) \tag{5.79c}$$

It then follows that the full amplitude contribution $f_L(\tilde{\underline{\rho}},\underline{\alpha},\underline{\beta})$ is obtained by (anti)-analytic continuation from real to complex moduli on the right-hand side of (5.78). A proof of this is sketched below.

Recall the expression for $\underline{\alpha}$ and $\underline{\beta}$ given by equation (4.85). Substituting from ρ according to the Mandelstam map (5.21) and using equation (5.69) leads to

$$\alpha_1 = 2\pi \sum_{r,m} (\pi p_r^+)(\mathrm{Im}\Pi)^{-1}_{1m} \; \mathrm{Im}\; u_m(z_r) \tag{5.80a}$$

$$\beta_1 = 2\pi \sum_r (\pi p_r^+)[-\mathrm{Re}\; u_1(z_r) + \sum_{m,n} (\mathrm{Re}\Pi)_{1m}(\mathrm{Im}\Pi)^{-1}_{mn} \; \mathrm{Im}\; u_n(z_r)] \tag{5.80b}$$

which may be rewritten as

$$i\beta_1 = -\; 2\pi i \sum_r (\pi p_r^+)\; u_1(z_r) + i \sum_m \Pi_{1m}\; \alpha_m \tag{5.81}$$

Furthermore, equations (5.69) and (5.80) imply that for given loopwidths $\underline{\alpha}$ the Mandelstam map has the form

$$\rho(z) = F_{\underline{\alpha}}(z) = \sum_{r=1}^{N} (\pi p_r^+) H(z,z_r) - i \sum_{1=1}^{g} (\mathrm{Re}\alpha_1) u_1(z_r) \tag{5.82}$$

Assume for the moment that $\underline{\alpha}$ is held fixed and $\underline{\beta} = 0$. Equation (5.81) then implies that the analytic moduli $\{\underline{w}, \underline{Z}_1, \underline{Z}_2, z_r)$ are constrained to lie on some analytic subspace M_o of moduli space. Moreover, equation (5.82) shows that $F_{\underline{\alpha}}$ is an analytic function of z and the moduli of M_o. Since the interaction points $\tilde{\underline{\rho}}$ are the solution to equation (5.23) we see that $\tilde{\underline{\rho}}$ is an analytic function of the moduli on M_o. Thus analytic continuation from $\tilde{\underline{\tau}}$ to $\tilde{\underline{\rho}}$ (on introduction of twists $\mathrm{Im}\underline{\rho}$) implies analytic continuation from real to complex moduli on M_o. Hence, for fixed $\underline{\alpha}$

$$\bar{\tilde{f}}_L(\underline{\rho},\underline{\alpha},\underline{\beta}=0) = \left[\left[\prod_p \bar{F}''_{\underline{\alpha}}\right]^{-D/48} \overline{g(\underline{w},\underline{z}_1,\underline{z}_2,z_r)}\right]\Bigg|_{M_o}$$

$$\times\left(\prod_{r=1}^{N} \pi p_r^+\right)^{-D/24} \tag{5.83}$$

Finally, if $\underline{\alpha}$ and $\underline{\beta}$ are now considered as free variables, with $\underline{\beta}$ non-zero, we have

$$\bar{\tilde{f}}_L(\underline{\rho},\underline{\alpha},\underline{\beta}) = \prod_{r=1}^{N} \left(\pi p_r^+\right)^{-D/24}\left(\prod_p \bar{F}''\right)^{-D/48} \times$$

$$\times\ \overline{g(\underline{w},\underline{z}_1,\underline{z}_2,z_r)} \tag{5.84}$$

where the moduli are now unconstrained and F is given by (5.21) once $\underline{\alpha}$ has been substituted for using (5.80a). To obtain the heterotic multi-loop amplitude it is necessary to combine f_L, for D = 16, with the corresponding contribution from the 10-dimensional right-moving superstring which is given by

$$f_s = (\det\partial_{\bar{\rho}}/\det'\Delta_\rho)^4 \exp\left[-\frac{1}{2}\sum_{r<s} P_r.P_s\ G(z_r,z_s)\right]$$

$$<\prod_p (2F''(\tilde{z}))^{-\frac{1}{2}}\ V_3{}^H[\partial_{\tilde{z}}\ \underline{X} + \frac{1}{2}\sum_r \underline{p}_r\ \tilde{G}'(z,z_r),$$

$$(2F''(\tilde{z}))^{-\frac{1}{2}}\left[\pi\sum_{r=1}^{N} \underline{\tilde{G}}'(z,z_r)p_r^+\ \theta_r + \sum_{i=1}^{g}\theta_i u_i'(\tilde{z})\right]]> \tag{5.85}$$

The product in (5.85) is over all interaction points P and the expectation value <...> is that already used in association with (5.17). In appendix XVI it is shown that the determinant factor defined on Σ_p satisfies

$$\left[\frac{\det \partial_{\bar\rho}}{\det'\Delta_\rho}\right]^4 = \left[\overline{AQ(\underline{w},\underline{z}_1,\underline{z}_2)}(\det \text{Im}\,\Pi)\right]^{-4} \qquad (5.86)$$

where A, is the conformal anomaly (5.72) and Q is the holomorphic factor in (5.74). Finally, combining equations (5.77), (5.79), (5.84), (5.85) and (5.86) the total expression for the heterotic multi-loop amplitude is thus

$$M_H = \left(\prod_{r=1}^{N} \pi p_r^+\right)^{-\frac{1}{2}} \int \prod_P d^2\tilde\rho \prod_{i=1}^{g} d\alpha_i \; d\beta_i \prod_{i=1}^{g} d^4\theta_i \left(\det \text{Im}\Pi\right)^{-4}$$

$$\int \prod_{r=1}^{N} d^4\theta_r \; \delta^4(\Sigma\, p_r^+\theta_r) \prod_{r=1}^{N} \Phi^{(r)}(\theta_r)$$

$$\prod_P (2F'')^{-\frac{1}{2}} \langle V_3{}^H\rangle \exp[-\tfrac{1}{2} \sum_{r<s} p_r\cdot p_s \; G(z_r,z_s)$$

$$\left[\prod_P \bar F'' \prod_{r=1}^{N} (\pi p_r^+)^2\right]^{-\frac{1}{2}} \overline{Q(\underline{w},\underline{z}_1,\underline{z}_2)}^{-12}$$

$$\overline{\theta(\underline{V}|\Gamma\otimes\Pi)} \exp\left[-\tfrac{1}{2} \sum_{r,s} \bar H(z_r,z_s)\tilde p_r\cdot\tilde p_s\right] \qquad (5.87)$$

It is important to check that M_H is invariant under the modular transformation (5.15). The non-trivial modular transformations of (5.87) are

$$(\det \mathrm{Im}\Pi)^{-4} \prod_i d^4\theta_i \to \det(C\bar{\pi}+D)^4 (\det \mathrm{Im}\Pi)^{-4} \prod_i d^4\theta_i \tag{5.88a}$$

$$\bar{\Theta}\, e^{-\frac{1}{2}\sum_{r,s} \tilde{\bar{p}}_r \tilde{p}_s H(z_r,z_s)} \to \det(C\bar{\pi}+D)^8 \times \bar{\Theta}\, e^{-\frac{1}{2}\sum_{r,s} \tilde{\bar{p}}_r \tilde{p}_s H(z_r,z_s)} \tag{5.88b}$$

$$Q(\underline{w},\underline{z}_1,\underline{z}_2)^{-12} \to \det(C\bar{\pi}+D)^{-12}\, Q(\underline{w},\underline{z}_1,\underline{z}_2) \tag{5.88c}$$

Equation (5.88b) follows from the properties of the generalised theta-function Θ [74] and the assumption that Γ is even, self-dual (see section 5.8). Equation (5.88c) is obtained by using equation (5.74) and noting that $\det'\Delta_z$ is itself modular invariant. The factors of $\det(C\bar{\pi}+D)$ in (5.88) cancel and hence M_H is modular invariant.

To complete our discussion of the heterotic multi-loop amplitude we consider equation (5.87) for the one-loop case and show that it agrees with results from the dual-resource model approach [75]. Firstly, observe that G and H can be expressed in terms of the first Jacobi theta function θ_1 [74]. That is, for $r\neq s$,

$$\exp[-G(z_r,z_s)] = \exp[-\pi(\mathrm{Im}\gamma_{sr})^2/\mathrm{Im}\tau] \left|\frac{\theta_1(\gamma_{sr}|\tau)}{\theta_1'(o|\tau)}\right| \tag{5.89a}$$

and

$$\exp[-H(z_r,z_s)] = \theta_1(\nu_{sr}\,|\tau)/\theta_1'(o|\tau) \tag{5.89b}$$

where $\nu_{sr}=\nu_s-\nu_r = (\ln z_s - \ln z_r)/2i\pi$, $\tau=\ln w/2i\pi$ and

$$\theta_1(\upsilon|\tau) = i \sum_{n=-\infty}^{\infty} (-1)^n \, e^{i\pi\tau(n-\frac{1}{2})^2} \, e^{i\pi\upsilon(2n-1)} \tag{5.89c}$$

$$\theta_1'(o|\tau) = \frac{\partial}{\partial\upsilon} \theta_1(\upsilon|\tau|_{\upsilon=o} \tag{5.89d}$$

There is also a non-singular contribution from the r=s terms in $\exp[-\frac{1}{2} \sum_{r,s} \tilde{p}_r \tilde{p}_s \bar{H}(z_r,z_s)$ of the form $\prod_{r=1}^{N} \bar{z}_r^{-1}$ (since $\tilde{p}_r^2 = 2$).

Under the short-string limit of section 5.5, the contribution of the vertex insertions $\langle V_3^H \rangle$, after integration over Grassmann variables, is

$$K_R (\Pi p_r^+)^{3/2} \, \gamma^{-4} \tag{5.90a}$$

where K_R is the right-moving kinematic factor of ref 65) and γ is the parameter of the Arfei mapping (5.43). Moreover, in the short-string limit, the determinental factor

$\left[\prod_P \bar{F}'' \prod_P (\pi p_r^+)^2 \right]^{-\frac{1}{2}}$ reduces to

$$\bar{\gamma}^{-4} \prod_{r=1}^{N} \bar{z}_r \, \prod_r (p_r^+ \pi)^{-\frac{1}{2}} \tag{5.91}$$

(see equations (5.45b) and (5.47a)). Finally combining equation (5.89)-(5.91) with the Jacobian

$$\frac{\partial(\rho,\alpha,\tilde{\beta})}{\partial(z_r,\tau)} = |\gamma|^8 \, |\mathrm{Im}\gamma|^2 \, \mathrm{Im}\tau \, \frac{\prod_r d^2 z_r}{|z_r|^2} \tag{5.92}$$

we obtain the one-loop expression

$$M_H = K_R \int \prod_{r=1}^{N-1} d^2\upsilon_r \int \frac{d^2\tau}{(\mathrm{Im}\tau)^5}\, Q(\tau)^{-12}$$

$$\prod_{r<s} \exp\left[\frac{-\pi(\mathrm{Im}\upsilon_{sr})^2}{\mathrm{Im}\tau}\right] \left|\frac{\theta_1(\upsilon_{sr}|\tau)}{\theta_1'(o|\tau)}\right|^{\frac{1}{2}P_r \cdot P_s}$$

$$\prod_{r<s}\left[\frac{\theta_1(\upsilon_{sr}|\tau)}{\theta_1'(o|\tau)}\right]^{\tilde{P}_r\cdot\tilde{P}_s} \sum_{\{n_i\}} \exp[i\pi\tau n_i \Gamma_{ij} n_j + 2\pi i n_i] \tag{5.93}$$

which is in exact agreement with dual-resonance model results [75]. Note that in deriving (5.93) we have used equation (5.52) and included an external wave-function normalisation given by $(\prod_{r=1}^{N} p_r^+)^{-1}$.

Finally we give the details of the heterotic graviton amplitudes in appendix XVIII.

Section 5.8: Modular Invariance of the Multi-loop Amplitudes of the Heterotic string.

The gauge-unfixed formlation of string theory is invariant under parametrisations of the world-sheet. For a consistent theory at the quantum level it is important to check that no anomalies arise breaking this reparametrisation invariance. In particular, the absence of global anomalies associated with global diffeomorphisms of the world-sheet requires that multi-loop amplitudes are invariant under the modular group Sp(2g,Z), (at one-loop the modular group is simply SL(2,Z)). Modular invariance is necessary for the finiteness and unitarity of the Polyakov string theory. For it permits the range of integration of the Polyakov multi-loop amplitudes to be defined over a single covering of moduli space.

In this section we show that the L.C. gauge multi-loop amplitudes of the heterotic string derived in the previous section are modular invariant, provided Γ is an even, self-dual lattice. We begin by discussing the one-loop case [65] and relate modular invariance to the range of integration of the variable τ, which parametrises the space of conformally inequivalent tori. This is then extended to higher loops. We end by discussing the significance of modular invariance in the L.C. gauge.

To discuss modular invariance at one-loop we consider in more detail the ranges of integration of the moduli τ and υ_r appearing in the amplitudes of equation (5.93). The range of integration for the variables υ_r is determined by noting that $\upsilon_r = \ln z_r/2\pi i$ with the z_r confined to the annulus given by $|\omega|\leq|z|\leq 1$, fig 13. Hence the υ_r are confined to the parallelogram of fig 15 with opposite sides identified

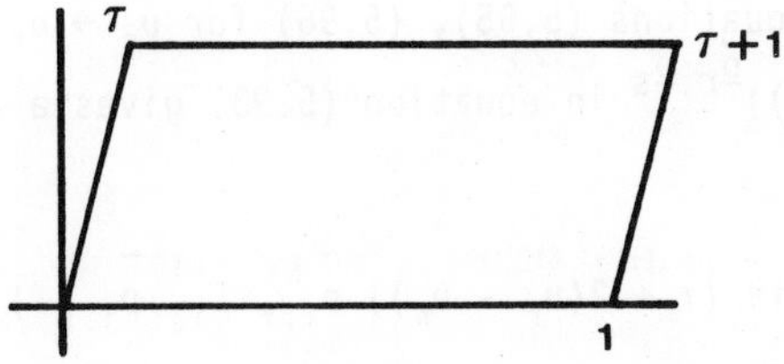

Fig 15. The integration region of the sources υ_r given in terms of a parallelogram with opposite sides identified to form a torus. The parameter τ specifies the conformal structure of the torus.

i.e.a. torus. The parallelogram is conformally equivalent to the surface obtained by cutting the annulus along the homology cycle δ. The vectors 1 and τ generate a lattice and the integration region of the υ_r can be taken to be any fundamental region of the lattice given by a displacement in integer multiples of 1 and τ of the above parallelogram. This corresponds to the fact that the integrand of (5.93) is a doubly periodic function of the υ_r, for fixed τ with periods 1 and τ. In other words, it is invariant under translations

$$\upsilon_r \longrightarrow \upsilon_r + 1 \ , \ \upsilon_r \longrightarrow \upsilon_r + \tau \tag{5.94}$$

To show this use the transformation formulae [74]

$$\theta_1(\upsilon + 1|\tau) = -\theta_1(\upsilon|\tau)$$

$$\theta_1(\upsilon + \tau|\tau) = -e^{-i\pi\tau - 2\pi i\upsilon}\theta_1(\upsilon|\tau) \tag{5.95}$$

and

$$\Theta(\underline{z} + \underline{m}|\Omega) = \Theta(\underline{z}|\Omega)$$

$$\Theta(\underline{z} + \Omega\underline{m}|\Omega) = -e^{-i\pi\underline{m}\Omega\underline{m} - 2\pi i\underline{m}.z}\Theta(\underline{z}|\Omega) \tag{5.96}$$

For example, using equations (5.95), (5.96) for $\upsilon_r \to \upsilon_r + \tau$, the term $\prod_{r<s}\{\theta_1(\upsilon_{sr}|\tau)\ \theta_1'(0|\tau)\}^{\tilde{\underline{p}}_r,\tilde{\underline{p}}_s}$ in equation (5.93) gives a factor

$$\prod_{\substack{s \\ s\neq r}} \exp\{-i\pi\ (\tau + 2(\upsilon_r - \upsilon_s))\ \tilde{p}_{i,r}\ \Gamma^{-1}_{ij}\ \tilde{p}_{j,s}\ \} \tag{5.97}$$

whereas the last term $\Theta(\underline{V}|\Gamma \otimes \tau)$ of equation (5.93) gives

$$\exp\{-i\pi\tau\tilde{p}_{i,r}\ \Gamma^{-1}_{ij}\ \tilde{p}_{j,r} - 2\pi i\ V_i\ \Gamma^{-1}_{ij}\ \tilde{p}_{j,r}) \tag{5.98}$$

Equation (5.98) is obtained by considering $\Theta(\underline{V}|\Gamma \otimes \tau)$ as an ordinary theta-function with 16-dimensional period matrix $\Omega_{ij} = \tau\Gamma_{ij}$ and noting that $\upsilon_r \to \upsilon_r + \tau$ induced the transformation $V_i \to V_i + P_{i,r}\tau$ (see equation (5.79c). Momentum conservation then implies that equations (5.97) and (5.98) cancel.

The parameter τ specifies the conformal structure of the torus constructed from the lattice $(1,\tau)$. However, more than a single τ corresponds to the same conformal structure. To see this, represent the points on the torus by

$$\upsilon = \sigma_1 + \sigma_2\ \tau \tag{5.99}$$

where $\sigma_1 + \sigma_2\ \tau \equiv \sigma_1 + m + (\sigma_2 + n)\tau$ for integer m and n. Then note that a different basis for the lattice could be chosen leading to the same torus with the same conformal structure. For example, the tori defined by $(1,\tau)$ and $(1,\tau+1)$ differ from each other by a global diffeomorphism or Dehn twist. This can be described in terms of cutting the torus along the γ-cycle, (σ_1), rotating in two boundary circles relative to one another by 2π, and gluing them back together.

Different choices of the lattice defining the same torus can be obtained from the group of 2 x 2 invertible integral matrices with unit determinant,

$$\begin{pmatrix} a & b \\ c & d \end{pmatrix} \quad a,b,c,d,\epsilon Z\ ,\ ad - bc = 1 \tag{5.100}$$

This is the modular group SL(2,Z). Under (5.100), τ transforms as

$$\tau \to (a\tau + b)/(c\tau + d) \tag{5.101}$$

The Dehn twists about the σ_1 and σ_2 directions correspond to the elements

$$\begin{pmatrix} 1 & 1 \\ 0 & 1 \end{pmatrix} \qquad \begin{pmatrix} 1 & 0 \\ 1 & 1 \end{pmatrix}$$

and these generate SL(2,Z). The homology cycles also transform under (5.100), since σ_1 and σ_2 describe translations along the cycles γ, δ respectively,

$$\begin{pmatrix} \gamma \\ \delta \end{pmatrix} \rightarrow \begin{pmatrix} d & c \\ b & a \end{pmatrix} \begin{pmatrix} \gamma \\ \delta \end{pmatrix} \tag{5.102}$$

Equations (5.101) and (5.102) then imply that

$$\upsilon_r \longrightarrow (c\tau + d)^{-1}\ \upsilon_r \tag{5.103}$$

since $\upsilon_r = \ln z_r/2\pi i = u(z_r)$.

We now show, following ref [65], that the one-loop amplitude of equation (5.93) is modular invariant, (invariant under SL(2,Z)). The presence of a global anomaly breaking modular invariance would be disastrous for the geometrical interpretation of string theory as a reparametrisation invariant theory, (since SL(2,Z) is isomorphic to the group of global diffeomorphisms of the torus). We shall need the following formula [76]

$$\frac{\theta_1\{\upsilon/(c\tau+d)\,|\,(a\tau+b)/(c\tau+d)\}}{\theta_1'\{0\,|\,(a\tau+b)/(c\tau+d)\}} = \frac{1}{(c\tau+d)}\quad e^{i\pi c v^2/(c\tau+d)}\,\frac{\theta_1(\upsilon|\tau)}{\theta_1'(0|\tau)} \tag{5.104}$$

and [74]

$$\theta\{(C\Omega+D)^{-1}\underline{z}|(A\Omega+B)(C\Omega+D)^{-1}\}$$

$$= \epsilon \det(C\Omega+D)^{\frac{1}{2}} \exp[i\pi\underline{z}^t(C\Omega+D)^{-1} C \underline{z}]\ \theta(\underline{z}|\Omega) \qquad (5.105)$$

where $\epsilon^8 = 1$ (the particular root ϵ depends on the choice of A,B,C,D). Equation (5.105) is the transformation law of a theta-function with an nxn period matric Ω and $\underline{z} \in R^n$, under the symplectic group Sp(2n,Z) consisting of the integral nxn matrices A,B,C,D, arranged as

$$M = \begin{pmatrix} A & B \\ C & D \end{pmatrix}, \ \det M = 1 \qquad (5.106)$$

Under (5.104), the left-moving Jacobi theta-function

$\prod_{r<s} \{\theta_1(\upsilon_{sr}|\tau)/\theta_1'(0|\tau)\}^{\tilde{\underline{p}}_r\cdot\tilde{\underline{p}}_s}$ of (5.93) transforms by a factor

$$(c\tau+d)^{-\sum_{r<s}(\tilde{\underline{p}}_r\cdot\tilde{\underline{p}}_s)} \prod_{r<s} \exp[i\pi c\upsilon_{rs}{}^2\, \tilde{\underline{p}}_r\cdot\tilde{\underline{p}}_s/(c\tau+d)] \qquad (5.107)$$

whereas the Jacobi theta-function associated with the uncompactified modes is invariant since $p_r^2 = 0$ for the ground state of the superstring. The exponential in (5.107) is cancelled by a similar constribution from the generalised theta-function. To show this we shall consider how $\theta(V|\Gamma\otimes\tau)$ transforms for the special cases $\tau \to \tau + 1$ and $\tau \to -1/\tau$. Together these two transformations generate the whole of SL(2,Z). It is easy to see from (5.79b) and (5.105) that under $\tau \to \tau + 1$

$$\theta(V|\Gamma\otimes\tau) \to \theta(V|\Gamma\otimes\tau)\exp(i\pi n_i\Gamma_{ij}n_j) \qquad (5.108)$$

If Γ is an even lattice, $n_i\Gamma_{ij}n_j = (\underline{e}_i)^2$ is even, then the exponential in (5.108) is unity. The Jacobi theta-functions are also invariant under $\tau \to \tau + 1$ as may be seen by setting c = 0, d = 1 in equation (5.107).

To analyse the case $\tau \to -1/\tau$ take θ to be an ordinary theta-function with d x d period matrix $\Omega_{ij} = \tau\Gamma_{ij}$ and use equation (5.105) with A = 0, B = 1, C = -1, D = 0. Then

$$\theta(-\Gamma^{-1} V/\tau|-\Gamma^{-1} \otimes \tau^{-1})$$

$$= \det(\Gamma\tau/i)^{\frac{1}{2}}\ \theta(V|\Gamma \otimes \tau)\ \exp(-i\pi V_i\ \Gamma^{-1}_{ij}\ V_j/\tau) \qquad (5.109)$$

The l.h.s. of (5.109) has the explicit form

$$\sum_{\{n_i\}} \exp\{-i\pi n_i\Gamma^{-1}_{ij}n_j/\tau - 2i\pi n_i\Gamma^{-1}_{ij}V_j/\tau\} \qquad (5.110)$$

Since Γ is assumed to be a self-dual lattice, the Cartan matrix satisfies the relation $\{\Gamma^{-1}n\} = \{n\}$, and (5.110) may be rewritten as

$$\sum_{\{n_i\}} \exp\{-i\pi n_i\Gamma_{ij}n_j/\tau - 2i\pi n_i V_i/\tau\} = \theta(-V/\tau|-\Gamma \otimes \tau^{-1}) \qquad (5.111)$$

Therefore the transformation law for θ under $\tau \to -1/\tau$ is

$$\theta(-V/\tau|-\Gamma \otimes \tau^{-1}) = (\tau)^8\ \theta(V|\Gamma \otimes \tau)$$

$$\exp(-i\pi \sum_{r,s} \tilde{p}_{i,r}\ \Gamma^{-1}_{ij}\ \tilde{p}_{j,s}\ \upsilon_r\ \upsilon_s/\tau) \qquad (5.112)$$

where we have used the fact that self-dual lattices are unimodular, so that $\det\Gamma = 1$, and where V_i has been replaced using equation (5.79c). The exponential in (5.112) cancels the exponential in (5.107) for $\tau \to -1/\tau$, after invoking momentum conservation. We conclude that under the general transformation (5.101)

$$\prod_{r<s} \overline{\{\theta_1(\upsilon_{sr}|\tau)/\theta_1'(0|\tau)\}}^{\tilde{p}_r\cdot\tilde{p}_s}\ \overline{\Theta(V|\Gamma\otimes\tau)} \longrightarrow \overline{(c\tau+d)}^{\sum_{r<s}\tilde{p}_r\cdot\tilde{p}_s}$$

$$\overline{(c\tau+d)}^{8} \prod_{r<s} \overline{\{\theta_1(\upsilon_{sr}|\tau)/\theta'(0|\tau)\}}^{\tilde{p}_r\cdot\tilde{p}_s}\ \overline{\Theta(V|\Gamma\otimes\tau)} \tag{5.113}$$

The other terms of equation (5.93) which transform under (5.101) are

$$\prod_{r=1}^{3} d^2\upsilon_r \quad\to |c\tau + d|^{-6}\ d^2\upsilon_r$$

$$d^2\tau/(\mathrm{Im}\tau)^5 \to |c\tau + d|^{6}\ d^2\tau(\mathrm{Im}\tau)^5$$

$$\overline{\theta(\tau)} \to \overline{(c\tau + d)}\ .\ \overline{\theta(\tau)} \tag{5.114}$$

Therefore combining equation (5.113) and (5.114) gives a net factor

$$\overline{(c\tau + d)}^{8}\ \overline{(c\tau + d)}^{-\sum_{r<s}\tilde{p}_r\tilde{p}_s}\ (c\tau + d)^{-12} \tag{5.115}$$

which is equal to unity since $\tilde{p}_r^2 = 2$ for all $r = 1,\ldots 4$ so that

$$\sum_{r<s} \{\tilde{p}_r \cdot \tilde{p}_s\} = - \sum_{r=1}^{4} \{\tilde{p}_r\}^2/2 = -4 \tag{5.116}$$

Therefore the one-loop amplitude of a string on a torus is modular invariant.

We briefly discuss the significance of this for the range of integration of τ. Naively the range of τ is $\mathrm{Im}\tau > 0$ and $-1/2 \leq \mathrm{Re}\tau \leq 1/2$, which follows from the relation $\tau = \ln\omega/2\pi i$ with $|\omega|=1$ and $0 \leq \mathrm{Arg}\omega \leq 2\pi$. However, the upper-half plane is mapped into itself by SL(2,Z), and hence divides into fundamental regions of the modular group. Each such region has the property that no two points in it are related by a non-trivial modular transformation, but any point in the upper-half plane may be carried to the region by a modular transformation. A standard region is the one defined by $\mathrm{Im}\tau > 0$, $-1/2 \leq \mathrm{Re}\tau \leq 1/2$, $|\tau|^2 > 1$ and corresponds to the shaded part of fig 16

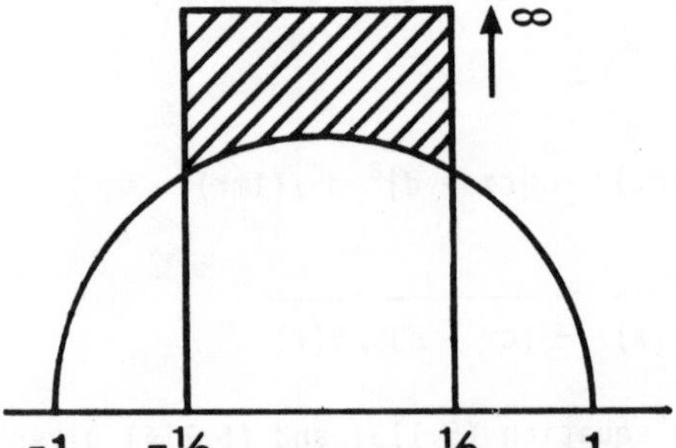

Fig 16. The fundamental region (shaded) of the modular group SL(2,Z).

Since the amplitude is modular invariant integrating over the whole strip $\mathrm{Im}\tau > 0$, $-1/2 \leq \mathrm{Re}\tau \leq 1/2$ would lead to infinite and non-unitary amplitudes (GWS) as there are an infinite number of fundamental regions within it. However, it has been shown by Giddings and Wolpert [53] that the natural ranges of integration for the L.C. diagram parameters $\tilde{\varrho}$, α, β determine a single covering of moduli space; that is, τ is integrated over a single fundamental region of the modular group. Modular invariance then implies that the amplitude is independent of which particular fundamental region is chosen as the range of integration for τ and hence the standard one of fig 16 may be taken as the region of integration for τ.

B. Modular Invariance at g Loops [77)]

Recall that a genus g Riemann surface has g pairs of canonical homology cycles. By 'canonical' we mean that

$$(\gamma_p,\gamma_q) = (\delta_p,\delta_q) = 0$$

$$(\gamma_p,\delta_q) = -(\delta_p,\gamma_q) = \delta_{pq} \tag{5.117}$$

where (,) is a quadratic form on the first homology group $H_1(\Sigma,Z) \cong Z^{2g}$ which counts the number of intersections of any two cycles, taking into account orientation, (see fig 17)

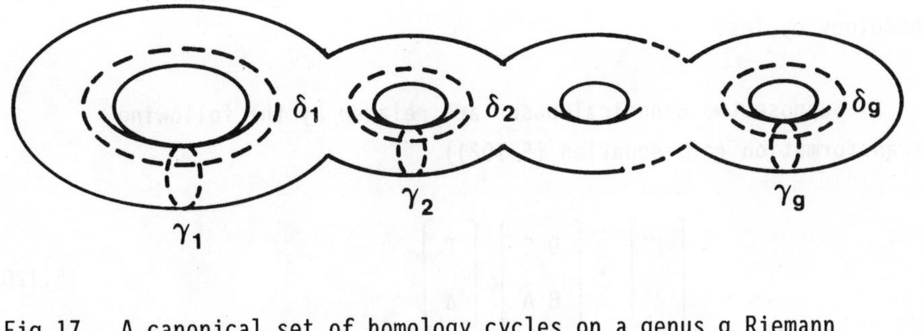

Fig 17. A canonical set of homology cycles on a genus g Riemann surface.

Using the canonical homology cycles as a basis, a general cycle can be written as $\mu = n_p\gamma_p + m_p\delta_p$, n_p , m_p integers such that

$$(\mu,\mu') = (n,m)\begin{pmatrix} 0 & I_g \\ -I_g & 0 \end{pmatrix}\begin{pmatrix} n' \\ m' \end{pmatrix} \tag{5.118}$$

where I_g is the g x g unit matrix.

Associated to a particular choice of canonical homology cycles are g Abelian differentials, equation (5.11), satisfying

$$\oint_{\gamma_q} du_p = \delta_{pq} \ , \ \oint_{\delta_q} du_p = \tau_{pq} \tag{5.119}$$

where τ_{pq} are the components of the period matrix Π. Using the Riemann bilinear relation (5.12) it is easy to show that Π is symmetric and that $\text{Im}\Pi > 0$. (see Appendix VI). The space of all period matrices denoted $K_g = \{\Pi;\ \Pi_{ij} = \Pi_{ji},\ \text{Im}\Pi > 0\}$ is a complex $g(g + 1)/2$ - dimensional space known as Siegel's upper-half plane. It can be shown that no two conformally inequivalent Riemann surfaces have the same Π, (Torelli's theorem), and hence K_g can be used to parametrise surfaces. However, in general the same surface can be described by a whole set of period matrices corresponding to the various choices of bases for the canonical homology cycles.

Suppose two canonical bases are related by the following transformation (cf. equation (5.102))

$$\begin{pmatrix} \Gamma' \\ \Delta' \end{pmatrix} \rightarrow \begin{pmatrix} D & C \\ B & A \end{pmatrix}\begin{pmatrix} \Gamma \\ \Delta \end{pmatrix} \tag{5.120}$$

where $\Gamma = (\gamma_1,\ldots,\gamma_g)$, $\Delta = (\delta_1,\ldots,\delta_g)$ and A,B,C,D are g x g integral matrices. Then for Γ', Δ' to describe a new canonical basis, the transformation (5.120) must preserve the symplectic form of equation (5.118),

$$\begin{pmatrix} D & C \\ B & A \end{pmatrix} \begin{pmatrix} 0 & I_g \\ -I_g & 0 \end{pmatrix} \begin{pmatrix} D & C \\ B & A \end{pmatrix}^t = \begin{pmatrix} 0 & I_g \\ -I_g & 0 \end{pmatrix}$$

Thus the matrix in (5.120) is an element of the symplectic modular group Sp(2g,Z). The change of u_p and Π under Sp(2g,Z) is then computed by imposing

$$\oint_{\gamma'_q} du'_p = \delta_{pq} \quad , \quad \oint_{\delta'_q} du'_p = \tau_{pq}$$

to give

$$du' = [(C\Pi + D)^{-1}]^t \, du \tag{5.121}$$

and

$$\Pi' = (A\Pi + B)(C\Pi + D)^{-1} \tag{5.122}$$

These formulae generalise the one-loop case for which τ is the 1 x 1 period matrix. We conclude that every Riemann surface Σ is represented by a point in the quotient space $A_g = K_g/Sp(2g,Z)$.

As discussed previously, invariance under the modular group at one-loop reflects the absence of global anomalies, since SL(2,Z) is isomorphic to the group of global diffeomorphisms of the torus. Moreover, the modular transformation properties of the parameter τ of the moduli space M_1, (the space of conformally inequivalent tori), is easily derived since τ can also be identified as the period matrix. In other words, $M_1 \cong A_1$. However, matters are more complicated at higher loops. For, in general, not every matrix of K_g corresponds to a Riemann surface and hence M_g is embedded in A_g as a proper subspace.

Consequently, the transformation properties of the moduli z_{1p}, z_{2p}, ω_p ($p = 1,\ldots g$) used to describe a genus g Riemann surface in Chapter 4 are non-trivial. Furthermore, Sp(2g,Z) is no longer equivalent to the group of global diffeomorphisms of the world-sheet.

Nevertheless, it turns out that it is sufficient to use invariance under Sp(2g,Z) as a check for absence of global anomalies. To understand this, note that the group of global diffeomorphisms is generated by Dehn twists just as at the one-loop level. In fact a complete set of generators is given by Dehn twists around curves which wrap around one or at most two handles of Σ in fig 17. Now suppose we have a marked Riemann surface, (a surface with homology basis specified). Since the intersection matrix of (5.117) is preserved by diffeomorphisms, the action of Dehn twists on the canonical homology cycles preserves the symplectic form and gives a representation of Sp(2g,Z). Although this representation is not faithful, the Dehn twists which leave the homology basis unchanged are harmless. For more details see [69),74)].

In light of the above, we now check that our multi-loop amplitude (5.87) is modular invariant, i.e. invariant under Sp(2g,Z). We shall consider each part of the amplitude separately. As noted above, the modular transformation properties of the moduli ω_p, z_{1p}, z_{2p} are highly non-trivial and it is therefore more useful to express the measure in terms of the L.C. diagram parameters $\underline{\rho}$, $\underline{\alpha}$ and $\underline{\beta}$. The complex times are themselves modular invariant as they are defined in terms of the real interaction times τ and twists σ which are independent of any particular choice of homology basis. The measure $\Pi\, d\alpha_p d\beta_p$ is also modular invariant as shown in section 5.2.

Next consider the terms involving the third Abelian integral H. Under the change of homology basis, equation (5.120), $H(z,z_r) \to H'(z,z_r)$ such that

$$\oint_{\gamma'_p} dH' = 0 \qquad \oint_{\delta'_p} dH' = 2\pi i\{u_p(z_N) - u_p(z_r)\} \tag{5.123}$$

As the singularities of H are unchanged we can set

$$H'(z,z_r) = H(z,z_r) + \sum_p y_p^{(r)} u_p(z) \tag{5.124}$$

Substituting (5.124) into (5.123) and using equations (5.119)-(5.120) gives

$$H'(z,z_r) = H(z,z_r) + \text{constant}$$

$$+ 2\pi i \sum_{p,q} u_p(z)[(C\Pi + D)^{-1}C]_{pq}[u_q(z_r) - u_q(z_N)] \tag{5.125}$$

Hence the terms in (5.87) involving $\overline{H}$ transform by a factor

$$\exp\{\pi i \sum_{r,s} \tilde{\underline{p}}_r \; \tilde{\underline{p}}_s \; \bar{u}_p(z_r)[C\Pi + D)^{-1} \; C]_{pq} \; \bar{u}_q(z_s)\} \tag{5.126}$$

The parts containing $u_p(z_N)$ can be dropped as a consequence of momentum conservation. The same applies to the constant in (5.125). A similar analysis shows that $\exp\{-\sum_{r<s} p_r \; p_s \; G(z_r,z_s)\}$ is modular invariant.

To obtain the transformation properties under Sp(2g,Z) of the generalised theta-function $\Theta(\underline{V}|\Gamma\otimes\Pi)$, equation (5.79b), we shall follow the one-loop case and consider specific elements of Sp(2g,Z). In particular, let

$$\begin{pmatrix} A & B \\ C & D \end{pmatrix} = \begin{pmatrix} 0 & -I_g \\ I_g & 0 \end{pmatrix} \tag{5.127}$$

so that

$$\theta(\underline{V}|\Gamma\otimes\Pi) \to \theta([I_{16} \otimes \Pi^{-1}]^t\ \underline{V}\ |- \Gamma \otimes \Pi^{-1}) \tag{5.128}$$

with I_{16} the 16 x 16 unit matrix. To determine the transformation law for θ under (5.127) reduce it to an ordinary theta function by taking $\Gamma \otimes \Pi$ to be a 16g x 16g matrix defined by

$$(\Gamma \otimes \Pi)_{16p+i,16q+j} = \Gamma_{i,j}\ \Pi_{p+1,q+1} \tag{5.129}$$

with $0 \le p,q \le g - 1$ and $1 \le i,j \le 16$. Also let $\underline{V}$ be a 16g column vector

$$(\underline{V})_{16p+i} = V_{i,p} = \sum_r \tilde{p}_{i,r}\ u_p(z_r) \tag{5.130}$$

Now embed Sp(2g,Z) into Sp(32g,Z) and use the standard formula, equation (5.105). Then

$$\theta([I \otimes \Pi^{-1}]^t\ \underline{V}\ |\ - \Gamma \otimes \Pi^{-1}) = \theta([\Gamma \otimes \Pi^{-1}]^t\ \underline{V}'|- [\Gamma^{-1} \otimes \Pi]^{-1})$$

$$= \{\det(\Gamma^{-1} \otimes \Pi/i)\}^{\frac{1}{2}} \exp\{i\pi(\underline{V}')^t\ [\Gamma^{-1} \otimes \Pi]^{-1}\underline{V}'\}$$

$$\theta(\underline{V}'|\Gamma^{-1} \otimes \Pi) \tag{5.131}$$

where $\underline{V}' = -(\Gamma^{-1} \otimes I_g)\ \underline{V}$. Also

$$\theta(\underline{V}'|\Gamma^{-1} \otimes \Pi) = \sum_{\{\underline{n}_i\}} \exp\{i\pi\underline{n}_i\Pi\underline{n}_j\Gamma^{-1}_{ij} + 2\pi i\underline{n}_i\Gamma^{-1}_{ij}\underline{V}_j\}$$

$$= \theta(\underline{V}|\Gamma \otimes \Pi) \tag{5.132}$$

since Γ is self-dual so that $\{\Gamma^{-1}n\} = \{n\}$. Furthermore, $\det(\Gamma^{-1}\otimes\Pi) = (\det\Pi)^8$ as $\det\Gamma = 1$. Therefore, using the definition of $\underline{V}$ we find

$$\theta(\underline{V}|\Gamma \otimes \Pi) \rightarrow \exp\{i\pi \Sigma\, \tilde{p}_{i,r} \tilde{p}_{j,s}\, \Gamma_{ij}\, u_p(z_r)\Pi^{-1}_{pq}\, u_q(z_s)\}$$

$$\theta(\underline{V}|\Gamma \otimes \Pi)\ (\det\Pi)^8 \tag{5.133}$$

The complex conjugate of the exponential in (5.133) cancels the exponential in (5.126) for C = 1, D = 0. If the above analysis is repeated for the cases

$$M = \begin{pmatrix} A & 0 \\ 0 & (A^t)^{-1} \end{pmatrix} \quad A \in GL(g,z) \tag{5.134a}$$

and

$$M = \begin{pmatrix} I & B \\ 0 & I \end{pmatrix} \quad \begin{matrix} B \text{ symmetric, even} \\ \text{diagonal} \end{matrix} \tag{5.134b}$$

it is found that θ is invariant provided the lattice Γ is even. The Abelian integral H is also invariant under (5.134). Together the three kinds of transformation, equations (5.129) and (5.134) generate all elements of Sp(2g,Z) [74]. Therefore, we conclude that if Γ is an even, self-dual lattice,

$$\overline{\theta(\underline{V}|\Gamma \otimes \Pi)}\ \exp\{-\sum_{r,s} \tilde{\underline{p}}_r\, \tilde{\underline{p}}_s\, \overline{H(z_r,z_s)}/2\} \rightarrow$$

$$\det(C\bar{\Pi} + D)^8\ \overline{\theta(V|\Gamma \otimes \Pi)}\ \exp\{-\sum_{r,s} \tilde{p}_r\, \tilde{p}_s\, \overline{H(z_r,z_s)}/2\} \tag{5.135}$$

The factor $\det(C\bar{\Pi}+D)^8$ in (5.135) must be combined with a factor of $\det(C\bar{\Pi}+D)^{-12}$ from $\overline{Q(\underline{\omega},\underline{z}_1,\underline{z}_2)}^{-12}$ and a factor of $\det(C\Pi+D)^4$ from the measure $(\det\mathrm{Im}\Pi)^{-4}\ \Pi d^4\theta_i$. (The transformation property of $\bar{Q}$ follows from equation (5.74) and the fact that det′Δ_z is itself modular invariant.) Hence the amplitude M_H, equation (5.87), is modular invariant if Γ is even, self-dual.

There still remains the question of the significance of modular invariance in the L.C. gauge. As stated above modular invariance reflects the absence of global anomolies associated with global diffeomorphisms of the world-sheet. However, this is a notion from the covariant Polyakov string [22] in which absence of anomalies is required to gauge-fix the theory properly. Modular invariance permits the range of integration of the Polyakov multi-loop amplitudes to be defined over a single covering of moduli space and is a necessary ingredient for the finiteness and unitarity of the theory. On the other hand, the L.C. gauge theory is gauge-fixed ab initio. Furthermore, as shown by Giddings and Wolpert [53], the range of integration of the L.C. parameters $\underset{\sim}{\tilde{\rho}}$, $\underset{\sim}{\alpha}$, $\underset{\sim}{\beta}$ determines a single covering of moduli space without the need to invoke modular invariance. Hence, finiteness of L.C. gauge amplitudes would appear to be independent of modular invariance. The same applies to unitarity, which is guaranteed by the existence of a second-quantised L.C. gauge field theory. Moreover, the construction of L.C. gauge amplitudes based on a second-quantised field theory, Chapters 3 and 4, seems to be possible for general lattices. Yet modular invariance limits the types of lattice constructions for string on tori to even, self-dual lattices.

We conclude that modular invariance is necessary for a consistent covariant formulation of string theory. Since the L.C. gauge is a gauge fixed version of this theory one expects this consistency requirement to be reflected within the L.C. formalism itself. (c.f vanishing of the conformal anomaly in Polyakov translating to closure of the Lorentz algebra in the L.C. gauge). Such a L.C. consistency requirement hasn't yet been found. This a one example of the remaining gap between L.C. and covariant formulations of string theory.

Section 5.9: Massive amplitudes.

Massive external states are usually disregarded in superstring calculations, since they correspond to unstable particles which can decay down to massless external particles which are stable. There is still intrinsic interest in such states, in spite of their instability since many elementary particles, such as the neutron, are unstable in certain environments (such as when free) but not in others (in a nucleus). However there is need to obtain scattering amplitudes for the graviton in the heterotic string in a manner identical to that for an external massive state, so our present discussion is doubly effective.

It is straightforward to construct the massive states of the closed superstring from the creation operators in the Fourier decomposition of the bosonic and fermionic variables given in section 2.2 [50]. The mode operators $\left\{ \alpha^I_m , \tilde{\alpha}^I_{-m} , Q^{\bar{A}}_{\pm m} , \tilde{Q}^{\bar{A}}_{\pm m} \right\}$ (m < 0) of (2.22) and (2.48) can be regarded as creation operators acting on the vacuum state

annihilated by the operators $\left\{ \alpha^I_m , \tilde{\alpha}^I_m , Q^{\bar{A}}_{\pm m} , \tilde{Q}^{\bar{A}}_{\pm m} \right\}$ (m > 0). From (2.24), (2.49) and (2.51) the massless states are thus obtained by acting with any combination of the $Q^{\bar{A}}_0$, $\tilde{Q}^{\bar{A}}_0$ on the vacuum | 0 > so as to give the 128 + 128 massless states with $N = \tilde{N} = 0$:

$$\prod_{r,s} Q_0^{\bar{A}_r} \tilde{Q}_0^{\bar{A}_s} |0\rangle \; \{|N=\tilde{N}=0\rangle\} \quad \text{with } 0 \le r,\ s \le 4 \qquad (5.136)$$

The set of states with $N = \tilde{N} = 1$ are obtained by decomposing the state

$$\{|N=\tilde{N}=1>\} = \left[\alpha^I_{-1} \oplus Q^{\bar{A}}_{-1} \oplus \epsilon^{ABCD} Q^{\bar{B}}_{-1} Q^{\bar{C}}_{-1} Q^{\bar{D}}_{-1} \right]$$

$$\otimes \left[\tilde{\alpha}^J_{-1} \oplus \tilde{Q}^{\bar{B}}_{-1} \oplus \epsilon^{ABCD} \tilde{Q}^{\bar{B}}_{-1} \tilde{Q}^{\bar{C}}_{-1} Q^{\bar{D}}_{-1} \right] \{|N=\tilde{N}=0>\} \qquad (5.137)$$

where there are $(256)^2$ states in all in (5.137). The next level with $N = \tilde{N} = 2$ can be obtained by decomposition of the states

$$\{|N=\tilde{N}=2>\} = \left[\alpha^I_{-2} \oplus R^A_{-2} \oplus Q^{\bar{A}}_{-2}\right] \otimes \left[\tilde{\alpha}^J_{-2} \oplus R^{\tilde{B}}_{-2} \oplus Q^{\tilde{\bar{B}}}_{-2}\right] \{N=\tilde{N}=0>\}$$

$$\oplus \left[\alpha^I_{-1} \oplus R^A_{-1} \oplus Q^{\bar{A}}_{-1}\right] \otimes \left[\tilde{\alpha}^J_{-1} \oplus \tilde{R}^B_{-1} \oplus \tilde{Q}^{\bar{B}}_{-1}\right] \{N=\tilde{N}=1>\}$$

$$\oplus \left[\alpha^I_{-2} \oplus R^A_{-2}\ Q^{\bar{A}}_{-2}\right] \otimes \left[\tilde{\alpha}^J_{-1} \oplus \tilde{R}^B_{-1} \oplus \tilde{Q}^{\bar{B}}_{-1}\right] \otimes \left[\tilde{\alpha}^K_{-1} \oplus \tilde{R}^C_{-1} + \tilde{Q}^{\bar{C}}_{-1}\right] \{N=\tilde{N}=0>\}$$

$$\oplus \left[\alpha^I_{-1} \oplus R^A_{-1} \oplus Q^{\bar{A}}_{-1}\right] \otimes \left[\alpha^J_{-1} \oplus R^B_{-1} \oplus Q^{\bar{B}}_{-1}\right] \otimes \left[\tilde{\alpha}^K_{-2} \oplus \tilde{R}^C_{-2} \oplus \tilde{Q}^{\bar{C}}_{-2}\right] \{N=\tilde{N}=0>\}$$

$$(5.138)$$

where $R^A_{-n} = \sum_{n_1+n_2+n_3=n} \epsilon^{ABCD} Q^{\bar{B}}_{n_1} Q^{\bar{C}}_{n_2} Q^{\bar{D}}_{n_3}$. Higher level (or mass) states can clearly be constructed similarly, and their SU(4) representations analysed; such analysis will not be crucial for the later construction of the massive amplitudes, though would be needed for a more detailed analysis required by a specific physical application.

We note that the state $<0|\bar{Q}^B_{+n}$ has N=n, where N is given by (2.51). Hence the conjugate massive states are given by reversing the sign of the suffices n in all of the $\bar{Q}^A_n$ operators in (5.137), (5.138).

Let us summarise the method developed earlier in this chapter to calculate the multi-loop amplitudes for closed superstrings, but now with emphasis on the massive modes. Each amplitude is obtained by reduction from the Feynman-Dyson perturbation expansion of the second-quantised L.C. gauge field theory. For external strings described by the transverse bosonic source $\underline{J}$, fermionic sources Q, $\tilde{Q}$ and energy p_r^- (for the rth external string, with $1 \le r \le N$) the amplitude is expressible as

$$\int \sum_{\alpha=1}^{2g+N-3} d\tilde{\rho}_\alpha \prod_{i=1}^{g} d\alpha_i \, d\beta_i \int D\underline{X} \, D\theta \, D\tilde{\theta} \, D\lambda \, D\tilde{\lambda} \exp\left[\frac{1}{\pi} \right.$$

$$\iint_\Sigma d\underline{X}^{\wedge *} d\underline{X} + \iiint \left[\lambda\partial_{\bar{\rho}}\theta + \tilde{\lambda}\partial_{\rho}\tilde{\theta}\right] + \iint_\Sigma (\underline{J}.\underline{X} + Q.\theta + \tilde{Q}\tilde{\theta}) -$$

$$\left. - \sum p_r^- \tau_r \right] \Pi \, V \left[\partial_{\bar{\rho}} X, \theta\right] \tilde{V} \left[\delta_{\rho} X, \tilde{\theta}\right] \left[\prod_{r=1}^{N} p_r^+\right] \qquad (5.139)$$

In (5.139) the variables $\tilde{\rho}_\alpha$ denote the interaction (fusion or splitting) points of the strings in the L.C. strip diagram (one of them being held fixed by overall translation invariance), α_i and β_i are the string-loop widths and twists, and τ_r are the times of the external string (to be taken to ± ∞ as appropriate). Moreover, $d\underline{X}$ is a set of (d-2) closed scalar 1-forms on the super-string world-sheet Σ, in terms of the variable $\rho = \tau - i\sigma$; λ, $\tilde{\lambda}$ and θ, $\tilde{\theta}$ are scalar and vector functions under the scale transformation $\rho \to k\rho$ (k real), as discussed more fully in section 5.1, 5.2. The sources $\underline{J}$, Q and $\tilde{Q}$ are defined so as to have support only at $\tau = \tau_r$ for the rth string, so that the extension of (5.14) for the source $\underline{J}$, in that case only containing zero modes, is

$$\underline{J} = \sum_{r=1}^{N} \underline{p}_r(\sigma)\, \delta(\tau - \tau_r) \tag{5.140}$$

with similar expressions for Q and $\tilde{Q}$. The expression (5.139) has to be folded with the appropriate external wave function $\psi(\{\underline{P}_r, Q_r, \tilde{Q}_r\})$ for the external string state being considered. Finally the vertex factors V, $\tilde{V}$, one for each vertex, are superfields in θ and $\tilde{\theta}$, and have the form (to within constant factors)

$$V\left[\partial_{\bar{\rho}} X, \theta\right] = \sqrt{\epsilon}\left[\partial_{\bar{\rho}} X^{7+i8} + \epsilon\, \partial_{\bar{\rho}} X^a\, \theta^{2a} + \epsilon^2\, \partial_{\bar{\rho}} X^{7-i8}\, \theta^4\right] \tag{5.141}$$

and a similar expression for $\tilde{V}$ with θ replaced by $\tilde{\theta}$, $\partial_{\bar{\rho}} X$ by $\partial_\rho X$ in (5.141), where $\epsilon = (\sigma - \sigma_p)$, and $\bar{\theta}^{2a} = \theta^A \theta^A \rho^a_{AB}$, with $1 \le a \le 6$; the limit $\sigma \to \sigma_p$ is ultimately to be taken in (5.141).

Integration over λ, $\tilde{\lambda}$ may be performed immediately in (5.141) leading to the constraints $\partial_{\bar{\rho}}\theta = \partial_\rho \tilde{\theta} = 0$. The former of these may be satisfied by

$$\theta(\rho) = \frac{1}{2\pi}\sum_r \int_{C_r} d\rho'_r \, \partial_\rho \, G(\rho,\rho'_r)\, \theta(\rho'_r) + \sum_{i=1}^{g} \theta_i u_i(\rho) \qquad (5.142)$$

where C_r is a small contour encircling the external string position ρ_r on Σ and θ_r is the boundary value of θ at ρ_r. The expression (5.142) is the higher mode extension of (5.18). Integration over $\underline{X}$ may be performed as before, by translation, after expressing $\underline{X}$ as:

$$\underline{X} = \underline{Y} + G^* \, \underline{J}, \; G^* \, \underline{J} = \frac{1}{2}\int d^2\rho' \; G(\rho,\rho')\underline{J}(\rho') \qquad (5.143)$$

to give finally for (5.139)

$$\int \Pi d\tilde{\rho}_\alpha \; d\alpha_i \; d\beta_i (\det\Delta_0)^{-4} \; (\det\partial_\rho \; \det_{\bar{\rho}})^4 (\det \mathrm{Im}\Pi)^{-4} \prod_{i=1}^{g} d^4\theta_i \; d^4\tilde{\theta}_i$$

$$\times \exp\left[-\frac{1}{2}\, \underline{J}.G^* \, \underline{J} - \sum p_r^- \tau_r\right] . \prod_P V(\partial_{\bar{\rho}} Y + \partial_{\bar{\rho}} G^* J, \; \theta)$$

$$\times \tilde{V} \; (\partial_\rho Y + \partial_\rho G^* J \; , \; \tilde{\theta}) \qquad (5.144)$$

There is a cancellation between the determinants as before.

Integration has yet to be performed in (5.144) over the Fourier modes θ_{rn}, $\tilde{\theta}_{rn}$ and P_{rn} of θ, $\tilde{\theta}$ and $\underline{J}$ after multiplication by the external wave functions, as well as over the inexact modes θ_i (where multiplication by the external wave functions and integration over Q and $\tilde{Q}$ in (5.139) have been implicitly performed, leaving only θ and $\tilde{\theta}$ to be integrated over in (5.144)). In order to consider the integration over

θ_{rn}, $\tilde{\theta}_{rn}$ and P_{rn} in detail let us turn first to analyse the bosonic momentum contributions.

We use again the Mandelstam mapping (5.21). On the rth string the angular variable η_r is introduced by

$$\rho = \tau + ip_r^+(\eta_r + \theta_r) \qquad (5.145)$$

where θ_r is an arbitrary origin, so that

$$i\eta_r = -\frac{\tau_r}{p_r^+} - i\theta_r + \frac{1}{p_r^+} F(z) \qquad (5.146)$$

Then for $z \sim z_r$, $d\eta \sim -i(z-z_r)^{-1}\,dz$, and for any smooth function of z,

$$\int_{C_r} d\eta\; f(z) = 2\pi f(z_r) \qquad (5.147)$$

Thus

$$\tau = (2\pi)^{-1} \sum_r p_r^+ \int_{C_r} d\eta_r'\; G(z,z')$$

Then, as for the tree case, the zero-th mode contribution to the term $-\frac{1}{2}\,\underline{J}.G^*\underline{J}$ in the exponent of (5.144) will be $\sum_{r,s} P_{ro}\,P_{so}\,G(z_r,z_s)$, so giving the usual Veneziano factor at g loops. The remaining factors can be evaluated in a similar manner as the tree level case treated in detail in [64]. In particular the massive modes arising from the non-zero Fourier modes of $\underline{J}$ in σ give a quadratic contribution to $\underline{J}.G^*\underline{J}$ equal to

$$\left(\frac{1}{2\pi}\right)^2 \sum_{r,s,n,m} \int_{C_r}\int_{C_s} d\eta_r\, d\eta'_r \left[\alpha_{rn}\, e^{in\eta_r} + \beta_{rn}\, e^{-in\eta_r}\right] \times$$

$$\times\, G(\rho,\rho') \left[\alpha_{sm}\, e^{im\eta_s} + \beta_{sm}\, e^{-im\eta_s}\right] \qquad (5.148)$$

where $\alpha_{rn} = \frac{1}{2}(P_{rn} - i\bar{P}_{rn})$, $\beta_{rn} = \frac{1}{2}(P_{rn} + i\bar{P}_{rn})$, $\underline{J} = \left(\frac{1}{2\pi}\right)\sum_{r,n\neq 0}$ $[\underline{P}_{rn}\sin n\sigma + \underline{\bar{P}}_{rn}\cos n\sigma + P_{ro}]\,\delta(\tau-\tau_r)$, so giving the coefficients of $\alpha_{rn}\,\alpha_{sm}$ or $\beta_{rn}\,\beta_{sm}$ in (5.146) as

$$C_{nrms} = -\frac{1}{4(2\pi)^2}\int_0^{2\pi} d\eta_r \int_0^{2\pi} d\eta'_s\; e^{i(n\eta_r + m\eta'_s)}\; G(\rho,\rho')$$

On partial integration, and in terms of the function $F_r = (e^F)^{1/p_r^+}$, with $F_r(s) \sim (z-z_r)$ as $z \sim z_r$, and $\xi_r = \tau_r/p_r^+$, C_{nrms} becomes proportional to

$$e^{-in\theta_r - im\theta_s - n\xi_r - m\xi_s}\,(mn)^{-1}\int_{C_r} dz \int_{C_s} ds'\, F_s^n\, F_s^m\, \partial_z \partial_{z'}\, G(z,z') \quad (5.149)$$

The coefficients C_{rnms} have similar properties as in the tree case, and in particular when r = s the only non-zero value is for n - -m, when the contribution is $-\frac{(2\pi)^2}{n}$. This latter may be regarded as the ground state wave function |0>, and α_{rn}, β_{rn} as creation and annihilation operators a^{+}_{rn} , $\tilde{a}^{+}_{rn}$, a_{rn}, $\tilde{a}_{rn}$ for the L and R movers where

$$\alpha_{rn} = \tilde{a}^{+}_{rn} - a_{rn}, \; \beta_{rn} = \tilde{a}_{rn} - a^{+}_{rn} \tag{5.150}$$

Then massive external wave functions are to be replaced by external states with dependence on the a^{+}_{rn}, $\tilde{a}^{+}_{rn}$ as described in the previous section. The remaining contribution to the exponent in (5.144) will be as for the tree level case as discussed in [55] and [64], of the form

$$\exp\left[\sum A_{nmrs}\,(a_{rn}\,a_{sm} + \tilde{a}_{rm}\,\tilde{a}_{sm})\right] \tag{5.151a}$$

where A_{nrms} is the double integral,

$$A_{nrms} = -\frac{1}{4(2\pi)^2}\,2\left[\frac{1}{nm}\right].\int_{C_r} dz\int_{C_s} ds'\; F^n_r\; F^m_s\; \partial_z\partial_{z'}\; G(z,z') \tag{5.151b}$$

and the other factors in (5.149) have been absorbed in the external state (for the θ's) and by the factor $\exp[-P^-_r\,\tau_r + p^2_r\,\frac{1}{4}\,G(\rho_r,\rho_r)]$ (for the ξ's)

Besides the above simple extension of the tree level factors for the massive bosonic modes to the multi-loop case, there are also further contributions containing the annihilation operators a_{rn}, $\tilde{a}_{rn}$ from the vertex factors V, $\tilde{V}$ in (5.144). These arise from the term $G'^{*}\underline{J}$, which has the expansion

$$\sum_r \left\{ G'(\tilde{z},z_r)\, \underline{P}_{ro} - \sum_{n\neq 0} \frac{1}{n} \int_{C_r} dz (F_r)^n\, \partial_z \partial_{\tilde{z}}\, G(\tilde{z},z)\, e^{-in\theta}\, r^{-n\xi}{}_r \underline{\gamma}_{rn} \right\} \tag{5.152}$$

and $\underline{\gamma}_{rn} = i[\theta(n)\, \underline{\alpha}_{rn} - \theta(-n)\, \underline{\beta}_{r-n}]$ is expressed using (5.150) in terms of the annihilation and creation operators a_n, $\tilde{a}_n$, a_n, $\tilde{a}_n^+$ acting on the initial and final states. Again, as for the tree case, only the annihilation operators are to be kept. It is to be noted that only $n<0$ enters in the sum in the last term of (5.152).

Similar expressions to (5.152) arise from the first term on the r.h.s. of (5.142) so that, using the mapping (5.145), with $\rho = F(z)$, $F'(z) = 0$, then

$$\sqrt{\epsilon}\, \theta(\tilde{\rho}) = [F''(\tilde{z})]^{-\frac{1}{2}} \left\{ \sum_{i=1}^{g} \theta_i u_i'(\tilde{z}) + \sum_{r=1}^{N} \left[G'(\tilde{z},z_r)\, p_r^+ \theta_{ro} \right.\right.$$

$$\left.\left. - \sum_{n\neq 0} \frac{p_r^+}{n} \int_{C_r} dz [F_r(z)]^n\, \partial_z \partial_{\tilde{z}}\, G(\tilde{z},z)\, e^{-in\theta_r - n\xi_r}\, \theta_{rn} \right] \right\} \tag{5.153}$$

As in the bosonic case, only $n<0$ enters in the sum in the last term of (5.153). However we notice that there is a factor $\epsilon\{p_r^+\}$ (the signature of p_r^+) different between θ_{rn} of (5.143) and the Fourier coefficients in (2.48):

$$\theta_{rn}^A = \bar{Q}_{rn}^A\, \epsilon(p_r^+)\ . \tag{5.154}$$

Since the conjugate (outgoing) states already have a reversal of the sign of n (as noted earlier in this section) then the further reversal

(5.154) leaves the signs of indices n in θ_{rn} unchanged in external states(although not at vertex factors). The factor $e^{-in\theta_r - n\xi_r}$ is absorbed again as in that case. The measure $D\theta$ will reduce, by the arguments given in [38] and appendix XV, to

$$(\det \mathrm{Im}\Pi)^{-2} \prod_{i=1}^{g} d^4\,\theta_i \prod_n \prod_{r=1}^{N} d^4\,\theta_{rn} . \delta^4 \left[\sum p_r^+ \theta_{ro} \right] \quad (5.155)$$

It is also necessary to take the Fourier transform of the massive states of section 2 with respect to their variables $Q_{rn}^{\bar{A}}$, $\tilde{Q}_{rn}^{\bar{A}}$ as remarked above, in order to have the correct external states to fold directly into (5.144). This Fourier transformation only changes the massless states to their complex conjugates [50], but will be highly relevant in the massive sector.

In conclusion, then, (5.150), (5.153) and (5.155) are to be used in the evaluation of the amplitude expression (5.144), along with the Fourier transforms of the massive (or massless) states. There is integration to be performed over all of the fermion modes θ_{rn}, and the inexact θ_i, whilst the bosonic modes P_{rn}, $\overline{P}_{rn}$ are to be interpreted as annihilation operators acting on the ground state.

We may extend the scattering amplitudes for external massless states in the heterotic string containing coloured gauge bosons, as constructed in section 5.7, to those containing gravitons by using the above formulae developed for non-zero mode scattering. In particular the change from section 5.7 is that the L-going modes are described by external states $\underset{\sim}{\zeta}_r . \tilde{\underline{a}}_{r-1}|0\rangle$, in the notation of (5.150), where $\underset{\sim}{\zeta}_r$ is the polarization vector associated with the rth L-going mode. Then the relevant factor for these latter modes may be read off directly from (5.151b) as $A_{-1r,-1s}$, which is proportional to $\partial_{z_r} \partial_{z_s} G(z_r, z_s)$. There will also be contributions from $A_{o,r'-1s}$ and $A_{-1r,os}$, which are proportional to $\partial_{z_s} G(z_r,z_s)$ and $\partial_{z_r} G(z_r,z_s)$ respectively. These terms lead to the usual factor $\exp [\frac{1}{2} \zeta_r \zeta_s \partial_{z_r}\partial_{z_s} G(z_r,z_s) + \zeta_r \cdot p_s \partial_{z_r} G(z_r,z_s)]$ for vector meson scattering in the bosonic string obtained by operator methods [65], and to amplitudes identical to those discussed in [78] at the one-loop level.

It is not possible to use the short string limit, developed in sections 5.3 - 5.6, for the calculation of massive amplitudes. This is because all massive amplitudes become divergent in this limit, although not the massless ones of the previous paragraph, to which the following remarks do not apply. This may be seen from the fact that the coefficients A_{rn} of the Fourier modes Υ_{rn} of (5.150) diverge in this limit. In detail, from (5.150)

$$A_{r,n}(\tilde{z}) = \int_{C_r} dz[F_r(z)]^n \, \partial_z \partial_{\tilde{z}} \, G(\tilde{z},z) = \int_{C_r} dz(z-z_r)^n \, H_{rn}(z) \quad (5.156a)$$

with

$$H_{rn}(z,\tilde{z}) = \psi(z,z_r)^n \prod_{s\neq r} [\phi(z,z_r]^{np_s^+/p_r^+} \, \partial_z \partial_{\tilde{z}} \, G(\tilde{z},z) \quad (5.156b)$$

where we are using the definitions

$$G(z,z_r) = \ln (z-z_r) + \ln\psi(z,z_r) = \ln\phi(z,z_r) \quad (5.156c)$$

Since $H_{nr}(z,\tilde{z})$ of (5.156b) is regular in z at z_r then, as remarked earlier, A_{rn} is non-zero only for $n < 0$, and with $n = -m > 0$,

$$A_{r,-m}(\tilde{z}) = 2\pi i\left[\left(\frac{\partial}{\partial z}\right)^{m-1} H_{r,-m}(z,\tilde{z})\right]_{z=z_r} \tag{5.157}$$

It is now possible to use the SSL (5.31) to evaluate the coefficient $A_{r,-m}(\tilde{z})$ at the external interaction vertex $\tilde{z}_r$ very close to z_r. Since, in this limit, from (5.156b)

$$G(\tilde{z},z) \sim \ell n(\tilde{z}-z) \tag{5.158}$$

then

$$H_{r,-m} \sim (z-\tilde{z})^{-2}$$

$$A_{r,-m}(z_r) \sim \sum_{\ell=0}^{(m+1)} a_{r\ell}^{(m)} (p_r^+)^{-\ell} \tag{5.159}$$

In (5.159) the coefficients $a_{rl}^{(m)}$ are functions of the external sources $z_1 \ldots z_N$ and of the moduli of the world-sheet Σ through the functions ψ and ϕ which may be determined from the first and third Abelian integrals on Σ [27].

The singularities in (5.159) involving increasingly high inverse powers of the vanishing parameters p_r^+ clearly give divergent values for the amplitudes, as claimed earlier. We might take this result to indicate that the SSL cannot be applied to obtain any amplitudes, but that sensible results including massive modes could be derived by some alternative method. We will justify the use of the SSL to obtain any amplitudes for superstrings in Chapter 7, where closure of the algebra guarantees that Lorentz transformations are allowed which affect the

SSL. Thus we must accept that massive amplitudes for superstrings (and strings) are infinite. Since the massive states are not in any case stable such an apparently disastrous result should not unduly disturb us; we will show in the next Chapter that amplitudes for the massless modes are, in fact, finite, so that all is well with the superstring.

CHAPTER 6: DIVERGENCE ANALYSIS

Section 6.1: The Origin of String Divergences

The expressions obtained in the previous chapter are integral representations which involve, as integration variables, the moduli of Riemann surfaces. If these latter are considered as punctured by the sources (the positions of the external strings), then integration is to be taken over the set of 3g-3+N complex moduli for the g-loop amplitude with N external sources.

By inspection of the multi-loop expressions in the last chapter (especially Sections 5.4, 5.5 and 5.6) it is clear that divergences may arise in them when two sources approach each other, $z_r \sim z_s$, say, since the integrand will aquire a behaviour, from the VVS factor, as $|z_r - z_s|^{-p_r p_s}$. Such singularities lead to poles and cuts in the amplitude as a function of the invariant masses $p_r p_s$, as is to be expected. However the presence of such divergences might lead one to suspect that there might also be divergences in the amplitudes when other moduli are integrated over. Due to the presence of the insertion factors in the $g \geq 2$ loop expressions of Section 5.6, the configuration $z_r \sim \tilde{z}_i$ or $\tilde{z}_i \sim \tilde{z}_j$, with close approach of an EIV and IIV or of two IIV's may be expected to lead to divergences in the integrand like $|z_r - \tilde{z}_i|^{-2}$ or $|\tilde{z}_i - \tilde{z}_j|^{-4}$. There may also be other sources of divergences due to approach to other boundary points of moduli space. Various possibilities are shown in figure 18.

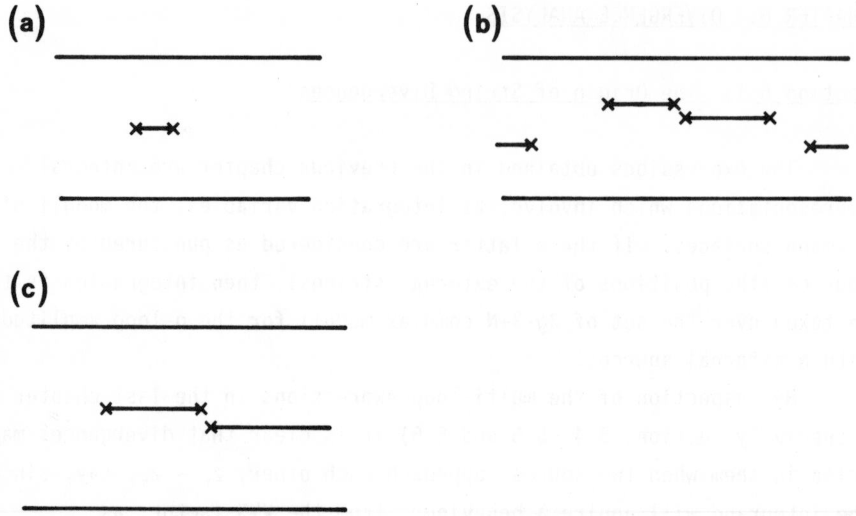

Fig. 18 (a) A vanishing loop in a one-loop L.C.D. diagram
(b) Two coinciding EIV's in a two-loop diagram.
(c) The L.C.D. of a coinciding EIV and IIV in a one-loop diagram.

The question of whether or not the multi-loop amplitudes for massless mode scattering of superstrings are finite or not is a crucial question to be answered. If a particular superstring model has finite amplitudes then it will give a perturbatively finite quantum gravity (where our attention will be focused on closed strings). Perturbative finiteness has proved of great importance in guiding the development of the standard gauge model of non-gravitational interaction [79], and clearly is a valuable property. If perturbative finiteness is not satisfied it would appear to be much harder to make any useful statements about such a theory without much more powerful tools of non-linear mathematical analysis than presently exists.

The problem of divergences for bosonic string amplitudes were analysed in the 70's [80], though with inconclusive results as to their renormalizability. It was already appreciated then that divergences did not arise from the ultra-violet regime of the constituent string modes, but rather from infra-red divergences due to the presence of a tachyon and of a massless scalar particle (the dilaton) with non-zero vacumm expectation value. The divergence of the planar one-loop open bosonic string amplitude can be recognised as arising from these sources [81]. The one-loop finiteness of the O(32) Chan-Paton mode [18), 24] has been argued [19] as being due to the absence of the tachyon and of dilaton tadpole-type divergences, due to supersymmetry annihilating the tachyon and causing the dilaton vacumm expectation value to vanish.

Attempts were made to give a detailed analysis of higher loop bosonic string amplitudes, especially in the last reference in ref. [80]. New developments, especially through the use of more sophisticated mathematical techniques, have allowed a more precise analysis to be made of divergences in the bosonic string multi-loop partition function. The tachyon and dilaton divergences are seen as arising from degenerations of the Riemann surface into one with the same number or fewer handles, with pinches at certain waists or dividing goedesics or at the handles. Such degenerations produce new zero modes which cannot be removed from laplacians on Σ. In particular the partition function $[\det'\Delta_0]^{-12}$ (in the bosonic case) is then divergent as a function of the length ℓ of a vanishing goedesic.

Initial analysis [82] of the bosonic string dilaton divergence by Selberg ζ- function techniques has more recently been superseded by use of the more powerful techniques of algebraic geometry [83]. These have allowed, in particular through the work of Belavin and Knizhnik [84], an elegant understanding of the dilaton and tachyon singularities in terms of a section of the appropriate holomorphic determinant line bundle of Σ. Poles in this section at points on the boundary of moduli space, obtained by adjoining stable curves with nodes, are the source of the above-mentioned divergences in bosonic string amplitudes.

Extension of the algebraic geometric approach to superstrings had been conjectured in the above refs. [83), 84]. Many papers have since appeared on this question (see, for example, the references quoted in [26] or [85]. There appear to be very difficult problems associated with the definition of the super-moduli space, although the structure of super-Teichmuller space has recently become clarified [86]. Even the split character of super-moduli space is yet uncertain, and integration over the even nilpotent elements introduces an ambiguity (discussed in [26] and [85]). A recent interesting attempt [87] uses the integration theory of [88]. This latter theory was only originally developed for split super-manifolds, so extension to the more general non-split case may be needed. The further interesting attempt in [85] requires use of a very special regularisation procedure, that of 'round' hypertubes, and only then works for the heterotic case.

The multi-loop superstring amplitudes constructed in the L.C. gauge by functional methods [32] involve the first and third Abelian differentials and the period matrix for the relevant Riemann surface Σ. The manner in which these depend on the parameters ℓ, etc. as these become zero near a degeneration of Σ can be determined by domain variational theory [89]. Such an approach leads to integral equations for the third Abelian differentials which may be solved by perturbation theory. Those solutions may then be used to analyse the possible singularities in the multi-loop amplitudes. The procedure had already been used for the bosonic string by Alessandrini and Amati [90] and by the authors earlier for the superstring [33], [91]. It is proposed to give a more detailed analysis of the divergence question in this Chapter using the above techniques.

We will do that briefly for the one-loop bosonic string and superstring in the next section before discussing divergences of the multi-loop amplitudes of sections 5.6 and 5.7 later in this Chapter.

Section 6.2: One-loop Superstring Divergences

Let us first consider the one-loop bosonic string amplitude, which may be obtained in a similar manner to that of Section 5.5 for the superstring, but now with neglect of the insertion factors. We shall follow closely the discussion of GSW, Chapter 8. The total expression for the lowest mass (tachyon) scalar mode scattering amplitude is (GSW, Chapter 8)

$$\int \prod_{r=1}^{N-1} \frac{d^2 z_r}{|z_r|^2} \exp \left[- \sum_{r<s} \frac{1}{2} p_r p_s G(z_r, z_s) \right] \frac{d^2 \tau}{\mathrm{Im}\tau} \left[\det\Delta_0 \right]^{-(D-2)/2} \tag{6.1}$$

where τ is the complex modulus (period matrix) of the one-loop surface, which is a torus, and D the dimension of the embedding space-time. Comparing (6.1) with the corresponding superstring amplitude (5.51) we see that the former has different powers of the period matrix $\mathrm{Im}\tau$ and also a contribution from the determinant of the bosonic Laplacian (which is not now cancelled). To analyse the divergences of the expression (6.1) it is necessary to use the explicit form of the one-loop Greens function $G(z_r, z_s)$ and also that of $\det\Delta_0$. The former of these is given by

$$\ell n\, \chi\, (z_r/z_s, \tau) \tag{6.2}$$

In terms of the modulus w of the annulus, fig 12, which is related to τ by

$$w = \exp\, [2\pi i \tau] \tag{6.3}$$

one may establish (see equations (5.69), (5.73), (5.74) and (5.75))

$$\chi(z,w) = \exp\, ((\ell n|z|)^2/2\ln|w|) \left| z^{-\frac{1}{2}}(1-z) \prod_{m=1}^{\infty} (1-w^m z)(1-w^m/z)(1-w^m)^{-2} \right| \tag{6.4}$$

whilst

$$(\det\Delta_0)^{-D/2} = |f(w)|^{-2(D-2)} (-4\pi/\ln|w|)^{(D-2)/2} \tag{6.5a}$$

with

$$f(w) = \prod_{n=1}^{\infty} (1-w^n)\ w^{1/24} \qquad (6.5b)$$

We do not wish to give a detailed analysis of the form of (6.1); our purpose here is to solely illuminate the divergence characteristics of that expression for the rest of this chapter. There are numerous symmetry properties of χ which may be helpful in understanding the nature of (6.1) further if that is desired; we refer the interested reader to the useful discussion in GSW Chapter 11.

The divergences in (6.1) can best be analysed by rewriting the expression in the form [65]

$$\int_F d^2\tau\ (\mathrm{Im}\tau)^{-2} C(\tau) F_v(\tau) \qquad (6.6a)$$

where C arises from the Laplacian (6.5a), and F_v from the external vertices, in D=26 (the critical dimension, as discussed in Section 2.3)

$$C(\tau) = 4(\tfrac{1}{2}\mathrm{Im}\tau)^{-12}\ \exp(4\pi \mathrm{Im}\tau).|f(w)|^{-48} \qquad (6.6b)$$

$$F_v(\tau) = (k\pi)^4\, \mathrm{Im}\tau \int \prod_{r=1}^{N-1} \frac{d^2 z_r}{|z_r|^2} \prod_{r<s} (\chi(z_r/z_s))^{-p_r p_s/2} \qquad (6.6c)$$

where $z_4 = 1$ in (6.6b). The measure $(\mathrm{Im}\tau)^{-2} d^2\tau$ and the factors C and F are separately modular invariant. The domain of integration is over a range which may be expressed succinctly in the variables

$$\nu_r = (\ln z_r)/2\pi i \quad (r=1,2,\ldots,N-1), \tag{6.7a}$$

as

$$0 \leq \mathrm{Im}\,\nu_r \leq \mathrm{Im}\tau \tag{6.7b}$$

The range of integration in τ then becomes a fundamental region of the modular group SL(2,Z),

$$-\frac{1}{2} \leq \mathrm{Re}\tau \leq \frac{1}{2}, \quad \mathrm{Im}\tau > 0, \ |\tau| \geq 1 \tag{6.7c}$$

which is illustrated in figure 16 of section 5.8.

As remarked in section (5.8), integrating τ over the whole strip $\mathrm{Im}\tau \geq 0$ and $-\frac{1}{2} \leq \mathrm{Re}\tau \leq \frac{1}{2}$, rather than the fundamental domain F, would definitely lead to an infinite, non-unitary amplitude. Roughly speaking, $\mathrm{Im}\tau$ can be viewed as the imaginary time that a string propagates in the loop. Thus restriction to the fundamental domain F eliminates the possibility of "ultraviolet" divergences associated with the small $\mathrm{Im}\tau$ region. However, this still leaves the possibility of "infrared" divergences associated with the large $\mathrm{Im}\tau$ region.

The various singularities of the amplitude (6.1) arise at the endpoints of the integration over the parameters ν_r and τ, i.e. at the boundaries of moduli space. It is important to distinguish between physical singularities, required by perturbative unitarity, and genuine divergences. The former corresponds to the production of on-shell intermediate physical states; they depend on external momenta for example, poles in the channel energies S_{IJ}, where

$$s_{IJ} = -\left(\sum_{r=I}^{J} p_r\right)^2$$

arise from the complex variables z_I, z_J converging together on the world-sheet. In this limit

$$\chi(\nu_{rs}, \tau) \sim 2\pi|\nu_{rs}| \tag{6.8}$$

where $\nu_{rs} = \nu_r - \nu_s \sim 0$. In order to exhibit the leading singularity it is convenient to introduce new variables $\epsilon\eta_r = \nu_r - \nu_J$ for r = I,I+1,...J-2 where the real variable ϵ is defined as $\epsilon\ e^{i\phi} = \nu_{J-1} - \nu_J$. The Jacobian for this transformation is

$$\prod_{r=I}^{J-1} d^2\nu_r = i\epsilon^{2(J-I)-1}\ d\epsilon\ d\phi \prod_{r=I}^{J-2} d^2\eta_r \tag{6.9a}$$

Using equation (6.8) and (6.9) we can expand the integrand of equation (6.1) in a power series in ϵ such that the most singular term in the ϵ-integral is of the form

$$\int_0^1 d\epsilon\ \epsilon^{2(J-I)-1} \prod_{I\leq r<s<J} \epsilon^{\frac{1}{2}p_r p_s} = \int_0^1 d\epsilon\ \epsilon^{-\frac{1}{4}s_{IJ}-3} \tag{6.9b}$$

where use has been made of the result

$$\sum_{I\leq r<s\leq J} p_r \cdot p_s = -\frac{1}{2}\ s_{IJ} - 4(J+1-I) \tag{6.9c}$$

for the tachyon mass $p_r^2 = 8$. The leading pole in (6.9b) occurs at the tachyon energy $s_{IJ} = -8$. Poles corresponding to higher mass states, $s_{IJ} = 0,8,\ldots$ arise at higher order terms in the ϵ expansion of the integrand. Also note that there are branch points in the integrand due to the vanishing of the z_r variables associated with internal propagators. It follows from equation (6.7) that such singularities occur in the limit $w \to 0$ (or $\text{Im}\tau \to \infty$). The branch points correspond to thresholds for producing pairs of on-shell imtermediate states.

The other type of divergence do not depend on special choices of the external momenta. There are a number of ways in which these "genuine" divergences may arise at one loop. One is associated with the limit in which all the ν_r's except one converge together. This corresponds to the insertion of a closed string loop on an external tachyon leg giving rise to an infinity due to an internal tachyon propagator evaluated on-shell (see fig 19).

Fig. 19 The insertion of the open-string loop on an external tachyon leg gives rise to an infinity due to an internal propagator evaluated on shell. This is an effect that is eliminated by a finite shift in the tachyon mass.

Such a divergence may be considered as a radiative correction to the tachyon mass and is not a crucial aspect of the finiteness of the theory. The more significant form of divergence arises from the convergence of all points ν_r on the torus (fig 20(a)).

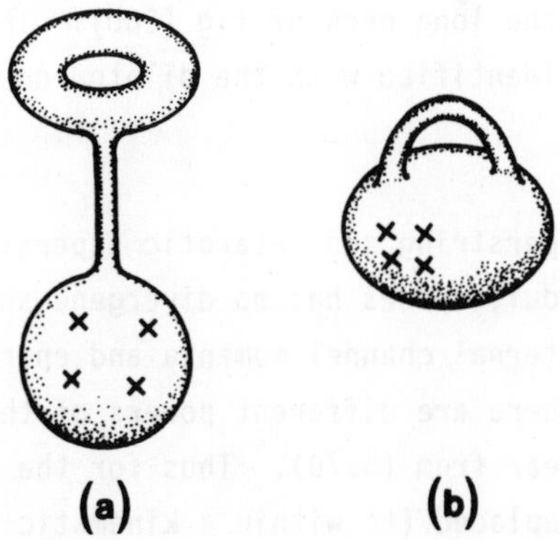

Fig. 20 A string world sheet with a long neck connecting a tree diagram to a torus is sketched in (a). The integral over the length of the neck diverges due to propagation of the tachyon. The world sheet of (a) can be conformally mapped into the picture sketched in (b) in which the world sheet has no long neck, and the external states are inserted very close together.

From equations (6.1), (6.8) and (6.9), with I = 1, J = M, the leading singular contribution to the amplitude is given by

$$\int_0^1 \frac{d\epsilon}{\epsilon^3}\, d\phi \left[\prod_{r=1}^{M-2} d^2\eta_r\right] \prod_{1\leq r<s\leq M-1} |\eta_r-\eta_s|^{k_r\cdot k_s/2}$$

$$\int_F \frac{d^2\tau}{\mathrm{Im}\tau}\, C(\tau) \qquad (6.10)$$

To interpret (6.10) it is useful to conformally map the worldsheet depicted in fig (20a) to that of fig (20b), which consists of a long, thin internal tube connecting a tree diagram to a torus without external state insertions. The divergence $\int d\epsilon/\epsilon^3$ in (6.10) corresponds to the propagation of a tachyon along the neck of fig (20b). The coefficient of this divergence factories into an M-point function on the sphere (the η_r integration) and a vacuum-to-vacuum amplitude on the torus (the τ integration). The next term in the ϵ expansion of the amplitude results in a divergence of the form $\int d\epsilon/\epsilon$ corresponding to the propagation of a massless dilaton along the long neck of fig (20b). The coefficient of this divergence can be identified with the dilaton one-loop expectation value.

For the closed superstring and heterotic superstring at one loop the integration over moduli spaces has no divergences arising independently of the external channel momenta and energies. This is because in both cases there are different powers of the moduli entering the amplitudes as is clear from (5.70). Thus for the type II superstring, (6.6) is replaced (to within a kinematic factor) by

$$\int_F d^2\tau(\mathrm{Im}\tau)^{-2}F_s(\tau)$$

$$F_s(\tau) = (\pi)^4(\mathrm{Im}\tau)^{-3}\int\prod_{i=1}^{3}\frac{d^2z_i}{|z_i|^2}\prod_{r<s}\left[\chi(z_r/z_s)\right]^{-p_rp_s/2} \qquad (6.11a)$$

We turn to discusss this question of divergences more generally for the superstring multi-loop amplitudes of Chapter 5 in the succeeding sections of this Chapter. In particular we propose to demonstrate the absence of divergences in the hetorotic amplitude from integration over the boundaries of moduli space. A similar analysis at one-loop to that for (6.6) shows there only to be poles and branch points from the intermediate channel states and no divergence from integration over the boundary points of moduli space. For example, consider the region in which all the ν_r's converge together. Introducing the variables η_r as in the case of the bosonic string and using the asymptotic estimate (6.8) gives the leading order approximation of (6.11a) as

$$\int_0^1 d\epsilon\ \epsilon^s\int_F\frac{d^2\tau}{(\mathrm{Im}\tau)^s} \qquad (6.11b)$$

which is convergent. This implies that the dilaton tadpole insertion is zero, which is to be expected since a non-zero value of the dilaton expectation value would violate supersymmetry. In the heterotic case, equation (5.93), we obtain in the same limit

$$\int_0^1 \epsilon d\epsilon\int_F\frac{d^2\tau}{(\mathrm{Im}\tau)^s}\ \frac{1}{\bar{\omega}}\ [f(\bar{\omega})]^{-24}\sum_{\{n_i\}}\exp[i\pi\tau n_i\Gamma_{ij}n_j] \qquad (6.11c)$$

Equation (6.11c) is finite, in spite of the apparant singularity in the limit $\bar{\omega}\to o$, on integration over the phase of ω.

Amplitudes for massless particle scattering by the heterotic and type II superstrings were calculated at multi-loop order in Chapter 5. The explicit calculation of the divergences at one loop order of this section showed that there are no momentum-independent divergences from integration at the boundary of moduli space in the superstring case. The space-time supersymmetry prevents the appearance of tachyons, dilatons or self-energy corrections. It is our purpose to extend those results to multi-loop diagrams by detailed analysis.

Section 6.3: Degeneration of Riemann Surfaces

The expressions for the multi-loop amplitudes, developed in Chapter 5, are of the general form

$$\int_M \prod_{\alpha} d^2\tilde{z}_\alpha \prod_{i=1}^{g} d^2\alpha_i \; I(\tilde{z}_\alpha, \alpha_i, z_r) \tag{6.12a}$$

where M is the moduli space of punctured Riemann surfaces, and the integrand I may be singular for some values of its variables. In more detail, and using the results of Sections 5.1 and 5.2, the amplitude for the type II superstring has form (to within external superfields and their associated Grassmann-valued measures)

$$\int \prod_{\alpha=2}^{2g+n-2} d^2\tilde{\rho}_\alpha \prod_{i=1}^{g} d^2\alpha_i \; \exp\{-\tfrac{1}{2} \sum_{r<s} p_r p_s G(z_r, z_s)\}(\det \mathrm{Im}\Pi)^{-4} \prod_{i=1}^{g} d^4\tilde{\theta}_i d^4\theta_i$$

$$\times \; \langle \prod_{\alpha} |2F''(\tilde{z}_\alpha)|^{-1} V\, (\partial_{\tilde{z}_\alpha} \underline{X} + \sum_{r=1}^{N} G'(\tilde{z}_\alpha, z_r)\underline{p}_r, (2F''(\tilde{z}_\alpha))^{-\frac{1}{2}}$$

$$\left[\sum_{r=1}^{N} G'(z_\alpha,\tilde{z}_r)p_r^+\theta_r + \sum \theta_i u_i(z_\alpha)\right])V(\partial_{\tilde{z}_\alpha}\underline{X} + \sum_{s=1}^{N} \bar{G}'(z_\alpha,\tilde{z}_r)\underline{p}_r{}',$$

$$[2\bar{F}''(\tilde{z}_\alpha)]^{-\frac{1}{2}}\left[\sum_{r=1} \bar{G}'(\tilde{z}_\alpha,z_r)p_r^+\tilde{\theta}_r + \Sigma\tilde{\theta}_i \bar{u}_i^1(\tilde{z}_\alpha)\right])> \tag{6.12b}$$

(where the contraction rules for < > are given in (2.68) and (2.71)), whilst that for the heterotic amplitude is

$$\int \prod_{\alpha=2}^{2g-N-2} d^2\tilde{\rho}_\alpha \prod_{i=1}^{g} d^2\alpha_i \exp\{-\frac{1}{2}\sum_{r<s} p_r p_s G(z_r,z_s)\}(\det\mathrm{Im}\Pi)^{-4}(\det\partial_z)^{-12}\mathrm{x}$$

$$\mathrm{x}\ \Theta(\underline{V}|\Gamma\otimes\Pi)\ \mathrm{x}\ \exp(-\frac{1}{2}\sum_{r,s}\bar{H}(z_r,z_s)\tilde{p}_r\cdot\tilde{p}_s)$$

$$\prod_{i=1}^{g} d^4\tilde{\theta}_i\,\mathrm{x}<\prod_\alpha[2\bar{F}''(\tilde{z}_\alpha)]^{-1}V(\partial_{\tilde{z}_\alpha}\underline{X}+\sum_{r=1}^{N}\tilde{G}'(z_\alpha,z_r)\underline{p}_r,[2\bar{F}''(\tilde{z}_\alpha)]^{-\frac{1}{2}}$$

$$\mathrm{x}\ (\sum_{r=1}^{N}\bar{G}'(\tilde{z}_\alpha,z_r)p_r^+\tilde{\theta}_r + \Sigma\tilde{\theta}_i\bar{u}_i(z_\alpha)))> \tag{6.12c}$$

Possibly dangerous configurations of these variables, which may lead to divergence of the integral, are (α) degenerations of the Riemann surface Σ, defined by a particular point of M. This degeneration may arise from the length of either some handle or dividing geodesic going to zero (so that the period Π of Σ becomes a degenerate matrix, or a new zero of Δ_0 appears and Π is a direct product of two matrices)

(β) two (or more) interaction points come together, $\tilde{z}_\alpha \sim \tilde{z}_\beta$, without a degeneration of the surface.
(γ) one (or more) source point approaches one (or more) interaction point, $z_r \sim \tilde{z}_\alpha$
(δ) two (or more) source points approach each other, $z_r \sim z_s$.
(ϵ) multiple degenerations of (α)-(δ) occur.

The degeneration of type (α) will be investigated in detail in Sections 6.4, 6.5, 6.6, 6.8, 6.9 those of type (β) in 6.7, and those of class (γ) and (δ) are considered in sub-section 6.10. Details of the behaviour of the mapping function F at some of these coalescences are given in appendix XI.

Section 6.4: Handle degenerations of Riemann Surfaces

In this sub-section the analysis of [89)] is developed using the notation of [92)]. Handle degeneration may be analysed by starting with a Riemann surface S, of genus g-1, and attaching a handle of small size to it to construct a new Riemann surface S^* of genus g. All quantities on the new surface are starred to distinguish them from the unstarred quantities on S. S^* is constructed from S by deleting the interiors of two parameter-discs on S, and identifying their boundaries.

To describe this in detail let (Γ,Δ) be a fixed one-dimensional homology on S, with $\Gamma = (\gamma_1,\dots,\gamma_{g-1})$, $\Delta = (\delta_1,\dots,\delta_{g-1})$ with only γ_i intersecting with δ_i. A basis of the first Abelian differentials on S is denoted by $(du_1,\dots,du_{g-1})$, with normalisation $\int_{\gamma_i} du_j = \delta_{ij}$. The period matrix Π_{ij} is defined by $\Pi_{ij} = \int_{\delta_i} du_j = \Pi_{ij}$, and has $\mathrm{Im}\,\Pi > 0$. A normal third Abelian integral η_{XY} has poles in $d\eta_{XY}$ with residue -1 at X and +1 at Y and is normalised so that $\int_{\gamma_i} d\eta_{XY} = 0$. For any two points P_0 and Q_0 of S not on (Γ,Δ), the parameter discs D_{P_0}, D_{Q_0} are defined about P_0 and Q_0 with boundaries C_{P_0} : $\mathrm{Re}\,\eta_{P_0Q_0}(P) = \log\epsilon$, C_{Q_0} :

$\mathrm{Re}\,\eta_{P_0Q_0}(P) = -\log\epsilon$. respectively. Points P′, Q′ on C_{P_0}, C_{Q_0} are identified if

$$\eta_{P_0Q_0}(p') - \eta_{P_0Q_0}(Q') = 2\log\epsilon + 2i\alpha \tag{6.13}$$

A Jordan curve γ is drawn from C_{P_0} to C_{Q_0} so as to have no intersection with (Γ,Δ), and $C_{P_0} = C_{Q_0}$ is denoted by δ. Then the surface S^* obtained by deleting the interiors of C_{P_0} and C_{Q_0} and taking $\Gamma^* = (\gamma_1, \ldots, \delta_{g-1}, \delta)$ as a Riemann surface of genus g.

The basic variational equation relating normal third Abelian integrals η_{XY} and η^*_{ZW}, in terms of the difference $\Delta\eta_{XY} = \eta^*_{XY} - \eta_{XY}$ is [92]

$$\Delta\eta_{XZ}(z) - \Delta\eta_{XY}(W) = \frac{1}{2\pi i}\int_{\delta s_0} \eta_{xy}\, d\eta^*_{zw} \tag{6.14}$$

where S_0 is the common domain of S and S^*, so is the whole of S outside C_{P_0} and C_{Q_0}. The integral equation (6.14) may be solved by expanding η_{XY} on the r.h.s. of (6.14) about P_0 and keeping only the lowest order term. From (6.13), $[2\ell n\epsilon + 2i\alpha]^{-1}\,\eta_{P_0Q_0}$ has the same discontinuity in W on S^* as $(2\pi i)^{-1}\int_\delta d\eta^*_{XY}$; the r.h.s. of (6.14) thereby reduces to an expression purely in terms of η to give

$$d\eta^*_{XY}(Z) = d\eta_{XY}(Z) - [2\ell n\epsilon + 2i\alpha]^{-1}[n_{XY}(P_0) - \eta_{XY}(Q_0)]\,d\eta_{P_0Q_0}(Z) + O(\epsilon) \tag{6.15}$$

Moreover the new first Abelian differential is therefore [64]

$$du^*_g(Z) = [2\ell n\epsilon + 2i\alpha]^{-1}\,d\eta_{P_0Q_0}(Z) + O(\epsilon) \tag{6.16}$$

The remaining first Abelian differentials are obviously $du_i^* = du_i$ $(i = 1,\ldots,g-1)$. Besides the first Abelian differenatials the object on Σ of crucial importance for the construction of the multi-loop amplitudes is the Greens function $\underline{G}$ defined from the first and third Abelian integrals. This is defined to have real part $G = \mathrm{Re}\,\underline{G}$ single-valued round (Γ,Δ), and may be given in terms of real parameters a_i^*, b_i^* $(1 \le i \le g)$. In terms of α_i^*, there is the explicit formula of [32]

$$\underline{G}^*_{P_1 P_N}(P) = \eta^*_{P_1 P_N}(P) + i \sum_{j=1}^{g} a_j^* u_j^*(P) \tag{6.17}$$

Then

$$a_j^* = \mathrm{Im}[\underline{G}^*_{P_1 P_N}(\gamma_i P) - \underline{G}^*_{P_1 P_N}(P)] \tag{6.18}$$

since η^* is single-valued and u_k^* has change δ_{jk} around γ_j. The variable b_i^* is then defined by

$$b_j^* = \mathrm{Im}[\underline{G}^*_{P_1 P_N}(\delta_j P) - \underline{G}^*_{P_1 P_N}(P)] \tag{6.19}$$

From the residue theorem

$$\int_{\delta_j^*} d\eta^*_{XY} = 2\pi i[u_j^*(Y) - u_j^*(X)]$$

and the definition of the period matrix gives

$$b_j^* = 2\pi.\mathrm{Re}[u_j^*(P_N) - u_j^*(P_1)] + \sum_i a_i \mathrm{Re}\Pi^*_{ij} \tag{6.20}$$

The condition of single-valuedness of $\mathrm{Re}\underline{G}^*_{P_1 P_N}$ on Σ leads directly to the condition

$$a_j^* = 2\pi(\mathrm{Im}\Pi^*)^{-1}_{jk}\ \mathrm{Im}[u_k^*(P_1)-u_k^*(P_N)] \tag{6.21}$$

(6.20) and (6.21) may be combined together as

$$b_j^* = 2\pi(u_j^*(P_N)-u_j^*(P_1)] + \sum_i a_i^*\Pi_{ij}^* \tag{6.22}$$

The notation being used here is that of [92], in which capital letters denote points on the surface and small latin letters denote their uniformizations. Then P_1 and P_N are usually denoted by the Koba-Nielsen source values z_1, z_N, and the mapping function for the string from the z to the ρ-plane is that given in (5.21). In the above notation

$$\rho = F^*(P) = \sum_{r=1}^{N} \underline{G}^*_{P_r P_N}(P).p_r^+ \tag{6.23}$$

Equations (6.15) and (6.16) may be used to give, to first non-trivial order, the functions and parameters on the surface S^* in terms of their values on S, provided that the period matrix Π^* is written explicitly in terms of Π. Using (6.16) and the definition of the period matrix this results in

$$\Pi_{ij}^* = \Pi_{ij} \quad (1 \le i,j \le (g-1))$$

$$\Pi_{gg}^* = 2\pi i[2\ell n\epsilon+2i\alpha))]^{-1}$$

$$\Pi_{gj}^* = 2\pi i[2\ell n\epsilon+2i\alpha]^{-1}[u_j(Q_0)-u_j(P_0)] \quad (1 \le j \le (g-1)] \tag{6.24}$$

with the associated inverse

$$(\mathrm{Im}\Pi^*)^{-1}_{ij} = (\mathrm{Im}\Pi^{-1})_{ij} \quad (1 \le i,j \le (g-1))$$

$$(\mathrm{Im}\Pi^*)^{-1}_{gg} = \pi^{-1}\ell n\epsilon$$

$$(\mathrm{Im}\Pi^*)^{-1}_{gj} = -\sum_{k=1}^{(g-1)} (\mathrm{Im}\Pi^*)^{-1}_{jk}\mathrm{Re}(u_k Q_0)-u_k(P_0)] \tag{6.25}$$

Using (6.15), (6.16), (6.20), (6.21), (6.23), (6.24) and (6.25) leads to

$$dF^*(P) = dF(P) + (2\ell n\epsilon)^{-1}\left[i\alpha^*_g - \sum_{r=1}^{N} p^+_r \eta_{P_r P_N}(P_0)\right.$$

$$\left. + \sum_{r=1}^{N} p^+_r \eta_{P_r P_N}(Q_0)\right] d\eta_{P_\beta P_N}(P) + O(\epsilon^2) \tag{6.26}$$

with the string width variables α^*_j of (4.85) given by

$$\alpha^*_j = \alpha_j = \sum_{r=1}^{N} 2\pi p^+_r (\mathrm{Im}\Pi)^{-1}_{jk}\mathrm{Im}[u_k(z_r)-u_k(P_N)] \quad (j < g) \tag{6.27}$$

$$\alpha^*_g = \sum_{r=1}^{N} p^+_r \mathrm{Im}\{[\eta_{P_r P_N}(P_0)-\eta_{P_r P_N}(Q_0)-2\pi i(u_k(Q_0)-u_k(P_0))](1-i\alpha(\ell n\epsilon)^{-1}]\}$$

$$\times\ (\mathrm{Im}\Pi)^{-1}_{kj} \times \mathrm{Im}(u_j(p_r)-u_j(P_N)] \tag{6.28}$$

and the string twist variables β_j^* of (4.85) by

$$\beta_j^* = \beta_j \quad (j < g)$$

$$\beta_g^* = +(\pi/\ell n\epsilon)\left\{\mathrm{Re}\sum_{r=1}^{N} p_r^+[\eta_{P_r P_N}(Q_0)-\eta_{P_r P_N}(P_0))(1-i\alpha/\ell n\epsilon)\right\}$$

$$+\ \alpha\alpha_g^*(\ell n\epsilon)^{-1} - \sum_{j=1}^{g-1}\alpha_j \mathrm{Re}(u_j(q_0)-u_j(P_0))(1-i\alpha(\ell n\epsilon)^{-1})]\} \qquad (6.29)$$

where terms of $O((\ell n\epsilon)^{-2})$ in (6.26) and $O((\ell n\epsilon)^{-1})$ in (6.28) have been dropped, though the terms $O(\alpha(\ell n\epsilon)^{-1}$, in the latter have been kept. The analysis of handle degeneration has now reached the point at which it can be applied to the specific analysis of the superstring amplitudes of Chapter 5 by solving for the interaction points $\tilde{P}$ for which

$$dF^*(\tilde{P}) = 0 \qquad (6.30)$$

This will be considered after the degeneration of a dividing geodesic is analysed in the next section.

Section 6.5: Dividing geodesic degeneration

Degeneration as the length ℓ of a dividing geodesic tends to zero has been considered in detail in [92] and also by Fay [93]. It will turn out that there is further work to do beyond a simple application of these references, since the degeneration parameter (ϵ of [92] or t of [93]) is not directly related to the length ℓ. Since the latter parameter enters the superstring amplitude the relation between these parameters ℓ and ϵ (or t) should be determined; that will be discussed at the end of this section.

The principles of the degeneration analysis of the last section can be used to go beyond the analysis of [92] to produce the equivalent to eqns. (6.22), (6.24), (6.26), (6.27) and (6.28). The degenerating surface S^* is obtained by gluing together two surfaces S_1, S_2 of genus g_1 and g_2 respectively at point A_1 and B_2. This can be achieved across the boundaries $C_P : t_1 = \exp[-\eta_{A_1, B_1, 1}(P)]$, $C_Q : t_2 = \exp[\eta_{A_2, B_2, 2}(Q)]$ on S_1 and S_2 respectively with $|t_1| = |t_2| = \epsilon$, where A_i, B_i are points on S_i and $\eta_{A_i, B_i, i}$ are normal third Abelian integrals S_i (i =1,2). The identification is by means of the condition

$$t_1 t_2 = t \tag{6.31}$$

whre t is a small parameter (in the notation of Fay [91]) equal to $\epsilon^2 \phi$ (in the notation of [92]) with $|\phi| = 1$. Following the arguments of [92] it is possible to obtain the equivalent integral equation to (6.14), in terms of $\Delta\eta_{XY} = \eta^*_{XY} - \tilde{\eta}_{XY}$, when X and Y are both in the interior of $S_1 \cap S^*$, with

$$\begin{aligned} \tilde{\eta}_{XY} &= \eta_{XY,1} \quad \text{on } S_1 \cap S^* \\ &= 0 \qquad\quad \text{on } S_2 \cap S^* \end{aligned} \tag{6.32}$$

The resulting integral equation is

$$\Delta\eta_{XY}(Z) - \Delta_{XY}(W) = \frac{1}{2\pi i} \eta_{XY,1}(A_1)\epsilon^2\phi \int_{C_{B_2}} t_2^{-1} d\eta^*_{ZW} + 0(\epsilon^4) \tag{6.33}$$

where t_2 is the local parameter on C_{C_2} given above. The standard relation

$$\frac{1}{2\pi i}\int_{\delta_1^*} d\eta^*_{XY} = [u^*_i(Y)-u^*_i(X)] \tag{6.34}$$

allows an evaluation of the r.h.s. of (6.34) using the value of $d\eta^*_{XY}$ on the l.h.s. given by (6.33), with that on the r.h.s. of (6.33), $d\eta^*_{ZW} = O(\epsilon^2)$ on S_2 if $Z,W \leftarrow S_1$, and $d\eta^*_{ZW} = [d\eta_{ZW,2}+O(\epsilon^2)]$ on S_2 if $Z,W \in S_2$. Then for $\delta_1^* \in S_1$, (6.34) leads, for $X \in S_1$, to

$$du^*_1(X) = du_{i1}(X) + tc_1 d_x \eta'_{XY,1}(A_1) \quad (1 \le i \le g_1) \tag{6.35a}$$

with $c_1 = (2\pi i)^{-1}\int_{C_{B_2}} t_2^{-1}\, d\eta_{ZW,2}$ as $\epsilon \sim 0$. Similarly if $\delta_1^* \in S_2$, then (6.33) and (6.34) lead, for $X \in S_1 \cap S^*$, by use of the residue theorem, to

$$du^*_i(X) = tu'_i(B_2) d_x \eta'_{XY,1}(A_1) \quad ((g_1+1) \le i \le g_2) \tag{6.35b}$$

A similar result follows for $X \in S_2 \cap S^*$, using the appropriate version of (6.33) with S_1 and S_2 interchanged, to give

$$du^*(X) =_i du_{i1}(X) + tc_2 d_x \eta'_{XY,2}(B_2)\ [(g_2+1) \le_1 i \le g_2)] \tag{6.35c}$$

with $C_2 = (2\pi i)^{-1}\int_{C_{A_1}} t_1^{-1} d\eta_{ZW,1}$, for $\delta^*_i \in S_2$. Finally for $\delta^*_i \in S_1$ and $X \in S_2 \cap S^*$ then

$$du^*_i(X) = t\, u'_i(A_1) d_X \eta'_{XY,2}(B_2) \quad (1 \le i \le g_1) \tag{6.35d}$$

Eqns. (6.35a) - (6.35d) agree with the values obtained by Fay [91] by a different method, on noting that $d_X\eta'_{XY,1}(A_1)$ is a differential of the second kind $W_1(X,A_1)$, on S_1, etc. These values may then be used, together with the definition of the period matrix to lead to [91]

$$\Pi^*_{ij} = \Pi_{1\,ij} \qquad 1 \le i,j \le g_1 \tag{6.36a}$$

$$= \Pi_{2\,ij} \qquad (g_1+1) \le i,j \le (g_1+g_2) \tag{6.36b}$$

$$= t\, u_i(A_1)u_j(B_2) \qquad 1\le i\le g_1,(g_1+1)\le j\le (g_1+g_2) \tag{6.36c}$$

It if finally possible to deduce from (6.33) that for $X,Y,Z \in S_1 \cap S^*$

$$d\eta^*_{XY}(Z) = d\eta_{XY,1}(Z) + O(t^2) \tag{6.37a}$$

and for $Z \in S_2 \cap S^*$, $X,Y \in S_1 \cap S^*$,

$$d\eta^*_{XY}(Z) = t\eta_{XY,1}(A_1)d_z\eta_{ZW,2}(B_2) \tag{6.37b}$$

The construction of $d\eta^*_{XY}(Z)$ for $X,Z \in S_1 \cap S^*$, $Y \in S_2 \cap S^*$ is simple, since to $O(t)$, $d\eta^*_{XY}(Z)$ only has a simple pole X, and so in this region

$$d\eta^*_{XY}(Z) = d_{Z_1}\eta_{Z_1,Z_2,1}(X)|_{Z_1=Z_2=Z} \tag{6.37c}$$

Simerlarly for $X \in S_1 \cap S^*$, $Y,Z \in S_2 \cap S^*$,

$$d\eta^*_{XY}(Z) = -d_{Z_1}\eta_{Z_1,Z_2,2}(Y)|_{Z_1=Z_2=Z} \tag{6.37d}$$

A source point P_0 is taken on S_1 in general position. The above equations lead to the values

$$dF^*(P) = \sum_{P_r \in S_1} p_r^+ d\eta_{P_r, P_0, 1}(P) + \sum_{P_r \in S_2} p_r^+ d_{Z_1} \eta_{Z_2, Z_1}(P)\Big|_{Z_1 = Z_2 = P_0}$$

$$+ i \sum_{j=1}^{g_1} \alpha_j^* du_{j1}(P) + O(t) \qquad (6.38a)$$

for $P \in S_1$, whilst for $P \in S_2$

$$dF^*(P) = \sum_{P_r \in S_2} p_r^+ d_{Z_1 Z_2}(P)\Big|_{Z_1 = Z_2 = P_r} + \sum_{j=g_1+1}^{g} \alpha_j^*{}_2(P) + O(t) \quad (6.38b)$$

Moreover

$$(\mathrm{Im}\Pi^*)^{-1}_{ij} = (\mathrm{Im}\Pi_1)^{-1}_{ij} \qquad 1 \le i,j \le g_1$$

$$= (\mathrm{Im}\Pi_2)^{-1}_{ij} \qquad g_1 < i,j \le g_1 + g + 2$$

$$= -t\mathrm{Im}[u_{i1}(A_1)u_{j2}(B_2)] \quad 1 \le i \le g_1, g_1 < j \le g_1 + g_2 \qquad (6.39)$$

whilst using the above formulae (6.35), (6.36), (6.37) and (4.85) (for the length-twist variable α_1^*, β_1^*) results in

$$\alpha_i^* = 2\pi \mathrm{Im} \left\{ \sum_{j=1}^{g_1} (\mathrm{Im}\Pi_1)^{-1}_{ij} \Big[\sum_{P_r \in S_1} p_r^+ (u_{j_1}(z_r) + \sum_{P_r \in S_2} t u'_{i1}(A_1) \eta'_{Z_r, Y, 2}(B_2) \Big] \right.$$

$$+\sum_{j=1}^{g_1}(-)t\ \mathrm{Im}(u_i(B_2)u_j(A_1)\sum_{P_r\in S_1}p_r^+u_{j1}(z_r)\Big\}\quad (\text{for } g_1<i\le g) \qquad (6.40a)$$

$$=2\pi\mathrm{Im}\Big\{\sum_{j=g_1+1}^{g}(\mathrm{Im}\Pi_2)^{-1}_{ij}\Big[\sum_{P_r\in S_2}p_r^+u_{j2}(z_r)+\sum_{P_r\in S_1}tu_i'(B_2)\eta'_{Z_r,Y,1}(A_1)\Big]$$

$$+\sum_{j=g_1+1}^{g}(-)t\ \mathrm{Im}(u_i(A_1)u_j(B_2))\sum_{P_r\in S_2}p_r^+u_{j2}(z_r)\Big\}(\text{for } 1\le\ \le g_1) \qquad (6.40b)$$

$$\beta_i^*=\sum_{j=1}^{g_1}(\mathrm{Re}\Pi_1)_{ij}\alpha_j^*+t\sum_{g=g_1+1}^{g}\mathrm{Re}[u_{i1}(A_1)u_{j2}(B_2)\alpha_j^*$$

$$-2\pi\Big[\sum_{P_r\in S_1}p_r^+\mathrm{Re}u_{i1}(z_r)+\sum_{P_r\in S_2}p_r^+t\ \mathrm{Re}(u_i'(A_1)\eta'_{Z_r Y,2}(B_2)\Big]\ (\text{for } 1\le 1\le g_1) \qquad (6.41a)$$

$$=\sum_{j=g_1+1}^{g}(\mathrm{Re}\Pi_2)_{ij}\alpha_j^*+t\sum_{j=1}^{g}\mathrm{Re}(u_{i2}(B_2)u_{j1}(A_1))\alpha_j^*$$

$$-2\pi\Big[\sum_{P_r\in S_2}p_r^+\mathrm{Re}u_{i2}(z_r)+\sum_{P_r\in S_1}p_r^+t\ \mathrm{Re}(u_i'(B_2)\eta'_{Z_r Y,1}(A_1)\Big](\text{for } g_1<i\le g) \qquad (6.41b)$$

This completes the analysis of the relevant functions on the surface associated with the dividing goedesic degeneration.

There is still the relationship between the vanishing parameters ℓ and t or ϵ. Upper and lower bounds on this relationship may be obtained by using the inequalities of Masur [94] relating the Poincare metric ρ^* on the annulus A_t^δ : $|t|(1-\delta)^{-1} < |z| < (1-\delta)$ around a uniformization of C_A, with the Poincare metric ρ on S*. For small enough t, Masur showed that there is a constant C so that on A_t^δ

$$\rho^* \geq \rho \geq C\rho^* \tag{6.42}$$

Moreover ρ^* is given explicitly by

$$\rho^*(z) = \pi|z| \log|t| \sin(\pi \log|z|/\log|t|)]^{-1} \tag{6.43}$$

As pointed out by Wolpert [95] the dividing geodesic has $|z| = |t|^{\frac{1}{2}}$ (by symmetry, as can be seen by uniformization in the strip with w=logz). Then from (6.42), (6.43) at $z|=|t|^{\frac{1}{2}}$

$$2\pi|t|^{\frac{1}{2}}\rho^*(|z|=|t|^{\frac{1}{2}}) \geq \ell \geq 2\pi C|t|^{\frac{1}{2}}\rho^*(|z|=|t|^{\frac{1}{2}})$$

or

$$2\pi^2(\log|t|^{-1})^{-1} \geq \ell \geq 2\pi^2 C(\log|t|^{-1})^{-1} \tag{6.44}$$

This proves that

$$\ell = O[(\log|t|)^{-1}] \tag{6.45}$$

in which also the left hand inequality in (6.44) was proved. The Weyl-Petersson measure $\prod_{i=1}^{3g-3} d\tau_i \wedge d\ell_i$ [94] (where the τ_i are the Fenchel-Nielsen twists associated with the geodesic length parameters ℓ_i), near a degeneration $\ell \sim 0$ has ℓ-dependent part $\ell d\ell$ d(argr). Use of (6.44) allows this to be rewritten as $d(\arg t)|t|^{-1}(\ell n|t|^{-3}d|t|$. This measure will be used later in analysis of the dividing geodesic degeneration in the heterotic superstring amplitude.

Section 6.6: Handle degeneration divergences

The analysis of possible divergences from a handle degeneration in either the type of II or heterotic multi-loop amplitudes of section 5.6 and 5.7 commences by solving the equation

$$dF^*(\tilde{z}) = 0 \tag{6.46}$$

where dF^* is given by (6.26). The term $O(\ell n\epsilon^{-1})$ on the r.h.s. of (6.26) may be neglected if $\tilde{P}$ is not close to P_0 or Q_0 so giving

$$dF(\tilde{z}) = 0. \tag{6.47}$$

There will be 2(g-1) solutions of (6.47). These correspond to the 2g-2 IV's of the degenerated surface in which the degenerated handle plays no role.

On the other hand $\tilde{P}$ may approach P_0 or Q_0 to $O(\ell n\epsilon^{-1})$. If P_0 and Q_0 are uniformised by p_+, p_- in the z-plane there will be two further solutions $\tilde{z}_\pm$ of (6.46) given by

$$\tilde{z}_\pm = p_\pm + a_\pm(\ell n\epsilon)^{-1} + O((\ell n\epsilon)^{-2}) \tag{6.48}$$

Since $d\eta_{P_0Q_0}(z_\pm) \sim \mp\, \ell n\epsilon(a_\pm)^{-1}$ then (6.46) will be satisfied provided

$$a_\pm = \pm\left[i\alpha^*_g + \sum_r \alpha_r(\eta_{P_rP_N}(Q_0) - \eta_{P_rP_N}(P_0))\right]F'(p_\pm)]^{-1} \tag{6.49}$$

To the order of $(\ell n\epsilon)^{-1}$ under consideration it is possible to write, from (6.28) and (6.29), that

$$\alpha_g^* = \alpha_g + \alpha(\ell n\epsilon)^{-1}\alpha_g'$$

$$\beta_g^* = (\ell n\epsilon)^{-1}\beta_g + \alpha(\ell n\epsilon)^{-2}\beta_g' \tag{6.50}$$

where α_g, α_g', β_g' are independent of α and ϵ, and may be read off directly from (6.28) and (6.29). Then

$$\frac{\partial(\alpha_g^*,\beta_g^*)}{\partial(\alpha,\ell n\epsilon)} = \alpha_g'\beta_g(\ell n\epsilon)^{-3} \tag{6.51}$$

It is clear from the analysis of section 6.5 that all of the functions entering the amplitudes in Chapter 5 have a finite, non-zero limit as $\epsilon \to 0$ except for $F^{*}{}''(\tilde{z}_\pm) = O(\ell n\epsilon)$ and $\det(\mathrm{Im}\Pi^*) = O(\ell n\epsilon)^{-1}$. For the type II string there is a total power of $(\det\mathrm{Im}\Pi^*)^{-4}$. Combined with the change of measure (6.51) to the variables α and $(\ell n\epsilon)$, these result in the asymptotic value of the integrand for the type II case as $\epsilon \to 0$ given at most by (with $x = (\ell n\epsilon)^{-1}$)

$$\int_0 x^3\,dx\int d\alpha \tag{6.52}$$

which is clearly finite. On taking account also of the factor $(\det\mathrm{Im}\Pi^*)$ in $\det'\Delta_0$ a similar analysis for the heterotic amplitude of section 5.7 gives the potentially dangerous term with divergence at most

$$\iint_0 x\;dx d\alpha \tag{6.53}$$

which is again finite. A similar analysis for multiple handle degeneration is expected to give the same result.

Section 6.7: Coinciding interaction points

It is clear that the degeneration of a handle (or any number of them) leads to a pair of interaction points $\tilde{z}_\pm$ converging to the resulting punctures P_0, Q_0 on the surface of Σ_{g-1}. Such a degeneration does not bring into play the potentially lethal Feynman ultra-violet-like divergences (short distance on the world-sheet) arising from the coincidence of two interaction points. Such a further degeneration can now be performed in a controlled manner by letting the points P_0, Q_0 approach each other. This control can be easily achieved with the handle degeneration occuring simultaneously. Thus $g > 1$ would seem to be required, since for $g = 1$ there is only one available modulus and two complex degrees of freedom are not available. However there is still the variable z_1, and a possible divergence from the above cause would also be expected here. Since one loop amplitudes are finite this potential divergence should not be present. This will be seen after the details of the singularity are determined.

The values of the parameters and functions given in section 6.4 can now be analysed further as the uniformising parameter δ between P_0 and Q_0 in the z-plane goes to zero. Some of the functions and parameters also vanish with δ and a careful analysis must be given. If $Q_0 = P_0 - \delta$ then the string parameters $\alpha_g, \alpha_g', \beta_g, \beta_g'$ in (6.50) are all of order δ, with

$$\alpha_g = \sum_r p_r^+ \operatorname{Im}[\partial\eta'_{P_r P_N}(P_0)] + 2\pi \operatorname{Re}[u_k'(P_0)\partial](\operatorname{Im}\Pi)^{-1}_{jk} \operatorname{Im}[u_j(P_r) - u_j(P_N)]$$

$$\alpha_g' = -\sum_r p_r^+ \operatorname{Re}[\partial\eta'_{P_r P_N}(P_0)] - 2\pi \operatorname{Im}[u_k'(P_0)\partial](\operatorname{Im}\Pi)^{-1}_{jk} \operatorname{Im}[u_j(P_r) - u_j(P_N)]$$

$$\beta_g = \pi\left\{\mathrm{Re}\sum_r p_r^+[\partial\eta'_{P_rP_N}(P_0)] + \sum_{j=1}^{g-1}\alpha_j \mathrm{Re}[u'_j(P_0)\partial]\right\}$$

$$\beta'_g = \pi\left\{\mathrm{Im}\sum_r p_r^+\partial\eta'_{P_rP_N}(P_0)] + \sum_{j=1}^{g-1}\alpha_j \mathrm{Im}[u'_j(P_0)\partial] + \alpha_g\right\}$$

$$a_\pm = \pm\left[i\alpha_g - \sum p_r^+\partial\eta'_{P_rP_N}(P_0)\right]\left[\sum p_r^+\eta'_{P_rP_N}(P_0) + i\sum_{j=1}^{g-1}\alpha_j u'_j(P_0)\right]^{-1} \qquad (6.54)$$

It is now possible to evaluate the singular functions

$$u'_g(\tilde{z}_\pm) \sim \mp (2a_\pm)^{-1} = O(\delta^{-1})$$

$$F''(\tilde{z}_\pm) \sim \ell n\epsilon(2a_\pm^2)^{-1}\left[i\alpha_g - \sum p_r^+\eta'_{P_rP_N}(P_0)\delta\right] = O(\ell n\epsilon.\delta^{-1})$$

$$\partial_{\tilde{z}_+}\partial_{\tilde{z}_-} G(\tilde{z}_+,\tilde{z}_-) \sim [\delta + (a_+ - a_-)(\ell n\epsilon)^{-1}]^{-2} = O(\delta^{-2})$$

$$\tilde{\rho}_+ - \tilde{\rho}_- \sim \delta\left[\sum_{r=1}^{N} p^+\eta'_{P_rP_N}(P_0) + i\sum_{j=1}^{g-1}\alpha_j u'_j(P_0)\right] = O(\delta) \qquad (6.55)$$

The jacobian J of (4.49) now becomes of order $\delta^2(\ell n\epsilon)^{-3}$.

One can now use the above estimates to determine the putative divergence character of the type II superstring. The relevant factor from that case

$$J.|u_i'(\tilde{z}_+)u_i'(\tilde{z}_-)|^4|f''(\tilde{z}_+)F''(\tilde{z}_-)|^{-3}|\partial_{\tilde{z}_+}\partial_{\tilde{z}_-}G(\tilde{z}_+,\tilde{z}_-)|^2 \qquad (6.56)$$

which reduces, using (6.55), to have order

$$\delta^{-4} \qquad (6.57)$$

(where ϵ-dependence is unchanged). Combining with the measure $d^2(\tilde{\rho}_+ - \tilde{\rho}_-) \sim d^2\delta$ leads to the aparent singularity

$$\int_0 |\delta|^{-3} d|\delta| = \infty^2 \qquad (6.58)$$

On the other hand the factor for the heterotic string replacing (5.11) is

$$J[u_i'(\tilde{z}_+)u_i'(\tilde{z}_-)]^2|F''(\tilde{z}_+)F''(\tilde{z}_-)|^{-2}.\partial_{\tilde{z}_+}\partial_{\tilde{z}_-}G(\tilde{z}_+,\tilde{z}_-) \qquad (6.59)$$

This now has order, instead of (5.12), equal to

$$\delta^0 \qquad (6.60)$$

and therefore (5.13) is replaced by the finite integral

$$\int_0 |\delta| d|\delta|. < \infty \qquad (6.61)$$

The heterotic string therefore does not suffer from the apparent divergence afflicting the type II superstring. When account is taken of the quartic terms, it will be shown in Chapter 7 that the leading singularity (6.58) is removed in the type II case, and the finite term

(6.61) in the heterotic case. Due to the difficulty experienced in Chapter 7 in attempting to close the 10-dimensional SUSY algebra for the type II case it is not possible, at the present, to consider cancellation of the non-leading order divergences beyond (6.58).

It is now necessary to consider the more general use of two interacting points approaching each other, $\tilde{z}_\alpha \sim \tilde{z}_\beta$, but the surface not degenerating (so that detImΠ does not tend to zero or infinity). That possibility was considered in a little detail, at one loop, in [91]. It corresponds to disentangling the simultaneous degenerations considered in this section.

Let us consider, first, the specific case of one loop. We use, as an example, a twice-punctured torus with punctures z_1, z_2. The Green's function is

$$G(z,z_1,z_2) = \ell n[\chi(\upsilon_1|\tau)/\chi(\upsilon_2|\tau)] \tag{6.62}$$

where $\nu_j = (2\pi i)^{-1}\ell n(z/z_j$ and $\tau = w'/w = e^{2\pi i w}$. The equation determining the two interaction points is

$$G' = 0 = z^{-1}(\ell n|z_2/z_1|(\ell n|w|/2\pi i)^{-1}$$

$$+ \partial_{\nu_1} \ell n\theta_1(\nu_1|\tau) - \partial_{\nu_2} \ell n\theta_1(\nu_2|\tau)] \tag{6.63}$$

(where θ_1 is one of the Jacobi θ-functions in the notation of Bateman [96])

whilst that for $\tilde{z}_1 = \tilde{z}_2$ is

$$G'' = 0 = \partial^2_{\nu_1} \ell n\theta_1(\nu_1|\tau) - \partial^2_{\nu_2} \ell n\theta_1(\nu_2|\tau) \tag{6.64}$$

This may be reduced, using $\partial^2 \ell n\theta(\nu|\tau) = \text{const.} + \text{const.}[\theta_0(\nu_1|\tau)/\theta_1(\nu_1|\tau)]^{-2}$ [96] to

$$[\theta_0(\nu_1|\tau)/\theta_1(\nu_1|\tau)]^2 = [\theta_0(\nu_2|\tau)\theta_1(\nu_2|\tau)]^2 \quad (6.65)$$

with solution $\nu_1+\nu_2 = n\tau$ or

$$(z^2/z_1z_2) = w^n. \quad (6.66)$$

Then (6.63) may be rewritten in terms of $\alpha = \nu_2-\nu_1$ as

$$\zeta(\omega\alpha) = \alpha\eta-\pi i\ \mathrm{Im}\alpha(2\omega\ \mathrm{Im}\ \tau)^{-1} \quad (6.67)$$

where $\eta = \zeta(1)$ and ζ is the Weierstrass ζ-function with poles at $\pm 2m\omega \pm 2n\omega'$ and quasi-periodicity [96]. Due to the pole and quasiperiodic structure ζ, Eq.(6.67) has a solution α for any τ. This may be seen to be specifically when, say $\omega' = i$, $\omega = 1$, when (6.67) reduces, to

$$\zeta(\alpha) = \alpha(\eta-\frac{1}{2}\pi).$$

This equation may be seen to have the solution

$$\alpha \sim 1.1i,$$

using the tabulated values of $\zeta(ix)$ [97]. In general, this solution may then be used with (20) to construct a value of $\tilde{z}$ as $(\tilde{z}/z_1) = (z_2/z_1)^{1/2}w^{n/2}$, where for $|w| < |z_1| < |w|^{1/2}$ we may take n = -1 and for $|w|^{1/2} < |z_1| < 1$ choose n = 0; each case leads to $|w| < |\tilde{z}| < 1$.

A smilar situation is to be expected for higher genus surfaces. In this case the mapping function F(z) is

$$F(z) = \sum_r p_r^+ .[\eta(z,z_r)+u_i(z)(\mathrm{Im}\Pi)^{-1}_{ij}\mathrm{Im}u_j(z_r)] \quad (6.68)$$

so the conditions $F' = F'' = 0$ for coincident roots become

$$\sum_r p_r^+ . [\eta'(z,z_r)+u_i'(z)(\mathrm{Im}\Pi)^{-1}_{ij}\mathrm{Im}u_j(z_r)] = 0 \tag{6.69}$$

$$\sum_r p_r^+ . [\eta''(z.z_r)+u_i''(z)(\mathrm{Im}\Pi)^{-1}_{ij}\mathrm{Im}u_j(z^r)] = 0 \tag{6.70}$$

For genus 1, the second term in (6.70) may be removed by choosing the local co-ordinate to be u(z), reducing (6.70) to (6.64). For g > 1 the same local co-ordinate may be used, without removal of the second term in (6.70), which is non-analytic in the moduli of the surface.

In spite of the property of non-analyticity just noticed, general continuity arguments indicate that there should be $\frac{1}{2}$ (2g+N-2)(2g+N-3) solutions of (6.69) and (6.70) for non-degenerate values of the period matrix Π, so that for det ImΠ ≠ 0. Even higher order coincidences $\tilde{z}_\alpha \sim \tilde{z}_\beta \sim \ldots \sim \tilde{z}_\delta$ are expected to occur without the latter degeneration, even up to the maximum possible of all (2g+N-2) interaction points coming together; we have already discussed the simplest case of that when g=1, N=2, and seen no degeneration of Π. Such a possibility is also natural in terms of the 3g+N-3 available moduli on the surface, since then the above maximum degeneration need only use up (2g+N-3) of the moduli. The handle or dividing geodesic degenerations can then arise from suitable limits in the remaining g moduli. There is exactly 1 component of the Deligne-Mumford compactification of the moduli space of stable curves of genus g, corresponding to a handle degeneration, which can occur in one of g possible ways, i.e. by letting any one (or more) of the $\alpha_i \to 0$.

It is possible, using the techniques of [53)], to give an explicit proof that the surface with any number of loops and co-inciding interaction points, but without handle or dividing goedesic degeneration, is indeed a Riemann surface. This will be done for three specific examples, the first for one loop, the other two for two loops.

For simplicity only two external strings will be considered, although it is straightforward to include any number of further external strings.

The L.C. strip diagram for the first example is shown in figure 21(a). The interaction points a,b,c,d,e,f are all to be identified, and the strips A,A′ to F,F′ are also to be identified in pairs.

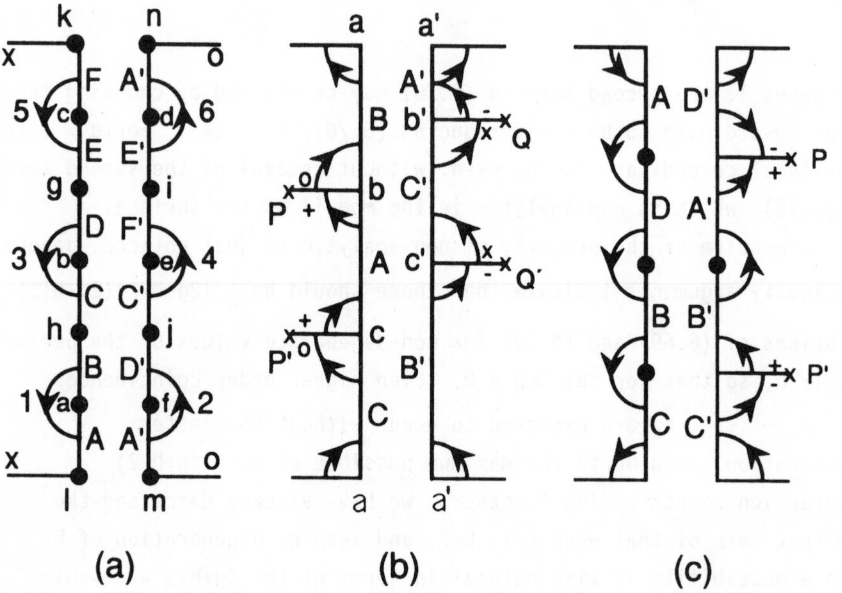

Fig. 21 The associated Riemann surface to fig. 8(a); the notation is given in the text
(b) Ditto for fig 8(b)
(c) Ditto for fig 8(c)

A complete circuit of the interaction point is shown on the figure, giving a total circulation angle of 6π. This implies, following [53], that

$$z_a = (\rho - a)^{\frac{1}{3}} \tag{6.71a}$$

so that

$$F'(z) \sim (z-\tilde{z})^2 \tag{6.71b}$$

and there is a double zero $F'(z)$ at $\tilde{z}$. It is exactly that case which was considered analytically earlier in this sub-section, by means of equations (6.63)-(6.67). The present argument shows very simply that the 1-loop surface with coincident interaction points is still a Riemann surface (without degeneration), as was proven earlier by analytic techniques. We note that the Euler characteristic χ = F-E+V is zero, since F=2, E=8 and V=6, where the two faces are I and II, the eight edges xx (identified), oo (identified), A,B,...,F and the six vertices are the points a=b=c=d=e=f,g,h,i,j and k=ℓ=m=n. Thus the surface is a torus, with g=1, as required.

The preceeding analysis, using the ideas of [53] can also be extended to the multi-loop case with coincident interaction points. Only the 2-loop case is considered here, although the method clearly generalises to arbitrary genus. The case of two coinciding points is shown in figure 21(b). These two points are denoted by a,...,d′, all of which points are to be identified, as are the strips A,A′,B,B′,C,C′ in pairs respectively. The non-coinciding interaction points are P,P′ (to be identified) and Q,Q′ (similarly identified). There are also identifications along the slits marked by -,+,x and O. On the figure is shown a complete 6π encircling of the coincident interaction points, so leading to the same behaviour (6.71).

Similar identifications are made in figure 21(c), which denotes three coincident points on a 2-loop surface, with similar identifications as in the previous figures. The neighbourhood of the common interaction point has a total angle of 8π, so that now (6.71) must be replaced by

$$z_a = (\rho - a)^{\frac{1}{4}}$$

$$F'(z) \sim (z - \tilde{z})^3$$

corresponding to a triple zero in F', as required by three coincident interaction points.

The above techniques thus gives a proof of the general result that interaction points may coalesce independently of degeneration of the Riemann surface as a handle or dividing goedesic degeneration occurs. The transversality of these different limiting configurations of the surface has already been shown analytically in the handle degeneration case by the analysis at the beginning of this section, in which the distance between P_0 and Q_0 may be allowed to go to zero independently of the radius of the handle. However the proof given by Figure 14 is more general (including the dividing geodesic degeneration), and thereby more satisfactory.

Let us now consider the contribution from the superstring multi-loop amplitudes from the sole degeneration $\tilde{z}_\alpha \sim \tilde{z}_\beta$ of type (β). This is an analysis independent of that done earlier, since it will use the configurations with $F'' \sim 0$ instead of $F'' \sim \infty$ as occured earlier in this section. The simplest case is the heterotic, since the insertion factors have no contribution from the L-movers. We need only work locally in $\nu = \tilde{z}_\alpha - \tilde{z}_\beta \sim 0$, and so have contributions to the amplitude as $\nu \sim 0$ (to leading and non-zero order) given by

$$9(F''^{\prime}_\alpha)^2 \iint |\nu|^4 d^2\nu 4F''_\alpha F''_\beta{}^{-1} (\nu)^{-2} [\tfrac{1}{3}(\theta^4_\alpha (2F''_\alpha)^{-2} + \theta^4_\beta (2F''_\beta)^{-2})$$

$$+ \theta^{2i}_\alpha \theta^{2i}_\beta (4F''_\alpha F''_\beta)^{-1}] \tag{6.72a}$$

If we use $F''_\alpha \sim -F''_\beta \sim -\nu F''^{\prime}_\alpha$ then (6.72a) reduces to be proportional to

$$\iint \nu^{-5}\bar{\nu}\, d^2\nu = -\frac{1}{2}\int_C \nu^{-2}\bar{\nu}\, d\bar{\nu} = \frac{1}{2}\int_C \nu^{-1} d\nu \qquad (6.72b)$$

where C is a small contour surrounding the origin and we have used $d\rho = -d\bar{\rho}$ on C, due to the fact that the contour C is initially to be chosen, before deformation, to have constant time τ on Σ. Thus $\bar{\nu}^2 d\bar{\nu} = -\nu^2 d\nu$ there. In order to remove the r.h.s. of (6.72b) we must include the quartic and quadratic field terms in the string field theory Hamiltonian to be discussed in the next Chapter. Inserting the correct constant of proportionality in (6.72b), equal to $(3/2\theta_\alpha^4(F_\alpha''))^{-2}$, the term (6.72b) will be shown in Chapter 7 cancelled by the contribution of H_4 (when $\tilde{z}_\alpha, \tilde{z}_\beta$ are on different handles) or of H_2 when $\tilde{z}_\alpha$, $\tilde{z}_\beta$ are on the same handle).

The expression (6.72b) is finite, so that its removal by the quartic (or quadratic) terms is not necessary to achieve finiteness of the multi-loop amplitudes. In other words the original proof of finiteness in the third paper of [33] was correct, though incomplete. The quartic contact interaction contributions remove the non-Lorentz covariant contribution (6.72b) (since the interaction positions are themselves non-covariant, as emphasized in [34] and [35]). The similar situation appears much more serious for the types I or II superstring, for which the corresponding boundary terms equivalent to (6.72b), are divergent, and hence have to be cancelled even to obtain finiteness. Thus the contact interactions play a more important role in that case. It has been shown in [98] that the leading singularity cancels in the type II case.

Finally we note that (6.72a) would be a logarithmically divergent integral (dilaton pole) if it were not initially regularized by removing a small disc, with boundary C, so as to represent it in the form (6.72a). The boundary contribution is then cancelled by the contact term, independently of the shape of the boundary of C. Such a regularisation is the natural way of combining the cubic and quartic contact terms so as to have a well-defined Lorentz-covariant expression. It may also be regularised to zero by phase integration as used extensively in [85], but independence of the shape of C is clearly to be preferred. Thus in this case, (6.72a) is replaced by

$$\iint |\nu|^{-6} d^2\nu = -\frac{1}{4}\int (\epsilon\bar{\epsilon}^2)^{-1} d\bar{\epsilon} \qquad (6.72c)$$

(3,1),(1,3) or (1,1), must also be considered. The leading order term (6.72c) is cancelled by the quartic or quadratic terms H_4, H_2 of the superstring field theory, exactly as in the heterotic case (though now cancelling a putative tachyon pole), and the terms with (n,m) = (3,1) or (1,3) may be set to zero by phase integration. The remaining term has a logarithmic divergence in it, corresponding to a putative dilaton pole. This does not appear to be able to be cancelled by terms in H_2 or H_4, and at tree level it may be seen to be explicitly cancelled between various terms of the amplitude (for details see [35]).

A similar analysis to the tree case can be performed at the multi-loop level. For four external massless states, a similar analysis to that in [55] may be performed, say for the external bosonic states, to remove any explicit dependence on the spatial components of the external momenta. The resulting expression for the amplitude depends in any non-covariant aspects only on the interaction point positions. In particular non-leading order singularities may arise from uncontracted products of $\partial_{\tilde{z}_1} X$ with $\partial_{\tilde{z}_2} X$, as $\tilde{z}_1 \sim \tilde{z}_2$. By the technique of [55] the particularly relevant factor in the amplitude containts the invariant energies explicitly as

$$(\partial_{\tilde{z}} X^I)^2 = tG'_{\tilde{z}1} G'_{\tilde{z}x} + s\, G'_{\tilde{z}0} G'_{\tilde{z}x} - (s+t) G'_{\tilde{z}1} G'_{\tilde{z}0} \qquad (6.72d)$$

where the limit $\tilde{z}_1 = \tilde{z}_2 = z$ has been taken in (6.72d). The factor (6.72d) is to be multiplied by the remaining factors in the insertion factor, in particular with the singular term

$$|F''_1F''_2|d^2x.(F''_1F''_2)^{-\frac{1}{2}}.\{\frac{\theta_1^4}{(2F''_1)^2}$$

$$\times \exp[-\tfrac{1}{2}(tG_{x1}+sG_{x0})] \qquad (6.72e)$$

The first part of (6.72e) arises from the measure $d^2\rho_{21}$, which has been transformed into the x-plane; the second factor is the conformal factor associated with $\overline{\partial_{\tilde{z}_1} X.\partial_{\tilde{z}_2} X}$, the third factor is that from the relevant θ-variables in the insertion factor at $\tilde{z}_1$ and $\tilde{z}_2$, where $\theta_i = \theta(\tilde{z}_i)$, i = 1,2; the final factor is that of Virasoro-Shapiro.

At tree level, the following identities hold, as may be shown by explicit computation (using that $\tilde{z}(\tilde{z}-x)(\tilde{z}-1)=(-p_1^+p_2^+p_3^+p_4^+/p_2^{+4})^{\frac{1}{2}}x(x-1)$):

$$G'_{\tilde{z},0}G'_{\tilde{z}x} - G'_{\tilde{z},1}G'_{\tilde{z},0} = cG'_{x,0}$$

$$G'_{\tilde{z},1}G'_{\tilde{z},x} - G'_{\tilde{z},1}G'_{\tilde{z},0} = cG'_{x,1} \qquad (6.72f)$$

with $c = (p_2^{+4}/p_1^+p_2^+p_3^+p_4^+)^{\frac{1}{2}}$. Let us assume that (6.72f) is true at multi-loop level. Then (6.72d) and (6.72e) together may be combined to replace (6.72d) by $\partial/\partial x$ acting on the exponential factor (6.72e). When combined with the remaining part of the singular curly bracket in (6.72e), the factor (6.72d) and the curly bracket in (6.72e) become

$$\frac{\theta_1^4}{(2F_1'')^2}\frac{\partial}{\partial x} - \frac{1}{2\tilde{z}_{21}^2}\left[\tfrac{1}{3}\left(\frac{\theta_1^4}{(2F_1'')^2}+\frac{\theta_2^4}{(2F_2'')^2}\right) - \frac{\theta_1^2\theta_2^2}{4F_1''F_2''}\right]$$

This may be further combined with the quartic contact interaction to be derived in Chapter 7 (which has an additional R-moving factor $\tilde{V}_3(P_1)\tilde{V}_3(P_2)$). Due to the similar form of the terms in (6.72g) to the

tree level case, we may show cancellation of the logarithm divergence present in the two terms. However the higher degenerations, when more than two interaction points coalesce, is much more difficult to analyse in this manner; it may be necessary to include higher order contact interactions to remove divergences arising in these cases. In the heterotic amplitude (6.12c) there is still a potential divergence from the vanishing of the partition function $P(\Sigma)$ at a degeneration. This will be discussed seperately in section 6.9 after the effect of dividing goedesic degeneration on the right-moving supersymmetric vertex factors has been analysed.
The special case of coalesence of interaction points at tree level for $p_1^+=-p_4^+$ is considered in appendix XII. this is a singular case indicating a surface deneration not present when $p_1^+\neq-p_4^+$. This case is considered in detail in [38].

Section 6.8: Dividing Goedesic Divergences

An analysis may now be performed of the possible divergences which could arise when a dividing geodesic length $\ell \to 0$ following the estimates of the various relevant functions in section 6.5. It appears necessary to consider separately the cases

(a) not all P_r's $\in S_1 \cap S^*$

(b) all P_r's $\in S_1 \cap S^*$

But in fact case (a) cannot arise, since it is clear from section 6.5 that none of the variables $\rho^*, \alpha_i, \beta_i$ become of $O(t)$ in that case. In other words the value $t = 0$ does not enter as a boundary point over the range of integration. Thus only the case (b) need be considered in detail.

The discussion of section 6.5 showed that there are g_1 IIV's P_α on $S_1 \cap S^*$ and g_2 on $S_2 \cap S^*$. Moreover the following estimates result:

$$F^{*''}(\tilde{P}_\alpha) = O(1) \qquad \tilde{P}_\alpha \in S^1 \cap S^* \tag{6.73a}$$

$$= O(t) \qquad \tilde{P}_\alpha \in S_2 \cap S^* \tag{6.73b}$$

$$\alpha^*_i,\ \beta^*_i \ = O(1) \qquad 1 \le i \le g_1 \tag{6.73c}$$

$$= O(t) \qquad g_1 < i \le g_1 + g_2 \tag{6.73d}$$

$$\tilde{\rho}^*_\alpha \ = O(1) \qquad \text{on } S_1 \cap S^* \tag{6.73e}$$

$$= O(t) \qquad \text{on } S_2 \cap S^* \tag{6.73f}$$

$$G^{*/}_{z_1 z_N}(z) = O(1) \qquad z, z^1 \in s_1 \cap s^* \tag{6.73g}$$

$$= O(1) \qquad z \in S_1 \cap S^*, z^1 \in S_2 \tag{6.73h}$$

$$= O(1) \qquad z, z^1 \in S_2 \cap S^* \tag{6.73j}$$

It is necessary to be more careful on the estimate (6.73h). In fact the results of section 6.5 of of Fay [93] that the second Abelian differential

$$w(x,y) = O(t) \qquad x \in S_1 \cap S^*,\ y \in S_2 \cap S^* \tag{6.74}$$

shows that

$$\eta'_{XY}(z) = \int_Y^X w(x',z)dx' = \int_{A_1}^X w(x',z)dx' + O(t) \tag{6.75}$$

Thus $\eta_{XY}(z)$ is independent of Y in the region (6.74). The line factors

$$\partial_{\tilde{z}_+} \partial_{\tilde{z}_-} G_{\tilde{z}-z_N}(\tilde{z}_+) = O(t) \text{ if } \tilde{z}_1 \in S_1 \cap S^*, \tilde{z}_- \in S_2 \cap S^* \tag{6.76}$$

Similarly the line factors

$$\sum_{s=1}^{N-1} G'_{z_N}(\tilde{z}_-, x_s) P_s = O(t) \text{ if all } z_s \in S_1 \cap S^*; \tag{6.77}$$

The contribution to these line factors for $z_s \in S_2 \cap S^*$ are also O(1), by (6.73j).

The more detailed estimates given by (6.73)-(6.77) can now be applied to the type II superstring amplitudes (6.76). Then the crucial factors in that amplitude can be written schematically as

$$\prod_{i>g_1} |u'_i(\tilde{P})|^8 \prod_{P_\alpha \in S_2 \cap S^*} |F''(\tilde{P}_\alpha)|^{-3} \Pi^2 d\tilde{\rho}^* \prod_{i>g_1} d\alpha_i d\beta_i \Pi |\partial_+ \partial_- G|^2 \tag{6.78}$$

The leading order in t contribution from the first factor is O(1), from the second $O((t)^{-6g_2})$, from $\Pi d^2\rho^*$ of $O(t^{4g_2})$, from the next factor is $O(t^{2g_2-1}dt)$ and the final factor contributes at least O(1) (where one $\tilde{\rho} \in S_1 \cap S^*$ has been fixed by translation invariance). The net factor of t that results is therefore at least going to zero as

$$t^{-1}dt \tag{6.79}$$

This is infinite and implies that the supersymmetric vertex factors give an unbounded contribution. The amplitude will have to be integrated by parts, with the boundary term cancelled by the contact terms of the next chapter. Since the contact term construction was incomplete for type II, that analysis will not be pursued further here.

In the case of the heterotic string, (6.78) is to be replaced by

$$\prod_{i>g_1} (u_i'(\tilde{P}))^4 \prod_{\tilde{P}_\alpha \in S_1 \cap S^*} [F''(\tilde{P}_\alpha)]^{-2} d\tilde{\rho}_\alpha \prod_{i>g_1} d\alpha_1 d\beta_i \Pi \partial_+ \partial_- G \tag{6.80}$$

The net factor t arising from this factor is now

$$t^{2g_2-1} dt \tag{6.81}$$

which is integrable, since $g_2 \geq 1$. Thus the heterotic string is finite from this degeneration.

Section 6.9: The Bosonic Partition Function

In the preceeding three sub-sections a careful analysis has been given of the possible divergences which may arise from the right-moving vertex factor contributions to the heterotic amplitude on degenerating Riemann surfaces as well as the companion left-moving vertex factor contributions to the type II superstring amplitudes. Putative divergences were discovered in the latter, the leading terms of which were shown to be absent on use of integration by parts, phase factor integration and inclusion of quartic field terms. There is still further analysis to perform on the heterotic amplitudes before it can be concluded that they are finite. In particular it is still necessary to analyse the putative tachyonic pole in the left-moving bosonic partition function $(\det\partial)^{-12}$ of (6.12c).

Removal of the non-holomorphic factor $[\det \mathrm{Im}\Pi]^{-4}$ from $(\det\partial_{\bar{\rho}}/\det'\Delta_0)^4$ has left the partition function $(\det\partial)^{-12}$ as a holomorphic section F of the determinant line bundle λ^{-12}, where λ is the Hodge line bundle on the moduli space M_g. The extension of this bundle to the stable compactification $\overline{M}_g$ [31] leads to a pole in F(y) of order one, as the explicit calculation in [84] shows. Thus

$$F(y) \sim \frac{1}{y_1} \text{ as } y_1 \sim 0 \tag{6.82}$$

where $y_1,\ldots,y_{3g-3}$ are the complex co-ordinates on M_g in which locally the curve $y_1 \to 0$ is the tranversal to the boundary $D = \overline{M}_g - M_g$. It is possible to relate this degeneration to the one of section 6.4 by taking, by suitable modular transformation,

$$\Pi_{12} = \frac{1}{2\pi i} \ell n y_1 \tag{6.83}$$

The handle degeneration analysis of section 6.4 can be related directly to (6.83) by the modular transformation

$$\Pi'_{11} = -(\Pi_{11})^{-1},\ \Pi'_{1i} = \Pi_{1i}/\Pi_{11}, \Pi'_{ij} = \Pi_{ij} \tag{6.84}$$

The dividing geodesic degeneration of section 6.5 cannot be so obtained, and will be discussed later.

The integration over the resulting singularity will then be of form

$$\int d^2 y_1 [y_1 |y_1|^2 (\ell n|y_1|)^1]^{-1} f_L(y_1, \ell n|y_1|) f_R(\bar{y}_1, \ell n|y_1|) \tag{6.85}$$

In (6.85), f_L, f_R are the contributions from the L and R moving modes at the degeneration, other then the y_1^{-1} arising from L-boson partition function. The measure $d^2 y_1 [|y_1|^2 . (\ell n|y_1|)^3]^{-1}$ comes from the identification $y_1 = (\epsilon e^{i\alpha})^2$ and the original measure $d(\ell n\epsilon)$ times suitable powers of $\ell n\epsilon$ from the partition function, as discussed earlier.

An alternative method of arriving at the measure in (6.85) multiplying f_L and f_R is by changing variables to the analytic ones, y_i, on Teichmüller space by means of the Jacobian, in the notation of [84]

$$j = (\det \mathrm{Im}\Pi)(\det \Delta_{-1})(\det \Delta_0)$$

and use of the work of Bost and Jolicoeur [82]

$$(\det \bar{\partial}_z)^{-12} = (\det \mathrm{Im}\Pi)^{-6} f_L(\bar{y})$$

giving the resultant measure $d\Omega(y)(\det \mathrm{Im}\Pi)^{-5}$. Near the degeneration $y_1 \sim 0$, $d\Omega \sim d^2y(|y|^2 \ell n|y|)^{-2}$, so that with (6.83) the value of the measure in (6.85) results.

The analysis of section 6.7 showed that $f_R(y_1, \ell n|y_1|)$ is bounded as $Y_1 \sim 0$ (and even as Y_1 ,$Y_2 \sim 0$ jointly with coinciding internal interaction vertices). Moreover the construction f_R from various products of derivative of Green's functions ensures that f_R is a function of y or $\ell n|y|$, but not explicitly of argy other than as in powers of y. This result is crucial to the analysis of (6.85), and follows from various features:

(i) all Feynman line contributions can be reduced to products of sums of third Abelian differentials $d\eta$ or of first Abelian terms $du_i(\mathrm{Im}\Pi)^{-1}_{ij}\mathrm{Im}u_j$ or $du_i(\mathrm{Im}\Pi)^{-1}_{ij}du_j$.

(ii) First and third Abelian differentials are analytic in the modular variables, and the first Abelian integrals are also analytic, at a degeneration $y_1 \sim 0$. This can be seen, for example, by use of the explicit Burnside expansions [69)], or by general arguments [66),99)]. Thus choosing y_1 to be the multiplier of the first generator γ_1, with fixed points 0 and ∞, the possible singularities in $du_i(z)$ can be seen explicitly not to be of form logw but always involve terms of O(w).

(iii) Only $\ell n|y|$ enters in $(\mathrm{Im}\Pi)^{-1}$, so that argy does not enter.

(iv) Factors F" are constructed only from the third Abelian differentials $d\eta$. The fact that f_R is a function of y_1 or $\ell n|y_1|$ then follows from the explicit expressions for the R-moving Feynman line contribution, together with (i)-(iv). The dependence of f_L solely on y_1 and $\ell n|y_1|$ as $y_1 \sim 0$ is also discernable by inspection. This is immediate for the H-function since it is constructed using Π and not $\mathrm{Im}\Pi$, and also in terms of the first Abelian differentials.

If $y_1 = re^{i\theta}$ the integral (6.85) can be written as

$$\int_0^{2\pi} dr[r^2(\ell nr)^7]^{-1}\int_0^{2\pi} e^{-i\theta}d\theta f_L(re^{i\theta},\ell nr)f_R(re^{-i\theta},\ell nr) \tag{6.86}$$

where as $r \sim 0$

$$f_L \sim c_L + c_L' re^{i\theta} + O(r^2)$$

$$f_R \sim c_R + c_R' re^{-i\theta} + O(r^2) \tag{6.87}$$

where the expressions on the r.h.s of (6.87) are to within powers of logr. Thus c_L, c_L', c_r, c_R are all functions of ℓnr. The function c_L', c_R' when regarded as depending on ℓnr, are bounded as $r \to 0$, since they both arise from powers $(\mathrm{Im}\Pi)^{-1} \sim (\ell nr)^{-1}$ associated with the first Abelian factors discussed in (i) above. The only possible singular term in (6.86), using (6.87), and integrating over the phase θ, is

$$\int_0 c_L' c_R' (\ell nr)^{-7} d(\ell nr) \tag{6.88}$$

The properties of c_L', c_R' noted above ensure that this integral is convergent. It is to be noted that such a detailed analysis is already necessary at 1 loop where the integral over the phase is crucial to remove the tachyon pole divergence; this completes the analysis of the divergences in [21)] and [100)].

The degeneration of a dividing geodesic seems easier to handle, since there is no singularity in $\mathrm{Im}\Pi$, so it is effectively constant near the degeneration $y_1 \to 0$. All of the other factors in f_R are bounded near this point, as shown in the previous section, and are analytic $y_1 \sim 0$ following the argument earlier for the handle geodesic degeneration but without powers of $\ell n|y_1|$ entering. A similar feature occurs for $f_L(y)$. The integral (6.85) still arises, but is now immediately seen to be finite without the additional remarks associated with c_R, etc. needed (since the quantities c_R, c_R', c_L, c_L' of (7.5) are now constants).

Section 6.10: Further degeneration analysis

The degenerations of type (δ) are well known at tree level to produce physical poles corresponding to the creation of real intermediate states. The only case which might be thought dangerous is when all of the z_r's tend together. However, if $z_r = z_0+\epsilon_r$ the momentum dependent factor $\prod_{r<s} |\epsilon_r-\epsilon_s|^{1/2p_rp_s}$ may be used to prevent any singularity. For the multi-loop case, and even at tree-level in the form of the amplitudes (6.12) care must be taken with the coincidence when some sources come together. For the degree (equal to the number of zeros minus the number of poles) of the Abelian differential $F'(z)$ is $2(g-1)$ and must be preserved under such analytic deformations as the coincidence of sources. Thus if n sources z_r coincide, (n-1) poles have been lost, and so (n-1) zeros (or interacting points) must also disappear, converging to the common z_r's, or to each other. The latter case has already been discussed, so let us consider the former. There will be no singularity in the amplitude arising from Y^A's, as can be seen from (6.12a), since as $\tilde{z}_\alpha \sim \tilde{z}_r$ the denominator factor $F''^{-1/2}$ cancels any singularity arising from $G'(\tilde{z}_\alpha,\tilde{z}_r)$.

The above discussion also applies mutatis mutandis to the degeneration of case (γ), where again the constancy of the degree of F' requires that (n-1) sources approach any set of n coinciding interaction points or to each other. Either of these cases have already been discussed above.

Finally we appeal to the expected transversality of the various degenerations (already shown in the multi-loop degeneration $\tilde{z}_\alpha \sim \tilde{z}_\beta$ above and in the joint handle degeneration and the $\tilde{z}_\alpha \sim \tilde{z}_\beta$ degeneration in section 6.6 or in section 5 of the third paper of [12]), to cover the case (ϵ), except of the following.

It is possible to take a handle degeneration simultaneously with one (or more) of the sources converging to the identified points P_0 or Q_0 at the end of the pinched handle (where P_0, Q_0 were defined in section 6.4). Co-ordinates for these sources may be taken as

$$z_r - z_+ = \epsilon^{x_r} e^{i\theta_r}$$

where $0 \leq \theta_r \leq 2\pi$ and x_r lies in a finite interval not containing 0. In this case the analysis of section 6.4 still applies, although that of 6.6 is modified in the following manner. In the expression for $F^{*\prime}$, the interaction points $\tilde{z}_+$ which previously converged $z_\pm$ (as in (6.48)) no longer need do so. This is due to the further factor of $\ell n\epsilon$ in the numerator of the second term on the r.h.s of (6.26), due to (6.89). Thus the previous estimate of $F''(\tilde{z}_\pm)$ is modified to be $O(1)$. However from (6.16) $u_g'(\tilde{z}_\pm)$ is now $O((\ell n\epsilon)^{-1})$. Finally, for the same reason as for the modification of estimates for $F^{*\prime}$, it is necessary to replace the Jacobian (6.51) by $O((\ell n\epsilon)^{-2})$. The net effect of these changes is to replace the measure in (6.52) by

$$(\ell n\epsilon\delta)^{-2} d(\ell n\epsilon) d\delta \tag{6.90}$$

There is now an external momentum-dependent factor in $\ell n\epsilon$ arising from the term proportional to $(\log\epsilon)^{-1}$ in $G^*(z_r, z_s)$ of (6.15) in the VVS factor (if z_r, z_s both approach z_+ as in (6.89)). From (6.15) this factor is proportional to

$$\exp[p_r p_s x_r x_s \ell n\epsilon] \tag{6.91}$$

Finally the integration over EIV's may be replaced by that over the set of z_r, with a factor proportional to $\ell n\epsilon$; the appropriate measure in z_r is given by $d^2 z_r . |z_r|^{-2} \sim d\, \ell n|z_r - z_+| d\, \mathrm{erg}(z_r - z_+)) \sim (\ell n\epsilon) dx_r d\theta_r$. thus the $\ell n\epsilon$-dependent expression in an amplitude is

$$\int \Pi dx_r d\theta_r \; d(\ell n\epsilon)\exp(p_r p_s \ell n\epsilon . x_r x_s).(\ell n\epsilon)^{M-2} \tag{6.92}$$

where there are supposed to be M external sources approaching P_+. This expression may be simplified to be proportional to

$$\int \Pi dx_r (p_r p_s x_r s_s)^{1-M} \tag{6.93}$$

This expression is finite, in general, where any divergence at the origin or co-ordinates x_r is not to be considered since the approximations used in the analysis breakdown since the source z_r does not then approach z_+. However it is necessary to consider if there might arise an infra-red divergence in (6.93) on integration over the external momenta to produce external states. This integration has measure

$$\prod_{r=1}^{M} \int \frac{d^9 p_r}{|\underline{p}_r|} \tag{6.94}$$

A scaling argument indicates that no divergence can arise from (6.93) and (6.94). However if M=N, and so only (N-1) independent four-momenta are integrated over in (6.94), one does find a logarithmic infra-red divergence in the four-dimensional situation. This has, however, to be analysed in more detail for the four dimensional case before such a conclusion can be fully justified.

CHAPTER 7: CLOSURE OF THE ALGEBRA

Section 7.1: Introduction

The light-cone gauge field theory approach to closed superstrings developed so far has the advantages over the covariant approach that in the former (a) multi-loop amplitudes are explicitly unitary (since they arise from a hermitian second quantised field theory) (b) there are no unphysical fields which have to be ultimately removed, and (c) there is no ambiguity arising from integration over the odd moduli of super-Riemann surfaces, since only moduli enter [101] (items (b) and (c) are closely related. These positive features of the light-cone gauge field theory have allowed a construction of multi-loop amplitudes to be made in Chapter 5 both for the type II and heterotic closed superstrings, the latter based on an extension [30],[31] of the original construction of [29]. This has further allowed an analysis of the divergences in these multi-loop amplitudes to be made in Chapter 6. No infinities were found in the heterotic case, and only those arising from partial integrations at coalescences of interaction points in the type II case. It is to be emphasized that the constructions and analysis of those chapters only used the cubic Hamiltonian of sections 3.3 and 4.6.

Contemporaneously to the analysis originally presented it was pointed out [34],[35] that a solely cubic Hamiltonian must violate the ten-dimensional super-Poincare algebra sP_{10}. This is because the SUSY sub-algebra of sP_{10} ensures positivity of the Hamiltonian whilst explicit analysis of the cubic Hamiltonian indicates that the latter is unbounded from below [34],[35]. This is clear from the fact that in the simple expression (x^2+gx^3) the cubic term dominates the quadratic one for large $|x|$, and so there is no lower bound. The conclusion that a quartic term was essential in the Hamiltonian, and hence in other generators of sP_{10}, was later drawn in [37] by different methods. Thus the field theory of Chapters 3 and 4 and hence the amplitudes of Chapter 5 and the divergence analysis of Chapter 6 are incomplete. The generators of the supersymmetry sub-algebra of sP_{10} had been obtained in detail in [50], and this construction extended to the whole of sP_{10} in [47]. In particular the crucial generators discovered in [47] were $J_3{}^{I-}$, $J_3{}^{+-}$ (where I= 1 to 8 and $\pm$ have the ususal L.C. definition, and the subscript 3 denotes the power of the fields in the generators). In order to construct the full multi-loop amplitutes for closed superstrings, and in particular to guarantee their invariance under sP_{10}, it is necessary to determine the quartic contributions to all of the generators of sP_{10}. Since the L.C. sub-algebra is linearly realised,it is thus required to determine $J_4{}^{+-}$, $J_4{}^{I-}$, $Q_4{}^{-A}$, $Q_4{}^{\bar{A}}$. The values of these for the heterotic superstring will now be presented, as will a discussion of the complete closure of sP_{10} at quartic level, obviating the need for any higher order generators. Comments on difficulties in the corresponding construction for type II superstrings will also be noted. If the algebra can be closed not only will the heterotic amplitudes proven finite in Chapter 6 have the correct ten-dimensional supersymmetry, but the short-string limit used in sections 5.3-5.7 will be justified (since this limit uses a Lorentz transform to go to the short-string frame).

The original methods used in [29] to close the SUSY algebra at the cubic level, and extended in [47] to $J_3{}^{+-}$, $J_3{}^{I-}$, were in terms of mode analysis. It has been claimed in many discussions on string field theory that such analysis is the only one which can be made reasonably rigorous, although already divergent sums arise in the algebra closure at the cubic level. The resulting manipulations and regularisations can be readily understood in functional form, as already noted in [50], and used extensively in [32],[33]. The functional approach has been outlined in sections 5.1 and 5.2, and used heavily since then. It was apparent already in [47] that mode methods lead to very hard summations, some being so difficult that not all of the identities for cubic closure in [47] were proven. Functional methods were used to attack the problem of cubic and quartic closure of the SUSY sub-algebra sP_{10} more directly in [36]. The resulting quartic heterotic Hamiltonian H_4 was used, with the cubic term H_3, to complete the construction of multi-loop heterotic amplitudes and analyse their finiteness [54].

The earlier situation on the functional approach to closure of the sP_{10} algebra, both for the heterotic and type II closed superstring (as well as a description of the amplitude construction and the finiteness analysis) was summarised in the review [38]. The results on the closure of sP_{10} presented there were incomplete, since in particular the quartic generators $J_4{}^{+-}$, $J_4{}^{I-}$ for the heterotic string were not analysed in detail, although a conjectured form for them was shown to give zero values for $[J_3,J_4]$ and $[J_4,J_4]$ contributions. Nor were the values of $\bar{Q}_4{}^A$, $\bar{Q}_4{}^{\bar{A}}$ obtained in that case, although zero values for the latter in the case of the heterotic superstring were found consistent with the SUSY sub-algebra of sP_{10}. Thus the situation was that full Lorentz invariance was not guaranteed for multi-loop amplitudes, though it was to be expected. Nor could the explicitly covariant multi-loop representation of [38] and [102] be justified, since the latter depended crucially on the use of such invariance.

The above defects for the heterotic superstring will be remedied in this section by constructing $J_4{}^{+-}$, $J_4{}^{I-}$ and completing the cubic closure of [47] by use of functional methods. In so doing we will justify the product form for these quartic terms raised as a possibility in [38], and hence also validate the covarant multi-loop representation and the finiteness results obtained in Chapters 5 and 6.

We do not attempt to analyse the type II superstring at the same level of detail. It has already been shown in [36],[38] for type II that Q_4^{-A} and $Q_4^{-\bar{A}}$ must be non-zero in order to close the SUSY sub-algebra of sP_{10}. However the singularities present for type II for the various commutators were noted to be of $O(\infty^2)$, and that the conjectured forms of $Q_4{}^A$, $Q_4{}^{\bar{A}}$ require use of special regularisation (called phase-integration, (in which $\int z^{-n}dz=0$ for n>1) in order to close the SUSY sub-algebra. This latter regularisation was also found essential to close the higher order commutators $[J_3, J_4]_-$, $[J_4, J_4]_-$. Such regularisation - dependence was absent from our analysis of the heterotic superstring algebra in [36),38)] since the terms being considered were all finite in that case. It will be seen by our present more complete analysis that these degrees of divergence of commutators persist at the quartic level in the sectors including J_4^{+-}, J_4^{I-} for both heterotic and type II.

Section 7.2: The General Approach

We will briefly summarise the functional approach here before discussing closing the algebra. As before, the superstring co-ordinate variables will be $\{\underline{X}(\sigma), \theta^{\bar{A}}(\sigma)\} \equiv \{Z(\sigma)\}$ (we only consider the fermionic L-movers explicitly in Z, although the R-movers will have to be present for type II). The corresponding field operator annihilating a string at Z is $\Phi_{p^+}(Z,\tau)$, at the (Euclideanised) L.C. time τ; the string parameter length is $\ell = 2\pi|p^+|$. The 1st - quantised generators of sP_{10} will be denoted g (it will also be a functional of the conjugate variable W to Z). The corresponding 2nd - quantised generators, at quadratic level, will be

$$G_2 = \int_0^\infty \ell d\ell \int DZ\, \Phi^+_{p^+} g \Phi_{p^+} \tag{7.1}$$

where conjugate variable W is replaced by the operator $\delta/\delta(iZ)$ on the field Φ_{p^+} in (7.1). The introduction of a cubic interaction H_3, corresponding to two strings splitting or fusing at an interaction point P, will lead also to cubic terms in the non- L.C. sub-algebra of sP_{10} ie to non-zero values of P^-, J_3^{+-}, J_3^{I-}, $Q_3^{-\bar{A}}$, $Q_3^{-\bar{A}}$. The general form of G_3 in these cases will then be, as already noted in (3.20),

$$G_3=\lambda\int_0^\infty \prod_{r=1}^{3} \ell_r d\ell_r dZ_r \delta(p_3^+-p_1^+-p_2^+) DZ_r \Delta(Z_1,Z_2;Z_3) V_3{}^G(P)[\Phi^+_{p_1}\Phi^+_{p_2}\Phi_{+p_3}+h.c] \quad (7.2)$$

In [9] the values of $V_3^{Q^{-\bar{A}}}$, $V_3^{Q^{-A}}$, V_3^H were obtained by closing the ten-dimensional SUSY algebra, (see (3.22))

$$V_3^{Q^{-\bar{A}}} = Y^{\bar{A}}(\tilde{z}), \quad V_3^{\bar{Q}^{-A}} = \frac{2}{3} Y^{3A}(\tilde{z})$$

$$V_3^H = [2F''(\tilde{z})]^{\frac{1}{2}}[2^{-\frac{1}{2}}\partial_z x^L(\tilde{z})+Y^{2i}\partial_z X^i+(\tfrac{\sqrt{2}}{3})\partial_z X^R Y^4] \quad (7.3)$$

with $\quad Y^{\bar{A}}= [2F''(\tilde{z})]^{-\frac{1}{2}} Q^A(\tilde{z}), \; Y^{3A} = \epsilon^{ABCD} Y^B Y^C Y^D.$

The expressions (7.3) are written in a form which arises on taking matrix elements between external string states. The function F(z) corresponds to the Mandelstam map $\rho=\tau-i\sigma=\Sigma_r \ell_r G(z_j,z_r)$ between the superstring world sheet co-ordinate ρ and a local uniformising variable z on the corresponding Riemann surface.

The method of [50] was to use perturbation theory in the string coupling constant λ in order to close the algebra of sP_{10} at succesively higher orders. In order to assess the need for quartic terms it was argued in [36] that matrix elements of commutators must also be taken in order to obtain the explicit (tho' generic) singularity structure. This is straightforward at cubic level, since matrix elements of fields can be reduced to first quantised (functional) form by well-known methods [28], as discussed in sections 3.4, 4.5 and 5.1.

However such reduction may not necessarily be valid for more general string interactions.

In order to proceed we will assumed that all higher order interactions and generators $G_n (n \geq 4)$ are local. This means that they have only one interaction point P, so will correspond to contact interactions. Moreover it will be assumed that these contact interactions may be obtained by taking limiting configurations of a world sheet with only cubic interaction points. The limit occurs when two (or more) of the interaction points coalesce.

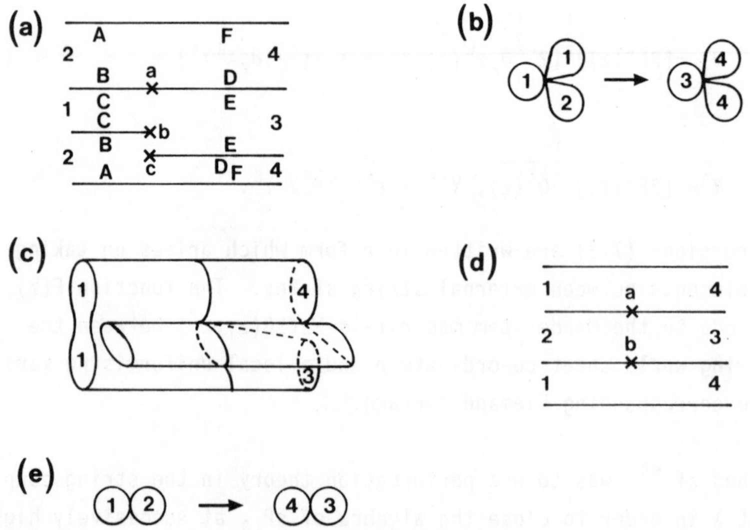

Fig.22 (a) The L.C. diagram arising when two interaction points p,p^1 coincide, in the evaluation of the matrix element $<12|[Q_3,Q_3]|34)$ for closed superstrings. The points a,b,c are to be identified, as are the edges marked with the same symbol. (b) As in (a), but a configuration space picture of the transformation from initial to final state, at the interaction time. (c) A three-dimensional picture of the time evolution of (a). (d) The degenerate case of (a) $p_1^+ = -p_4^+$ for the open string. (e) The degenerate case of (a) when $p_1^+ = -p_4^+$ for the closed string.

These limiting situations have been shown earlier to correspond to a Riemann surface. Moreover before the coalescence occurs, matrix elements involving these contact terms may be calculated by reduction to functional form, as in the cubic case. The resulting expressions in the coalescing limit, will then be regarded as defining the corresponding contact interactions. It is clear from [36),38)] that only such contact interactions seem needed to close the algebra at quartic level, since only such terms survive in calculations of various brackets $[G_3, G_3']$. Since the algebra sP_{10} will be shown to close for the heterotic string with only such contact terms the above assumptions appear to have been justified. Finally it may be interesting to note that a differential geometry of string fields based only on contact interactions, as string field products has been developed elsewhere [121)]. This is a natural product structure if there were to be any chance of direct gauge fixing to the light-cone contact products considered above.

The methods of [32),33),38)] may now be used to express the massless mode matrix elements of any commutator $[G_m, G_n']$ (for m,n>2) as a sum of functional integrals of the form

$$\Pi d\alpha_i\, d\beta_i\, \Pi d^2\tilde{\rho}_\alpha \; \exp[-\tfrac{1}{2} \sum_{r<s} p_r p_s G(z_r, z_s)\}\langle \Pi V_3^H(\tilde{\rho}_\alpha). \quad [V_m^G(P) V_n^{G'}(P') - V_m^G(P') V_n^{G'}(P)]\rangle \tag{7.4}$$

where the expectation value < > is defined by taking all possible contraction of variables X, λ and θ in (7.4) entering in the insertion factors V_m, V_n, with a single contraction being given by (see section 5.2)

$$\overbrace{X^I(\tilde{z}_\alpha) X^J}(\tilde{z}_\beta) = -\tfrac{1}{2} G(\tilde{z}_\alpha, \tilde{z}_\beta)\delta^{IJ} \tag{7.5}$$

$$\overline{\tilde{\lambda}(z_\alpha)\tilde{\theta}(z_\beta)} = -\frac{1}{\pi}\partial_{\tilde{z}_\alpha} G(\tilde{z}_\alpha,\tilde{z}_\beta)$$

and V^G_m, $V^G_n{}'$ are assumed to be functions only of $\partial_\rho x^I$, θ^A and λ at P or P′, for m,n≥3. In the expression (7.4) the variables α_i and β_i are the internal loop widths and twists, $\tilde{\rho}$ are the interaction positions (excluding the points P and P′). Ultimately $Y^{\bar{A}}z$ is to replaced by its value obtained by integrating out the conjugate variables λ^A to $\theta^{\bar{A}}$ so as to lead to the solution $\partial_{\bar{\rho}}\theta=0$ from section (5.2), as

$$Y^{\bar{A}}(\tilde{z}) = [2F''(\tilde{z})]^{-\frac{1}{2}}\left[\sum_{r=1}^{N}\partial_z G(\tilde{z},z_r)\theta_r p_r^+ + \sum_{i=1}^{g}\theta^i U_i'(\tilde{z})\right] \qquad (7.6)$$

with θ_r being the value of $\theta(z)$ at the external string position z_r with momentum p_r, and u_i' are the g first Abelian differentials on t the Riemann surface of genus g and θ^i are the zero modes of θ (external state wave functions should also be folded in to (7.4)). As assumed earlier, $V^G_m(P)$ or $V^G_n{}'$ (P′) are to be constructed by taking coalescences on (m-2) or (n-2) interaction points respectively. The manner in which these limits will be taken will be determined in the succeeding sections (see also [36),38)]).

The structure of matrix elements of $[G_2, G_m]$ takes a different form from (7.4). This is due to the fact that each G_2 has the expression at 1st quantised level [47)]

$$g_2 = \int d\sigma(g_2(\sigma)+f.\partial/\partial\ell) \qquad (7.7)$$

In (7.7), g_2 is a function of λ^A, $\theta^{\bar{A}}$ and $\partial_{\tilde{\rho}}x^I$, f a function of the zero modes ℓ or p^+. Thus it is clear that care must be taken in the evaluation of matrix elements of $[G_2, G_m]$ by the integration over the variables λ^A. It was already noted in [50] that evaluation of the field commutator brackets in $[G_2, G_m]$ leads to a modified vertex factor in which g_2 for each string acts so that at the contact interaction point P in G_m one has the insertion factor

$$\left[\sum g_{2r} \right] V_m^G(P) \tag{7.8}$$

By string continuity (7.8) leads in (7.9) to the expression

$$\oint d\sigma \left[g_2(\sigma) + \sum f_r \partial/\partial \ell_r \right] V_m^G(P) \tag{7.9}$$

where the contour integral $\oint$ encircles the contact point P. The expression (7.9) now replaces the antisymmetrised product of $V_m^G(P)\, V_n^G{}'(P')$ in the matrix elements (2.4) in order to obtain matrix elements of $[G_2, G_m]$. The manner in which the contributions from different coalescing interaction points are calculated will be discussed in detail in section 7.5.

It is the purpose of this Chapter to close the super-Poincaré algebra sP_{10}, which has generators which will be denoted generically by J,Q,P. We use the notation that the bracket [,] denotes a graded commutator. To simplify the notation we will not use explicitly the coefficients of the sP_{10}, but instead we denote by $\cong$ equality when the sP_{10} coefficients are included. Those are given in appendix IX. We intend to use the graded Jacobi indentities, valid for graded commutators, to reduce the number of independent commutators defining sP_{10} to a smaller set. In the following two sections non-linear realisations at cubic and quartic level will then be constructed of this smaller set of commutators. The assumptions to be made in this section are the following:

(i) $[J,J] \cong J$ (7.9a)

$[J,Q^{+}] \cong Q$ (7.9b)

(ii) $[Q^{+A},Q^{+B}] = 0$ (7.9c)

$[Q^{+\bar{A}},Q^{+\bar{B}}] = 0$ (7.9d)

$[Q^{+\bar{A}},Q^{+B}] = 2\delta^{\bar{A}B}p^{+}$ (7.9e)

$[P^{+},Q^{-A}] = 0$ (7.9f)

where Q^{-A} is defined by the l.h.s. of (7.9b) (the coefficients on the r.h.s. of (7.9b) being given in detail in Appendix IX).
(iii) Jacobi identities are valid for the J and Q^{+} operators
Assumption (iii) is satisfied if the bracket [,] is a graded commutator, though its detailed validity at quartic level will be discussed in section 7.5 (where the coalescence process has to be specified). The assumption is certainly valid at quadratic and cubic level. It is the purpose of this section to deduce the remaining set of brackets for sP_{10}, viz:

(a) $[J, Q^{-A}] \cong Q$ (7.10a)

$[J, Q^{-\bar{A}}] \cong Q$ (7.10b)

(b) $[Q, Q] = 0 = [\bar{Q}, \bar{Q}]$ (7.10c)

$[Q, \bar{Q}] = P$ (7.10d)

(c) $[J, P] = P$ (7.10e)

(d) $[P, Q] = 0$ (7.10f)

(e) $[P, P] = 0$ (7.10g)

To prove (a), let J^{α} denote any of J^{IJ}, J^{+-}, J^{-I} or J^{+I}. If (7.10b) is used to define $Q^{-A} \cong [J^{L-},Q^{+A}]$, we then have

$$[J^{\alpha},Q^{-A}] \cong [J^{\alpha},[J^{L-},Q^{+A}]] \cong [[J^{\alpha},J^{L-}],Q^{+A}]$$

$$+[[J^{\alpha},Q^{+A}],J^{L-}]\cong[J,Q^{+A}]+[Q^{+},J^{L-}]\cong Q \qquad (7.11)$$

where (i) has been used in the last step of (7.11). and the coefficient of Q arising on the r.h.s. of (7.11) may then be checked using the details of the coefficients in Appendix IX to give the known results for $[J^{\alpha}, Q^{-A}]$. Thus in detail for $J^{\alpha}=J^{+-}$,

$$[J^{+-},Q^{-A}] =-i[J^{+-},[J^{L-},Q^{+A}]]$$

$$=-i[[J^{+-},J^{L-}],Q^{+A}]+i\ [[+J^{+},Q^{+A}],J^{L}]$$

$$=-[J^{L-},Q^{+A}]\ -\frac{1}{2}[Q^{+A},J^{L-}]$$

$$=-\frac{1}{2}\ Q^{-A} \qquad (7.12)$$

and agrees with the coefficients in Appendix IX for this bracket. For J's other than J^{+-}, say J^{L-}

$$[J^{L-},Q^{-A}]\cong[J^{L-},[J^{i-},Q^{+\bar{B}}]\rho^{iBA}\cong\{[[J^{L-},J^{i-}],Q^{+B}]-$$

$$[[J^{L-},Q^{+\bar{B}}],J^{i-}]\}\rho^{iBA} = 0$$

using the detailed coeffients in (i). Similarly

$$[J^{L-},Q^{-\bar{A}}]\cong[J^{L-},[J^{R-},Q^{+\bar{A}}]]\cong[[J^{L-},J^{R-}],Q^{+\bar{A}}]$$

$$-[[J^{L-},Q^{+\bar{A}}],J^{R-}]=0$$

$$[J^{R-},Q^{-A}]\cong[J^{R-},[J^{L-},Q^{+A}]]=0$$

$$[J^{R-},Q^{-\bar{A}}]\cong[J^{R-},[J^{i-},Q^{+A}]]=0$$

$$[J^{i-},Q^{-A}]\cong[J^{i-},[J^{L-},Q^{+A}]]\cong[[J^{i-},Q^{+A}],J^{L-}]$$

$$\cong \rho^{iAB}[Q^{-\bar{B}},J^{L-}]=0$$

$$[J^{i-},Q^{-\bar{A}}]\cong[J^{i-},[J^{R-},Q^{+\bar{A}}]]\cong[[J^{i-},Q^{+\bar{A}}],J^{R-}]$$

$$\cong\rho^{i\overline{AB}}[Q^{-B},J^{R-}]=0$$

so that $$[J^{I-},Q^{-A}]=[J^{I-},Q^{-\bar{A}}]=0 \qquad (7.13)$$

This proves (a).
To prove (b), consider first (using standard commutator or anticommutator notation [,]$\pm$)

$$\begin{aligned}&[Q^{-B}, Q^{+A}]_+= i[[J^{L-}, Q^{+B}]_-, Q^{+A}]_+\\ &= i[[J^{L-}, Q^{+A}]_-, Q^{+B}]_+ +i\ [[Q^{+B}, Q^{+A}]_+, J^{L-}]_-\\ &= - [Q^{-A}, Q^{+B}]_+\end{aligned} \qquad (7.14)$$

From (7.14) it follows that each side is symmetric in A, B, so that

$$[Q^{-B}, Q^{+A}]_+= -\surd 2\ \rho^{jAB}P^j \qquad (7.15)$$

so defining P^j. Note that we used the graded Jacobi identity in (7.14).
A similar analysis shows that

$$[Q^{+\bar{A}}, Q^{-\bar{B}}]_+= \surd 2\rho^{j\overline{AB}}\bar{P}^j \qquad (7.16)$$

for some $\bar{P}^j$. To show that P^j and $\bar{P}^j$ are identical one may proceed as follows:

$$2^{-\frac{1}{2}}\ i\rho^{i\bar{C}\bar{B}}[Q^{-B},Q^{+A}]_+ = [[J^{i-},Q^{+\bar{C}}]_-, Q^{+A}]_+$$

$$= {}_{-}[[J^{i-},Q^{+A}]_{-},\ Q^{+\bar{C}}]_{+}-[[Q^{+\bar{C}},Q^{+A}]_{+},J^{i-}].$$

$$= 2^{-\frac{1}{2}}\ i\rho{=}^{iAD}[Q^{-\bar{D}},Q^{+\bar{C}}]_{+}\ -\ 2\delta^{A\bar{C}}[P^{+},J^{i-}] \tag{7.17}$$

so that using (7.15) and (7.16) in (7.17)

$$-i\rho^{i\bar{C}\bar{B}}\rho^{jAB}P^{j} = i\rho^{iAD}\rho^{j\bar{C}\bar{D}}P^{j}\ -\ 2\delta^{A\bar{C}}[P^{+},J^{i-}].$$

or

$$\frac{1}{2}\ i\ \rho^{ij\bar{C}A}\ (P^{j}-P^{\bar{j}}) + \delta^{\bar{C}A}\ (---) = 0 \tag{7.18}$$

where the matrices ρ^{ij} were introduced in (A.6) of [50]. Independence of the two separate terms on the l.h.s. of (7.18) leads to the desired result.

Next, one may take

$$[Q^{-\bar{B}},Q^{+A}]_{+} \cong [[J^{R-},Q^{+\bar{B}}],\ Q^{+A}]$$

$$\cong [[J^{R-},Q^{+A}],\ Q^{+\bar{B}}] + [[Q^{+\bar{B}},Q^{+A}],\ J^{R-}]$$

$$\cong \delta^{A\bar{B}}(---)$$

Therefore P^{R} may be defined by

$$[Q^{-\bar{B}},Q^{+A}]_{+} = 2\ \delta^{A\bar{B}}P^{R} \tag{7.19}$$

Similarly

$$[Q^{-B},\ Q^{+\bar{A}}] \cong [[J^{L-},Q^{+B}],Q^{+\bar{A}}]$$

$$\cong [[Q^{+\bar{A}},Q^{+B},], J^{L-}] \cong \delta^{B\bar{A}}(---)$$

so that P^L is defined by

$$[Q^{-B},Q^{+\bar{A}}]_+ = 2\delta^{B\bar{A}} \; P^L \tag{7.20}$$

Furthermore

$$[Q^{-A},Q^{-B}]_+ \cong [[J^{L-},Q^{+A}], Q^{-B}] \cong [[J^{L-},Q^{-B}], Q^{+A}] + [[Q^{-B},Q^{+A}], J^{L-}] \tag{7.21}$$

The r.h.s. of (7.21) is antisymmetric in A,B, by (7.15), whilst the ι.h.s. of (7.21) is symmetric in A, B, so that

$$[Q^{-A},Q^{-B}]_+ = 0 \tag{7.22}$$

One may show similarly that

$$[Q^{-\bar{A}},Q^{-\bar{B}}]_+ = 0 \tag{7.23}$$

Finally in (b),

$$[Q^{-\bar{B}},Q^{-A}]_+ \cong [[J^{R-},Q^{+\bar{B}}], Q^{-A}] \cong [[Q^{-A},Q^{+\bar{B}}], J^{R-}]$$

$$\cong \delta^{A\bar{B}} \; [P^L,J^{R-}]$$

so that $H = P^-$ may be defined by

$$[Q^{-A},Q^{+\bar{B}}]_+ = 2\delta^{A\bar{B}}H \tag{7.24}$$

To prove (c),

$$[J,P] \cong [J,[Q,Q]] \cong [[JQ]Q] \cong [Q,Q] \cong P \tag{7.25}$$

Here the successive ≅'s needed the use of (b), (iii), (a) and (b) respectively. One may prove (d) as follows. Firstly

$$2\delta^{A\bar{B}}\ [P^{+},Q^{+C}]_{-} = [[Q^{+A},Q^{+\bar{B}}]_{+},Q^{+C}]_{-}$$

$$= -\ [[Q^{+\bar{B}},Q^{+C}]_{+},\ Q^{+A}] = -2\delta^{\bar{C}B}[P^{+},Q^{+A}]_{-} \qquad (7.26)$$

and one may easily see from (7.26) that

$$[P^{+},Q^{+C}]_{-}=0$$

In a similar manner it may be shown that

$$[P^{+},Q^{+C}]_{-}\ = 0$$

Next one has

$$[[Q^{+A},Q^{-\bar{B}}]_{+},Q^{+C}]_{-}\ = 2\delta^{A\bar{B}}[P^{R},Q^{+C}]_{-}= -[[Q^{-\bar{B}},Q^{+C}]_{+},Q^{+}]_{-}$$

$$= -2\delta^{C\bar{B}}[P^{R},Q^{+A}]_{-}$$

and again results that

$$[P^{R},Q^{+A}]_{-}=0$$

Similarly

$$[P^{L},Q^{+\bar{A}}]_{-}=0$$

Finally

$$2\delta^{A\bar{B}}[P^{L},Q^{+C}]_{-}=[[Q^{+\bar{A}},Q^{-B}]_{+},Q^{+C}].$$

$$=-[[Q^{+\bar{A}},Q^{+C}]_{+},Q^{-B}]_{-}\ -[[Q^{-B},Q^{+C}]_{+}Q^{+\bar{A}}]_{-}$$

$$=-2\delta^{C\bar{A}}[P^+,Q^{-B}] + \sqrt{2}\rho^{jCB}[P^j,Q^{+\bar{A}}]_- \tag{7.27}$$

If $[P^L,Q^{+C}]_-$ is denoted by M^C, $[P^+,Q^{-B}]_-$ by N^B, then symmetrisation of (7.27) in B,C gives

$$\delta^{\bar{A}(B}M^{C)} + \delta^{\bar{A}(C}N^{B)} = 0 \tag{7.28}$$

and on taking the trace of (7.28) in $\bar{A}$ and B one obtains

$$M^C = - N^C \tag{7.29}$$

But from (7.9f), $N^B=0$, so by (7.29), $M^C=0$, and from (7.27)

$$\rho^{jCB}[P^j,Q^{+A}]_-=0 \tag{7.30}$$

One may use

$$\sqrt{2}\rho^{j\bar{A}\bar{B}}[P^j,Q^{+\bar{C}}] = [[Q^{+\bar{A}},Q^{-\bar{B}}]_+,Q^{+\bar{C}}]_-$$

$$= - [Q^{-\bar{B}},Q^{+\bar{C}}]_+,Q^{+\bar{A}}]_-=-\sqrt{2}\rho^{j\bar{C}\bar{B}}[P^j,Q^{+\bar{A}}]_-$$

to deduce that, with $L^{j\bar{C}}=[P^j,Q^{+\bar{C}}]_-$,

$$\rho^{j\bar{A}\bar{B}}L^{j\bar{C}} + \rho^{j\bar{C}\bar{B}}L^{j\bar{A}} = 0 \tag{7.31}$$

From (7.30) and (7.31) follows that

$$[P^j,Q^{+\bar{A}}]_- = 0$$

One may similarly prove that

$$[P^R, Q^{+C}]_- = [P^j, Q^{+\bar{A}}]_- = 0$$

and from

$$[J^{RI}, Q^{-A}] \cong \rho^{iA\bar{B}} Q^{-B}$$

that

$$[P^+, Q^{-\bar{B}}] = 0$$

Finally one may use (c), and in detail the relation

$$iP^I = [J^{I-}, P^+]_-$$

to obtain

$$[P^I, Q^{-A}]_- = [P^I, Q^{-\bar{A}}]_- = 0$$

and use that

$$[J^{I-}, P^K] = i\,\delta^{IK} H$$

to obtain that

$$[H, Q] = 0$$

for Q^+ and Q^-.

(e) may then be obtained from

$$[P,P] \cong [P,[QQ]] \cong [[PQ]Q] = 0$$

It has thus been shown that the relations (i) and (ii), together with the Jacobi identity, leads to the construction of the whole of the algebra sP_{10}. This feature allows for a considerable reduction in the calculations required to close the algebra of sP_{10}. What has been done is to put the onus of the closure on the [J,J] commutators, since from [47] those latter relations are clearly the most difficult. The action of $J_2{}^{I-}$ had not been attempted to be understood in [36],[38] in the functional approach. This question will be considered in detail in the following sections.

The result of this section appears to be new, and of independent interest. Presumably something similar to it is also valid in other dimensions, although the dimensionality of space-time may appear to have been used in the split of the SUSY generators into the L.C subset $Q^{+A}, Q^{+\bar{A}}$ and the non L.C. subset $Q^{-A}, Q^{-\bar{A}}$.

It would appear that this split may also be done in general, although the presence of triality in ten dimensions may be relevant.

The extension of the above result to including R-going modes, with generators $\tilde{Q}$, may be performed in an obvious manner, and need not be given in any detail.

There is further reduction on the analysis by noticing that it is enough to construct J^{R-} satisfying the following relations

$$[J^{Li}, J^{R-}] = i\delta^{LR}\ J^{i-} \tag{7.32a}$$

which defines J^{i-},

$$[J^{Ri}, J^{R-}] = 0 \tag{7.32b}$$

$$[J^{ji}, J^{-R}] = 0 \tag{7.32c}$$

$$[J^{RL}, J^{R-}] = -\ i\delta^{LR}\ J^{R-} \tag{7.32d}$$

$$[J^{R-}, J^{i+}] = iJ^{Ri} \tag{7.32e}$$

$$[J^{R-}, J^{L+}] = -i\delta^{RL}\ J^{+-} - iJ^{RL}, \tag{7.32f}$$

which defines J^{+-},

$$[J^{R-}, J^{R+}] = 0 \tag{7.32g}$$

$$[J^{Li}, J^{j-}] = -i\delta^{ij}\, J^{L-}, \tag{7.32h}$$

which defines J^{L-},

$$[J^{Li}, J^{+-}] = 0 \tag{7.32i}$$

$$[J^{R-}, J^{i-}] = 0, \tag{7.32j}$$

to ensure the full closure of the

$$[J,J] \cong J$$

commutation relations involving J^{K-} and J^{+-}, K = L,R,k. This statement may be shown by following the same spproach we have used to show (7.10). In the next section we discuss the cubic closure which was first obtained in [47] (we get complete agreement). In the following section we obtain the closure of quartic contributions of the algebra and show there are no higher than quartic terms in the generators of sP_{10} for the heterotic superstring in 10-dim.

Section 7.3: Cubic Closure

The cubic closure is concerned, following the results of the previous section, with the brackets

$$[J_2, J_3] \cong J_3 \tag{7.33}$$

where J_3 is one of the generators J^{+-}, J^{i-}, J^{L-}, J^{R-} in the notation of Appendix IX, with $1 \leq i \leq 6$ and $A^{L,R} = \frac{1}{\sqrt{2}}(A^7 \pm iA^8)$ for any 8-vector. There is also a non-trivial consistency condition arising from (7.24), since Q^{-A}, for example, arises on the r.h.s of both

$$[J_3^{R-}, Q_2^{+\bar{A}}]_- = iQ_3^{-\bar{A}} \tag{7.33a}$$

$$[J^{i-}_{3}, Q^{+A}_{2}] = -2^{-\frac{1}{2}} i\rho^{iAB} Q_3^{-\bar{B}} \tag{7.33b}$$

This consistency, and a similar one for $Q_3^{\bar{A}}$, is checked straightforwardly once (7.33) is satisfied, so the latter is the crucial relation to satisfy. The more expanded form of (7.33) is

$$[J_2^{\alpha}, J_3^{\beta}] + [J_3^{\alpha}, J_2^{\beta}] \cong J_3 \tag{7.34}$$

where α and β are pairs of space-time indices involving the - direction. Thus the left hand side of (7.34) required the evaluation of the effect of $J_2{}^{I-}$ or $J_2{}^{+-}$ on a cubic interaction vertex. This evaluation was discussed briefly in section 7.2. It is necessary to turn first to the action of the second term of (7.8) on a cubic insertion factor before evaluating the effect of the contour integral of (7.8) on that factor. It is this former calculation which was not done in functional form in [36],[38], nor, as far as we know, anywhere else.

The insertion factors are both explicit and implicit functions of ρ_r^+, the latter feature being through their dependence on the interaction point position $\tilde{z}$. Therefore from the vanishing of $F'(\tilde{z})$ follows that

$$(\partial \tilde{z}/\partial l_s)\, F''(\tilde{z}) + G'(\tilde{z}, z_s) = 0$$

so that

$$\frac{\partial \tilde{z}}{\partial l_s} = -\frac{G'(\tilde{z}, z_s)}{F''(\tilde{z})} \tag{7.35}$$

Let us define

$$\delta = i\sum_{r=1}^{3} \ell_r.\partial/\partial\ell_r + \frac{1}{2}\, i\oint idz\theta^{\bar{A}}\lambda^{A} \qquad (7.36)$$

and evaluate the action of δ on objects associated with the cubic insertion factors of section 3.3. From the definition of the mapping F in section 2 and the vanishing of $F'(\tilde{z})$ it follows that for any function M of $\tilde{z}$ alone,

$$\sum_r \ell_r\partial/\partial\ell_r M = 0 \qquad (7.37)$$

Moreover it then follows that

$$\sum_r i\ell_r\ \partial/\partial\ell_r\ (F")^a = ia(F")^a \qquad (7.38)$$

The action of the first part of δ on Y^A of (7.3) is $-\frac{1}{2}\, i^{Y\bar{A}}$; by contraction, with the value

$$\overline{\lambda^A(z_1)\theta}{}^{\bar{B}}(z_2) = \frac{1}{2\pi}\,\delta^{A\bar{B}}\ (z_1-z_2)^{-1} \qquad (7.39)$$

one obtains the opposite contribution from the second part of δ on $Y^{\bar{A}}$, resulting in

$$\delta Y^{\bar{A}} = 0 \qquad (7.40)$$

Let us include the L-mode determinental contribution $(\Pi_p F"_p)^{-\frac{1}{2}}$ [30),31),38),62)] in the cubic Hamiltonian insertion factor by defining

$$W_3^H = V_3^H(\bar{F}'')^{-\frac{1}{2}} \tag{7.41}$$

Then by (7.38) applied also to $\bar{F}''$ one obtains

$$\delta W_3^H = -iW_3^H$$

Again by contractions and the definition of $Q^{+A}, Q^{+\bar{A}}$ [2] it is clear that

$$\delta Q^{+A} = \frac{1}{2}\, iQ^{+A} \tag{7.42a}$$

$$\delta Q^{+\bar{A}} = \frac{1}{2}\, i\, Q^{+\bar{A}} \tag{7.42b}$$

(with $Q^{+\bar{A}} = (i\sqrt{2/\pi})\int dz\theta^A(z)$, $Q^{+A} = -i\sqrt{2}\int dzF'(z)\lambda^A(z)$). Then it follows from (7.3) that

$$\delta Q^{-A} = -\frac{1}{2} iQ^{-A},\quad \delta Q^{-\bar{A}} = -\frac{1}{2} iQ^{-\bar{A}} \tag{7.43}$$

When the co-ordinates $Z=(\underline{X},\theta)$ are considered in the z-plane, then no explicit θ_s-dependence occurs, so that

$$(\partial/\partial \ell_s)\underline{X}(\tilde{z}) = (\partial\tilde{z}/\partial \ell_s)\partial_z\underline{X}(\tilde{z}) \tag{7.44}$$

and similarly for X^+ and θ. In particular, with $X^+ = 2\mathrm{Re}F(\tilde{z})$,

$$(\partial/\partial \ell_s)X^+(\tilde{z}) = 2(\partial\tilde{z}/\partial \ell_s)\partial_z F(\tilde{z}) + \mathrm{H.C.} = 0 \tag{7.45}$$

Finally

$$\sum_s p_s^k (\partial/\partial \ell_s)(V_3(\tilde{z})(\bar{\tilde{F}}'')^{-\frac{1}{2}}) = \sum p_s^k \,[\partial/\partial \ell_s) V_3].(\bar{\tilde{F}}'')^{-\frac{1}{2}} \qquad (7.46)$$

since

$$\sum p_s^k (\partial/\partial \ell_s)(\bar{F}''(z))^{-\frac{1}{2}} = -\frac{1}{2}\sum p^k \bar{G}''(z,z_s).[\bar{F}''(z)]^{-3/2}$$

$$= -\frac{1}{2}\,\partial_z^2 X^k .\,[\bar{F}''(\tilde{z})]^{-3/2} = 0 \qquad (7.47)$$

when only right-movers are considered in insertion factors, as in the heterotic. Let us assume that H_3 is known. Then at cubic order,

$$[J_2{}^{+-}, H_3] + [J_3{}^{+-}, H_2] = -iH_3 \qquad (7.48)$$

has to be solved. Using from [10] that $J_2{}^{+-} = \delta + X^+ h$ and taking matrix elements, denoted generically by < > as before, we have, considering only insertion factor on the r.h.s.

$$\langle J_2{}^{+-}, H_3\rangle = \delta W_3(\tilde{z}) + \langle X^+ h, \tilde{V}_3{}^H\rangle = -i\tilde{W}_3{}^H + \langle X^+ h, \tilde{V}_3{}^H\rangle \qquad (7.49)$$

where $\tilde{W}_3$ denotes the insertion factor (7.42) at $\tilde{z}$. Thus from (7.49) and (7.50) is required that

$$\langle X^+ h\tilde{V}_3{}^H\rangle = \langle h\tilde{J}_3{}^{+-}\rangle$$

$$\text{or}\quad \langle h, \tilde{X}^+\tilde{V}_3{}^H - J_3{}^{+-}\rangle = 0 \qquad (7.50)$$

This relation, together with

$$<Q^{+A},(\tilde{J}_3{}^{+-}-\tilde{X}^+V_3^H)>=<Q^{+\bar{A}},(\tilde{J}_3{}^{+-}-\tilde{X}^+\tilde{V}_3^H)> = 0 \qquad (7.51)$$

suggest $\tilde{J}_3{}^{+-}=\tilde{X}^+\tilde{V}_3{}^H$. In the notations of [36),38)] this reads as

$$V_3^{J+-} = X^+V_3^H \qquad (7.52)$$

(where the extra factor $(\bar{F}")^{-\frac{1}{2}}$ in (7.41) arises in (7.51) on taking matrix elements) as given in [47)]. From the relation

$$[J^{I-},P^K]_3 = i\delta^{IK}H_3$$

one expects

$$V_3^{JI-} = X^I V_3^H + F_3{}^{I-} \qquad (7.53)$$

where $F_3{}^{I-}$ has no dependence on X. This form also occurs in [47)]; we may calculate $F_3{}^{I-}$ from the relation for $[J^{I-},J^{K-}]$. After checking that $[J^{I-},J^{K-}]=0$ and $[J,Q^+]\cong Q$ at cubic level, the rest of the cubic level of the algebra of sP_{10} follows, as was shown in section 7.3.

In order to be able to acquire a satisfactory level of expertise for the quartic calculations of the next section let us determine the purely bosonic part involving ∂X^L of the cubic term in $[J^{[i-},J^{k]-}]$. The expression to be evaluated is therefore

$$<X^{[i}h,\tilde{X}^{k]}\tilde{\hat{V}}_3^H> + 2i\sum_r p_r{}^{[i}\partial/\partial\ell_r<\tilde{X}^{k]}\hat{\tilde{W}}_3> + <h,\tilde{X}^{[i}>\tilde{X}^{k]}W_3^H \qquad (7.54)$$

where we have defined

$$\hat{W}_3 = i(2|F"|^2)^{-\frac{1}{2}}\partial_z X^L$$

The first and third terms in (7.54) become

$$2|\Pi\tilde{F}''|^{-1} \oint dz[F'(z)]^{-1}X^{[i}(z)2\partial X^L \partial X^R(z).\quad \partial\tilde{X}^L\ \tilde{X}^{k]} \tag{7.55}$$

In (7.55) no contraction should be taken on the variable $X^i(z)$, since cancellation occurs between such contractions and those arising between point-split variables $X'(z)$ and $h(z)$ [103]. Using the contractions of (7.5) and a clockwise circulation around $\tilde{z}$, with $F'(z)=(z-\tilde{z})g(z)$, one obtains for (7.55)

$$-i\ (2|\tilde{F}''|)^{-1}\ \partial_z^2(\bar{g}^1 X^{[i}\partial X^L)_{z=\tilde{z}}\ \tilde{X}^{k]} \tag{7.56}$$

The second term in (7.54) may be evaluated using the identity

$$(\partial/\partial p_r^{+}\ \overset{1}{)}\ ((F'')^{-2}\partial_z X^L) = -(i/2)G''(z,z_r)(F'')^{-2/3}\partial_z X^L$$

to obtain

$$(i/2)(F''^{-1}\bar{F}''^{-\frac{1}{2}})\tilde{X}^{[k}\partial_z^2[(F'')^{\overset{1}{-2}}X^{i]}\partial X^L]|_{z=\tilde{z}} \tag{7.57}$$

Now using that

$$\partial(\tilde{F}'')^{-\frac{1}{2}}=--(\tilde{F}''')(\tilde{F}'')^{-3/2},\partial_z(g^{-1})|_{z=\tilde{z}}=-(1/2)\tilde{F}'''(\tilde{F}'')^{-2}$$

one obtains that (7.57) exactly cancels (7.56), so (7.54) is zero, as required for closure at the cubic level. The above analysis can be extended straightforwardly to the fermionic terms to obtain agreement with the result of [47]

For completeness we present the closure of $[J^{+-}J^{R-}]$ at cubic order. According to the argument presented earlier this closure must hold automatically. That result will now be derived. Closure of the relation for J^{+-},J^{R-} at cubic level requires the satisfaction of

$$[J_2{}^{+-},J_3{}^{R-}]_- - [J_2{}^{R-},J_3{}^{+-}] = -iJ_3{}^{R-} \tag{7.58}$$

Since $\delta J_3{}^{R-} = -iJ_3{}^{R-}$ then (7.58) reduces in matrix elements

$$\langle X^+h,J_3{}^{R-}\rangle - \langle J_2{}^{R-},X^+V_3{}^{H}\rangle = 0 \tag{7.59}$$

We use (7.52), (7.53) for $J_3{}^{R-},J_3{}^{+-}$, so that the purely bosonic part of $\langle J_2{}^{R-},J_3{}^{+-}\rangle$ becomes

$$\langle \overset{\sqcap}{X^R} h,\tilde{V}_3^H\rangle\tilde{X}^+ + \langle h,\overset{\sqcap}{\tilde{X}^R}\tilde{V}_3^H\rangle X^+ \tag{7.60}$$

The left hand side of (7.59) is therefore, using (7.60)

$$\langle X^+h,\tilde{X}^R\tilde{V}_3^H\rangle + \langle X^+h,F_3{}^{R-}\rangle - \langle h,\overset{\sqcap}{\tilde{X}^R}\tilde{V}_3^H\rangle X^+ - \langle \tilde{X}^R h,\overset{\sqcap}{\tilde{V}_3^H}\rangle\tilde{X}^+$$

$$-\langle F_2{}^{R-},\tilde{X}^+\tilde{V}_3^H\rangle - 2i\Sigma\langle p_r^R(\partial/\partial\ell_r),X^+\tilde{V}_3^H\rangle \tag{7.61}$$

In terms of Z^I defined by

$$(\tilde{F}")^{-\frac{1}{2}}\partial_z X^I(\tilde{z}) = Z^I$$

then

$$2i\sum p_r{}^R\partial/\partial\ell_r .\ Z^I = -({}^1/_2)(\tilde{F}")^{-1}\partial_z{}^2\tilde{X}^R.Z^I - (\tilde{F}")^{-1}\partial_z X^R.\partial_{\tilde{z}}Z^I \tag{7.62}$$

Since V_3^H is linear in Z^I, and considering only bosonic parts,

$$\langle X^R h, \tilde{V}_3^H \rangle = \langle X^R h, \tilde{Z}^I \rangle (\partial \tilde{W}_3^H / \partial z^I)$$

and one may use above techniques to arrive at

$$\langle X^R h, \tilde{Z}^I \rangle = -(1/2)\{\partial_z^2 [g^{-1} \partial_z X^I] . \tilde{X}^R + 2\partial_z [g^{-1} \partial_z X^R$$

$$+ g^{-1} \partial_z X^I \partial_z^2 X^R\} C \quad (7.63)$$

where $C = (F")^{-\frac{1}{2}}$ for I=L,i,R respectively.

The first term on the r.h.s. of (7.63) exactly cancels $\langle h, \tilde{V}_3{}^H \rangle \tilde{X}^R$, so that there results the relation

$$\langle h\ ,\ \tilde{V}_3{}^H \rangle\ \tilde{X}^+ \tilde{X}^R\ -\ \langle X^R h, \tilde{V}_3{}^H \rangle\ \tilde{X}^+ - 2i \sum_r p_r^R [(\partial/\partial \ell_r) Z^I] . \frac{\partial \tilde{W}}{\partial Z^I}{}^3 . \tilde{X}^+ = 0 \quad (7.64)$$

This proves (7.59) except for the fermionic terms. These can also be obtained by similar but more lengthy analysis.

Section 7.4: Quartic Closure

The discussion of section 7.2 allows representations of generators of the sP_{10} algebra in terms of insertion factors which in many aspects is similar to the description of operator algebras in terms of symbols and the star product of symbols [104), 105), 106)]. The latter description provides a functional representation of the algebra which does not depend on a given basis in the Hilbert space of states. In particular it codes in a systematic way the use of contracted and non-contracted terms coming from matrix representations of the generators.

To each generator we associate a functional, which in this representation will essentially be the insertion of the corresponding generator as has been described in section 7.2 and in [38]. We call this functional the associated symbol. To the product of two operators we associate a product of symbols which we denote by a *. It is non-commutative. We then define the commutation of two symbols in terms of the * product.

Given two symbols $A(x,\lambda,\theta)$ and $B(x,\lambda,\theta)$, where the dependence is on x,λ,θ and the z-derivatives, we define a time ordered * product by

$$A*B = \overline{A_1 B_2} + A_1 B_2 \tag{7.65}$$

We will also denote it as $A_1 * B_2$ when necessary. The subindices 1 and 2 denote evaluation of X,θ,λ and their z-derivatives at points z_1 and z_2. Points z_1 and z_2 are ordered according to light-cone time by using the Mandelstam map $\rho = F(z)$ as described in section 5.1. This definition is related to the functional regularization used earlier. In our approach we will then define the limit $1 \to 2$, but we will not give further specification

of points 1 and 2. The contracted term $\overline{A_1 B_2}$ denotes all possible contractions between the arguments at 1 and 2 according to the definitions

$$\overline{\partial X_1 \ \partial X_2} = \frac{1}{4(z_1-z_2)^2} \quad , \quad \overline{\lambda_1 \ \theta_2} = \frac{1}{2\pi(z_1-z_2)} \tag{7.66}$$

$A_1 B_2$ denotes the product of the two functionals, and are the non-contracted terms of the product of operators (in the quantum operator language).

The * product is associative and non-commutative. In fact

$$(A*B)*C = \overbracket{\overbracket{A_1 B_2} C_3} + \overbracket{A_1 \overbracket{B_2 C_3}} + \overbracket{A_1 B_2 C_3} + A_1 \overbracket{B_2 C_3} + \overbracket{A_1 B_2} C_3 +$$

$$+ A_1 B_2 C_3 = \overbracket{A_1 \overbracket{B_2 C_3}} + \overbracket{A_1 \overbracket{B_2} C_3} + \overbracket{A_1 B_2 C_3} + \overbracket{A_1 B_2} C_3 +$$

$$+ A_1 \left(\overbracket{B_2 C_3} + B_2 C_3 \right) = A*(B*C)$$

$$A*B = \overbracket{A_1 \ B_2} + A_1 B_2 \neq \overbracket{B_1 A_2} + B_1 A_2 = B*A \tag{7.67}$$

since in general

$$\overbracket{A_1 B_2} \neq \overbracket{B_1 A_2} \quad , \quad A_1 B_2 \neq B_1 A_2 \tag{7.68}$$

as we will explicitly see.

We associate to the product of operators the * product of symbols, and to the graded commutator of two operators the graded * commutator defined by

$$[A,B]^* = A*B \mp B*A, \tag{7.69}$$

(with the + sign in (7.69) for odd symbols). The [,]* satisfy the Jacobi identity. This arises from the following relation (where we take the time ordering 1,2,3 and add a subindex to each symbol to be more precise):

$$[C,[A,B]_*]_* = C_1 * (A_2 * B_3 - B_2 * A_3) - (A_1 * B_2 - B_1 * A_2)*C_3$$

$$= A_1 * (C_2 * B_3 - B_2 * C_3) - (C_1 * B_2 - B_1 * C_2) A_3 +$$

$$+ (C_1 * A_2 - A_1 * C_2)*B_3 - B_1 * (C_2 * A_3 - A_2 * C_3) =$$

$$= [A,[C,B]_*]_* + [[C,A]_*,B]_* \tag{7.70}$$

To the Hamiltonian we associate the symbols

$$\hat{H}_2 \to \tilde{h} = \int_C h \ , \ \tilde{h} = -\frac{2i}{\pi}\int_C dz \left[\frac{(\partial_z x)^2}{F'} - \pi \frac{\theta^A}{F'} \partial_z \lambda^A \right] \tag{7.71}$$

$$\hat{H}_3 \to H = \frac{1}{[2\bar{F}'']^{\frac{1}{2}}} \left[\frac{1}{\sqrt{2}} Z^L - Z^i Y^{2i} + (\sqrt{2}/3)\, Z^R\, Y^4 \right] \tag{7.72}$$

$$\hat{H}_4 \to H_4 \text{ to be determined} \tag{7.73}$$

where $Z^I = 2i \dfrac{\partial_z X^I}{(2F'')^{\frac{1}{2}}}$, $Y^{\bar{A}} = \dfrac{\theta^{\bar{A}}}{(2F'')^{\frac{1}{2}}}$, C is defined as a contour of constant τ on the Riemann surface. τ is the light cone gauge time, and is a conformal invariant definition. $\hat{H}_2, \hat{H}_3, \hat{H}_4$ are the quadratic, cubic and quartic contributions to the Hamiltonian in terms of field operators.

We then have for the * product of cubic symbols

$$H{*}H = \overline{H_1 H_2} + H_1 H_2 = \frac{1}{(4\bar{F}_1''\bar{F}_2'')^{\frac{1}{2}}} \Big[\frac{1}{3} \overline{Z_1^L Z_2^R}\, Y_2^{\ 4} + \overline{Z_1^i Z_2^j}\, Y_1^{\ 2i} Y_2^{\ 2j} +$$

$$+ \frac{1}{3} \overline{Z_1^R Z_2^L}\, Y_1^{\ 4} \Big] + H_1 H_2 = \frac{1}{(4\bar{F}_1''\bar{F}_2'')^{\frac{1}{2}}} \cdot \frac{1}{(4F_1''F_2'')^{\frac{1}{2}}} \cdot \frac{-1}{z_{12}^{\ 2}}$$

$$\left[\frac{1}{3} Y_2^{\ 4} + \frac{1}{3} Y_1^{\ 4} + Y_1^{\ 2i}\, Y_2^{\ 2i} \right] + H_1 H_2 \tag{7.74}$$

For the (quadratic * cubic) terms we obtain

$$\tilde{h} * H = \int_{C_1} h * H_2 = \int_{C_2} dz\,(\overline{h\,H_2}) + \int_{C_1} dz\; h\,.\,H_2$$

$$[\tilde{h},H]^* = \oint_C \overline{h\,H_2} + \oint_C h.H_2 \qquad (7.75)$$

where contractions are to be performed before integration. The symbol $\tilde{h}$ acting on a product yields

$$[\tilde{h}\ ,\ A{*}B] = \oint_C \Big(\overline{h_1\ \overline{A_1 B_2}} + \overline{h\,\overline{A_1}\ B_2} +$$

$$+ \overline{h\,A_1 B_2} + \overline{h\,A_1} B_2 + h\,\overline{A_1 B_2} + h\,A_1 A_2 \Big) =$$

$$= \oint_C \Big(\overline{h\ \overline{A_1 B_2}} + \overline{h\ \overline{A_1} B_2} + \overline{h\ A_1 B_2} + \overline{h\ A_1} B_2 \Big) + \oint_C h\ .\ \overline{A_1 B_2} + \oint_C .A_1 B_2 \qquad (7.76)$$

where C encircles the points 1 and 2. (7.76) may be rewritten in the following way, which we will extensively use in our calculations,

$$[\tilde{h}\ ,\ A{*}B]_* =$$

$$= \Big[\oint_{C_1} \Big(\overline{h\,A_1} + h\,A_1 \Big)\Big].\ B_2 + \Big[\oint_{C_1} \Big(\overline{h\,A_1} + h\,A_1\Big)\Big]\ B_2 +$$

$$+\left[\oint_{C_2}\left(\overbrace{\overbrace{h\,B_2} + h\,B_2\Big]\Big]\cdot A_1}\right.\right. + \left[\oint_{C_2}\left[\overbrace{h\,B_2} + h\,B_2\right]\right]\cdot A_1 \qquad (7.77)$$

where in the first and third terms of the right hand number of (7.77) we first do the integrals around point 1 excluding 2 and around 2 excluding 1, respectively, and do all possible contractions between terms in 1 and terms in 2.

It can be seen that expressions (7.76) and (7.77) are equal, which is in other words the relation

$$[\hbar\ ,\ A*B]_* = [\hbar, A]*B \quad + \quad A*\,[\hbar, B]_* \qquad (7.78)$$

which is valid for all the symbols we have defined. The function F(z) which appears in all symbols is constructed from the associated Greens function of Light Cone diagrams. The only relevant property for our construction is the singular behaviour of this function at interacting points, which are associated to our points 1 and 2. This means that our construction of the sP_{10} generators will be independent of any dynamics. This is the right property to have for the kinematic construction.

To construct F(z) we consider a punctured Riemann Surface of any genus. We then construct the Green function $G(z,\tilde{z},a,b)$ associated to it and the third abelian function G(z) whose real part is G. We define in local coordinates the usual Mandelstam map

$$F(z) = \sum_i l_r\ G(z,z_i)\quad ,\quad \sum l_r = 0$$

where z_r (the punctures) are so chosen as to have $F'(z_1) = F'(z_2) = 0$. We are interested in the limit situation $z_1 \to z_2$. This can always be done without having a degeneration of the surface as we discussed in section 6.7. We also are going to use the following notation

$$F'(z) = (z-z_1)(z-z_2)f(z).$$

We now define another symbol which is relevant to our construction. It is intrinsic for quartic constructions, but has no meaning for cubic symbol. We denote by Δh,

$$[\Delta h\ ,\ A*B]_* = \frac{1}{2}\Delta_{12}\ [\tilde{h},A]_* * B + \frac{1}{2}\Delta_{21}\ A* [\tilde{h},B]_* \tag{7.79}$$

where $\Delta_{12} = F(z_1)-F(z_2) = -\Delta_{21}$.

It would have been possible to use the symbol $\Delta_F h$ defined by

$$[\Delta_F h\ ,\ A]_* = \oint_{C_1} (F(z)-F_o)\ (\overbrace{h\ A_1} + h\ A_1)\ ,\ F_o = \frac{F(z_2)+F(z_1)}{2} \tag{7.80}$$

The construction we give here is in terms only of Δh. The procedure is exactly the same in both cases, but Δh has simpler * commutation rules than $\Delta_F h$. The presence of factors Δ_{12} in (7.80) are necessary for the closure of the algebra. It is related to the δ-weight of each generator. In fact consider the δ operator of section 3 acting on symbols,

$$\delta.\equiv\ il_r\frac{\partial}{\partial l_r} - \frac{1}{2}\oint dz\ \theta^{\bar{A}}\lambda^A.,$$

$$\delta.A = il_r \frac{\partial A}{\partial l_r} - \frac{1}{2} \oint dz \, \overline{\theta}^A \lambda^A A \tag{7.81}$$

We then have

$$\delta h = -i\tilde{h} \quad , \quad \delta H = -iH$$

$$\delta \Delta_{12} = +i\Delta_{12}$$

$$\delta(\Delta h) = 0 \quad , \quad \delta(\Delta_{12} H) = 0 \; , \; \delta(\Delta_{12} H^* H) = -i(\Delta_{12} \; H^* H) \tag{7.82}$$

We have thus to introduce such factors if we are trying to construct quartic symbols from cubic ones but preserving the δ-weights. We come back to this point later.

We now give a geometrical interpretation of Δh. Let us consider

$$\Delta_{12} \; [\tilde{h} \; , \; \partial x_1^I]_* \; = \; \Delta_{12} \oint_{C_1} \tilde{h} \, \partial x_1^I \; + \; \Delta_{12} \oint_{C_1} h \; . \; \partial x_1^I ,$$

$$\oint_{C_1} h \partial x_1^I \; = \; - \, \partial \partial \left(\frac{\delta x^I}{(z-z_1)} \right) \Bigg|_{z=z_1} \; = \; -2 \, \frac{\delta x_2^I}{z_{12}^3 f_2} + O(1)$$

$$\oint_{C_1} h = O(z_{12}^{\;2}),$$

We then have

$$\lim_{1\to 2} \Delta_{12} [\tilde{h} , \partial x_1^I]_* = \lim_{1\to 2} \left[\frac{-2\Delta_{12}}{z_{12}^{\ 3} f_2} \partial x_2^I + O(z_{12}^{\ 2})\right] = + \frac{1}{2} \partial x_2^I \tag{7.83}$$

In the same way

$$\lim_{1\to 2} \Delta_{12} [\tilde{h} , \bar{\theta}_1^A] = + \frac{1}{3} \bar{\theta}_2^A \tag{7.84}$$

We thus obtain by applying $\Delta_{12}[\tilde{h}, .]$ that symbols at point 1 are

translated to point 2 and vice-versa, with a factor in this case of 1/3. We obtain the general expression for (7.79),

$$\Delta_{12} [\tilde{h},A]_* = -\Delta_{12} \partial\partial\left[\frac{\partial x^I}{(z-z_2)}\right]\Bigg|_1 \frac{\delta A}{\delta x^I} -\Delta_{12} \partial\partial\left[\frac{\bar{\theta}^A}{(z-z_2)f}\right]\Bigg|_1 \frac{\delta A}{\delta \bar{\theta}^A}$$

$$- 4\Delta_{12} \frac{\partial x_1 \partial x_1}{z_{12} f_1} A_1 + 4\pi \Delta_{12} \frac{\theta_1 \partial \lambda_1}{z_{12} f_1} A_1 \tag{7.85}$$

which again at highest order in $\dfrac{1}{z_{12}}$ reduces to

$$= \frac{1}{3} \partial x_2^I \frac{\delta A}{\delta x^I} + \frac{1}{3} \bar{\theta}_2^A \frac{\delta A}{\delta \bar{\theta}^A} + O(z_{12}) \tag{7.86}$$

One has to be very careful, however, with expressions like (7.86) because if we again apply $\Delta_{12}[\tilde{h},.]_*$, that is if we consider

$$\Delta_{12}\ [\tilde{h}\ ,\ \Delta_{12}[\tilde{h},A]_*]_* \tag{7.87}$$

then the highest order contribution to (7.87) is not obtained by applying $\Delta_{12}\ [\tilde{h}, .]_*$ to the highest order terms in (7.86). There are also contributions coming from $O(z_{12})$. Instead expression (7.85) is exact so we can apply $\Delta_{12}[h,.]_*$ to it.
The symbols for $\hat{J}^{R-}$ were obtained in [47]. The symbol associated with the cubic generator, which we are going to use extensively, is

$$\hat{J}^{R-}_s \longrightarrow J^{R-} = x^R H + F^R,$$

$$F^R = \frac{i\pi}{\sqrt{2}} \frac{Y^{\bar{A}}\lambda^A}{(2\bar{F}'')} \tag{7.88}$$

The Δh symbol commutes with the quadratic symbol j^{R-} associated to $J_2{}^{R-}$,

$$[\Delta\tilde{h}\ ,\ j^{R-}]_* = 0 \tag{7.89}$$

and also with $\tilde{h}$,

$$[\Delta h\ ,\ \tilde{h}]_* = 0 \tag{7.90}$$

These properties follow from (7.79).
Let us introduce, for a given symbol G,

$$K_{12}(G) = \sum_{i=0}^{n} a_i \underbrace{[\Delta h, [\ \Delta h,[\ldots\ldots,G]_* \ldots]_*}_{i} \tag{7.91a}$$

$$K(G) := \lim_{1,2\to 0} \int d\Delta_{12}\ K_{12}\ (G\ ), \tag{7.91b}$$

where a_i, $i = 0,\ldots n$ are constant coefficients which may depend on $f(z_1)$, $f(z_2)$ and their derivatives at 1 and 2. A suitable choice for n will be determined later. The point $z_0 = \dfrac{z_1+z_2}{2}$.

We define

$$G_{12}{}^{R} = \Delta_{12}\ (H_1 * J_2{}^{R-} + J_1{}^{R-} * H_2) \tag{7.92a}$$

$$J_4^{R-} = K(G_{12}^{R}) \tag{7.92b}$$

We first show that (7.92) satisfy (7.32) for any a_i, i=0,...,. For (7.32b) we obtain

$$[J^{Ri}, G^{R}_{12}]_* = \Delta_{12}\ ([J^{Ri}, H]_* * J^{R-} + H * [J^{Ri}, J^{R-}]* +$$

$$+ [J^{Ri}, J^{R-}]_* * H + J^{R-} * [J^{Ri}, H]) = 0, \tag{7.93}$$

from the known * commutations of cubic symbols.

We now use that

$$[\Delta h, J^{IK}]_* = 0 \tag{7.94}$$

to obtain

$$[J^{Ri}\ , K(G^{R}_{12})]_* = 0,$$

and

$$[J^{Ri}\ , J_4{}^{R-}]_* = 0 \tag{7.95}$$

In the same way we obtain (7.32c) and (7.32d).
From (7.32a) we get $J_4{}^{i-}$,

$$J_4^{i-} = K(G_{12}^{\ i}),$$

$$G_{12}^{\ i} = \Delta_{12}\ (H_1 *\ J_2^{\ i-} + J_1^{\ i-} * H_2) \tag{7.96}$$

We may now use

$$[J^{I+},[\Delta h, A*B]_*]_* = -i[\Delta P^I, A*B]_*\ {}_*[\Delta h,[J^{I+},\Delta * B]],$$

where

$$[\Delta P^I, A*B] = \frac{1}{2}\Delta_n\ [P^I,A]*B + \frac{1}{2}\ \Delta_{21}\ A*[P^I,B],$$

to show that

$$[J_4^{\ R-}, J^{R+}] = 0,$$

$$[J_4^{\ R-}, J^{R+}] = 0. \tag{7.97}$$

Additionally, we obtain from (7.32f)

$$G_{12}^{+-} = \Delta_{12}\ (H_1 * J_2^{\ +-} + J_1^{\ +-} * H_2)$$

$$J_4^{+-} = K(G_{12}^{+-}) \tag{7.98}$$

(7.32h,i) follows in the same way using (7.94). We get for $J_4^{\ L-}$

$$G_{12}^{\ L} = \Delta_{12}\ (H_1 * J_2^{\ L-} + J_1^{\ L-} * H_2)$$

$$J_4^{L-} = k(G_n^L) \tag{7.99}$$

The only restrictions arising from (7.20) are

$$[J_4^{i-},Q^{+\bar{A}}]_* - \frac{1}{\sqrt{2}} p^{i\bar{A}\bar{B}} [J_4^{L-},Q^{+A}]_*,$$

$$[J_4^{i-},Q^{+A}]_* = -\frac{1}{\sqrt{2}} p^{iAB} [J_4^{R-},Q^{+\bar{A}}]_* \tag{7.100}$$

They are satisfied using

$$[Q^{+\bar{A}},\Delta h]_* = [Q^{+A},\Delta h]_* = 0 \tag{7.101}$$

and relations (7.101) for cubic symbols.
Condition (7.10f) yields

$$[P^+,[J_4{}^{L-},Q^{+A}]] = 0 \tag{7.102}$$

which is again trivially satisfied.
We are thus left with commutation relation (7.32j). The closure of the sP_{10} algebra reduces then to proving that there exist coefficients a_i, $i=0,\dots,n$ which allow (7.32j) to be satisfied.
First we are going to show that

$$[h,J_4{}^{R-}]_* + [H,J^{R-}]_* + [H_4,j^{R-}]_* = 0 \tag{7.103}$$

and then we prove that (7.103) implies

$$[j^{R-},J_4{}^{i-}]_* + [J^{R-},J^{i-}]_* + [J_4{}^{R-},j^{i-}]_* = 0 \tag{7.104}$$

H_4 is obtained from

$$[J_4{}^{R-},P^L] = iH_4,$$

we thus get

$$H_4 = k(G_{12}^{H})$$

$$G_{12}^{H} = 2\Delta_{12}\ H_1 * H_2$$

We have from (7.89)

$$[G_{12}^{H}, j^{R-}]_* = 2\Delta_{12}\ ([H_1, j^{R-}]_* * H_2 + H_1 * [H_2, j^{R-}]_*) =$$

$$= 2\Delta_{12}\ ([J_1^{R+}, \tilde{h}]_* * H_2 + H_1 * [J_2^{R-}, \tilde{h}]_*),$$

$$[H_4, j^{R-}]_* = K(2L_{\eta 2}([J_1^{R-}, \tilde{h}]_* 8H_2 + H_1 * [J_2^{L+} \tilde{h}]*)). \qquad (7.105)$$

The first and third terms of (7.103) then combine to give

$$[H, J^{R-}]_* + K([\Delta h, [H, J^{R+}{}_*]_*) = 0 \qquad (7.106)$$

We have now to prove the existence of $a_i, i=0,\ldots,n$ which satisfy (7.106).
We first explicitly show all different forms that arise in (7.106).
Consider all the terms contained in $[\Delta h, H]_* J$,

$$\text{i)}\ \frac{\partial x \partial x}{F'} H_1\ (x_2{}^R H_2) \qquad ,\text{ii)}\ \frac{\partial x \partial x}{F'} H_1\ (x_2^R H_2)\ ,$$

$$\text{iii)}\ \frac{\partial x \partial x}{F'} H_1\ F_2{}^R \qquad ,\text{iv)}\ \frac{\partial x \partial x}{F'} H_1\ (x_2{}^R H_2 + F_2{}^R)\ ,$$

$$\text{v)}\ \frac{\theta\partial_z\lambda}{F'} H_1\ x_2{}^R H_2 \quad ,\text{vi)}\ \frac{\theta\partial_z\lambda}{F'} H_1\ x_2{}^R H_2 \quad ,\text{vii)}\ \frac{\theta\partial_z\lambda}{F'} H_1\ \theta_2\ \lambda_2\ ,$$

$$\text{viii)}\ \frac{\theta\partial_z\lambda}{F'} H_1\ x_2{}^R H_2 \quad ,\text{xi)}\ \frac{\theta\partial_z\lambda}{F'} H_1\ \theta_2\ \lambda_2 \quad ,\ \text{x)}\ \frac{\theta\partial_z\lambda}{F'} H_1\ \theta_2\ \lambda_2\ ,(7.107)$$

$$\text{xi)}\ \frac{\partial x_1\partial x_1}{z_{12}f_1} H_1\ x_2{}^R H_2 \quad ,\text{xii)}\ \frac{\partial x_1\partial x_1}{z_{12}f_1} H_1\ x_2{}^R H_2,$$

$$\text{xiii)}\ \frac{\partial x_1\partial x_1}{z_{12}f_1} H_1\ x_2{}^R H_2 \quad ,\text{xiv)}\ \frac{\partial x_1\partial x_1}{z_{12}f_1} H_1\ x_2{}^R H_2\ ,\ \text{xv)}\ \frac{\partial x_1\partial x_1}{z_{12}f_1} H_1\ x_2{}^R H_2,$$

$$\text{xvi)}\ \frac{\partial x_1\partial x_1}{z_{12}f_1} H_1\ x_2{}^R H_2 \quad ,\text{xvii)}\frac{\partial x_1\partial x_1}{z_{12}f_1} H_1\ x_2{}^R H_2\ ,\text{xviii)}\frac{\partial x_1\partial x_1}{z_{12}f_1} H_1\ x_2{}^R H_2,$$

$$\text{xix)}\ \frac{\partial x_1\partial x_1}{z_{12}f_i} H_1\ x_2{}^R$$

$$\text{xx)}\ \frac{\partial x_1\partial x_1}{z_{12}f_1} H_1\ \theta_1\lambda_2 \quad ,\text{xxi)}\ \frac{\partial x_1\partial x_1}{z_{12}f_1} H_1\ \theta_2\lambda_2,$$

xxii) $\dfrac{\theta_1 \partial_z \lambda_1}{z_{12} f_1} H_1 \;\; x_2{}^R H_2$,xxiii) $\dfrac{\theta_1 \partial_z \lambda_1}{z_{12} f_1} H_1 \;\; x_2{}^R H_2$

xxiv) $\dfrac{\theta_1 \partial_z \lambda_1}{z_{12} f_1} H_1 \;\; x_2{}^R H_2$,xxv $\dfrac{\theta_1 \partial_z \lambda_1}{z_{12} f_1} H_1 \;\; x_2{}^R H_2$,xxvi) $\dfrac{\theta_1 \partial_z \lambda_1}{z_{12} f_1} H_1 \;\; x_2{}^R H_2$,

xxvii) $\dfrac{\theta_1 \partial_2 \lambda_1}{z_{12} f_1} H_1 \;\; x_2{}^R H_2$,xxiii) $\dfrac{\theta_1 \partial_z \lambda_1}{z_{12} f_1} H_1 \;\; \theta_2 \lambda_2$,xxix) $\dfrac{\theta_1 \partial_z \lambda_1}{z_{12} f_1} H_1 \;\; \theta_2 \lambda_2$,

xxx) $\dfrac{\theta_1 \partial_2 \lambda_1}{z_{12} f_1} H_1 \;\; \theta_2 \lambda_2$,xxxi) $\dfrac{\theta_1 \partial_z \lambda_1}{z_{12} f_1} H_1 \;\; \theta_2 \lambda_2$,xxxii) $\dfrac{\theta_1 \partial_2 \lambda_1}{z_{12} f_1} H_1 \;\; \theta_2 \lambda_2$,

xxxiii) $\dfrac{\theta_1 \partial_z \lambda_1}{z_{12} f_1} H_1 \;\; \theta_2 \lambda_2$ (7.108)

The different terms we get from (7.108) are

$\partial x^R \theta^4$, $\partial\partial x^i \theta^4$, $\partial x^i (\theta\partial\theta)^i$, ∂x^L

$\partial x^R \theta^4 (\partial\theta\lambda)$, $\partial x^i \; \theta^{21} \; \theta \; \lambda)$,

$\partial\partial\partial x^R \theta^4$, $\partial x^R \partial\theta \; \partial\theta \; \theta\theta$, $\partial\partial x^R \partial\theta \; \theta\theta\theta$, $\partial x^R \partial\partial\theta \; \theta^3$,

$$\partial x^R \ (\partial x \ \partial x \ \theta^4) \tag{7.109}$$

where ∂x^R and θ factors are evaluated at points 1 and 2. Let us consider the term $\partial x^R \theta^4$ which has a coefficient behaving as $\frac{1}{z_{12}^2}$ as $z_{12} \to 0$; all other terms are of order O(1) as $z_{12} \to 0$. Explicit evaluation shows that the terms from (7.108) contributing to the term $O(1/z_{12})$ are

$$\text{ii)} \quad \text{iii)} \quad \text{vi)} \quad \text{vii)} \quad \text{ix)} \quad \text{xxii)} \quad \text{xxiii)} \quad \text{xxix)} \tag{7.110}$$

Explicit expressions for those contributions are

$$\text{ii)} \quad = - \partial_1 \partial_1 \left[\frac{1}{4z_{12}^{\ 4} f_1}\right] \lambda(\theta_1, \theta_2) x_2^{\ R}$$

$$\text{iii)} \quad = \frac{8}{5} \ \frac{\partial_2 x^R}{z_{12}^{\ 4} f_2} \left[\frac{\theta_2^{\ \bar{A}} \theta_4^{\ \bar{B}} \theta_1^{\ \bar{C}} \theta_1^{\ \bar{D}}}{(2F_1'')^2}\right]$$

$$\text{vi)} \quad = \frac{2}{\partial_{12}^{\ 3} f_2} \ \frac{x_2^{\ R}}{4z_{12}^{\ 2}} \left[\frac{4}{3} \ \frac{\theta_2^{\ \bar{A}} \theta_1^{\ \bar{B}} \theta_1^{\ \bar{C}} \theta_1^{\ \bar{D}}}{(2F_1'')^2} \epsilon_{\overline{ABCD}} - 4 \ \frac{\theta_2^{\ \bar{A}} \theta_1^{\ \bar{B}} \theta_2^{\ \bar{C}} \theta_2^{\ \bar{D}} \epsilon_{\overline{ABCD}}}{(2F_1'')(2F_2'')}\right]$$

$$\text{vii)} \quad = \frac{8}{z_{12}^{\ 4} f_2} \cdot \frac{\partial x_1^{\ R} \theta_2^{\ \bar{A}} \theta_2^{\ \bar{B}} \theta_1^{\ \bar{C}} \theta_1^{\ \bar{D}}}{(2F_1'')^2} \epsilon_{\overline{ABCD}}$$

$$\text{ix)} \quad = \frac{\partial_1\partial_1}{2}\left[\frac{1}{z_{12}^{\;2} f_1}\right] \cdot \partial x_1^{\;R}\ \frac{8}{3}\ \frac{\theta_2^{\;\bar{A}}\theta_1^{\;\bar{B}}\theta_1^{\;\bar{C}}\theta_1^{\;\bar{D}}}{F_1''^{2}}$$

$$\text{xii)} \quad = -2 \cdot \frac{1}{4z_{12}^{\;4} f_1}\ \partial x_1^{\;R}\ \lambda(\theta_1,\theta_2)$$

$$\text{xxii)} = \frac{1}{z_{12}^{\;4} f_1}\ \frac{2}{3}\ \partial x_2^{\;R}\ \theta_1^{\;\bar{A}}\theta_2^{\;\bar{B}}\theta_2^{\;\bar{C}}\theta_2^{\;\bar{D}}\ \epsilon_{\overline{ABCD}}$$

$$\text{xxiii)} = \frac{-4\pi}{z_{12} f_1} \cdot \frac{1}{4z_{12}^{\;2}} \cdot \frac{-1}{2\pi z_{12}^{\;2}} \partial x_2^{\;R} \epsilon_{\overline{ABCD}} \left[\frac{4}{3}\frac{\theta_1^{\;\bar{A}}\theta_2^{\;\bar{B}}\theta_2^{\;\bar{C}}\theta_2^{\;\bar{D}}}{(2F_2'')^2} - 4\frac{\theta_1^{\;\bar{A}}\theta_2^{\;\bar{B}}\theta_1^{\;\bar{C}}\theta^{1\bar{D}}}{(2F_1'')(2F_2'')}\right]$$

$$\text{xxix)} \quad = \frac{8}{z_{12}^{\;4} f_1} \cdot \partial x_1^{\;R} \left[\frac{\theta_1^{\;4}}{3(2F_1'')^2}\right] \tag{7.111}$$

where we have denoted

$$\lambda(\theta_1,\theta_2) = \frac{1}{4}\left[\frac{\theta_1^{\;4}}{3F_1''^{2}} + \frac{\theta_2^{\;''}}{3F_2''^{2}} + \frac{\theta_1^{\;2}\theta_2^{\;2}}{F_1''F_2''}\right]$$

All terms in (7.111) give, as $z_1 \to z_0$, $z_2 \to z_0$, $z_0 = \dfrac{z_1+z_2}{2}$, a maximum contribution of order $\dfrac{1}{z_{12}^{\;2}}$, when all z_{12} factors are considered, and also O(1) contributions for terms $\partial\partial\partial x^R\theta^4$, $\partial x^R\theta^4(\partial x\partial x)$, $\partial x^R\partial\partial\theta\ \theta^3$. In the same way one can show that (7.109) are all the forms of order $\frac{O(1)}{z_{12}^2}$ and O(1) which appear in K)[Δh,[H,J^{R-}]$_*$]$_*$). The explicit coefficients are not however required to find a_i, i= 1,...,n which solve (7.106), as we shall see. The contributions from the term [H,J] in (7.106) are given by

$$\overline{H_1 H_2}\ (x_1{}^R - x_2{}^R) = \frac{-4}{\sqrt{2F''_1}\sqrt{F''_2}}\ \frac{1}{z_{12}^{\ 2}} \left(\frac{1}{3}Y_1{}^4 + \frac{1}{3}Y_2{}^4 - Y_1{}^{2i}Y_2{}^{2i}\right)\left(x_1{}^R - x_2{}^R\right),$$

$$\overline{H_2 H_1 (x_1}{}^R - x_2{}^R) = i\ \frac{1}{\sqrt{2}}\ \frac{1}{2}\ \frac{1}{z_{21}}\left(\frac{H_1}{\sqrt{2F''_2}} + \frac{H_2}{\sqrt{2F''_1}}\right),$$

$$\frac{i\pi}{\sqrt{2}}\ \lambda_2{}^A \overline{\bar{Y}_2{}^A\ H_1} = -\ \frac{i\pi}{2\sqrt{2}}\ \frac{1}{z_{21}}\ \bar{Y}_2{}^A\ \frac{1}{\sqrt{F''_2}}\ \frac{\partial}{\partial y_2^A}\ H_2\ ,$$

$$-\frac{i\pi}{\sqrt{2}}\ \lambda_2{}^A\ \overline{\bar{Y}_2{}^A\ H_1} = +\ \frac{i\pi}{2\sqrt{2}}\ \frac{1}{z_{21}}\ \bar{Y}_2{}^A\ \frac{1}{\sqrt{F''_1}}\ \frac{\partial}{\partial y_1^A}\ H_2\ ,$$

$$H_2\ (x^R H_1 + F_1^R) - H_1\ (x^R H_2 + F_2^R)\ , \tag{7.112}$$

where the last term in (7.112) correspond to the non-contracted term in the * product. The * commutator [H,J^{R-}]$_*$ is equal to the sum of the terms listed in (7.112). By considering the different contributions in

(7.103) one can check that the term of order $\frac{1}{z_{12}^2}$ are again of the form $\partial x^R \theta^4$ while the contributions of order correspond exactly to the rest of terms in (7.109). Having shown the structure of the equation (7.106) we are now going to consider the limit procedure 1,2,→0. We first evaluate the relevant numerical factors for it. Let us consider the operation of Δh^* over the different terms ∂x^I, $\partial\partial x^I$, $\partial\partial\partial x^I$, $\theta^{\bar{A}}$, $\partial\theta^{\bar{A}}$, $\partial\partial\theta^{\bar{A}}$ which are the factors involved in (7.109).

We denote by x_o or θ_o the $(x(z_o), \theta(z_0))$ respectively where $z_o = \frac{z_1+z_2}{2}$

We obtain

$$\Delta h^* \partial x_0^I = -8\frac{\Delta_{10}}{z_{12}^3}\left[\frac{\partial x_1^I}{f_1} + \frac{\partial x_2^I}{f_2}\right] - 4\frac{\Delta_{10}}{z_{12}^{-}}\left[\left[\frac{\partial x \partial x}{f}\right]_1 + \left[\frac{\partial x \partial x}{f}\right]_2\right]\partial$$

$$= \frac{4}{3}\partial x_0^I + \frac{1}{6}\partial\partial\partial x_0^I z_{12}^2 + \frac{2}{3}(\partial x \partial x)_0\, \partial x_0^I\, z_{12}^2 + \ldots.$$

$$\Delta h^* \theta_0^{\bar{A}} = -\frac{\Delta_{10}}{z_{12}^3}\left[\frac{\theta_1^{\bar{A}}}{f_1} + \frac{\theta_2^{\bar{A}}}{f_2}\right] - 4\frac{\Delta_{10}}{z_{12}^3}\left[\left[\frac{\theta\partial z\lambda}{f}\right]_1 + \left[\frac{\theta\partial z\lambda}{f}\right]_2\right]\theta_0^{\bar{A}} z_1$$

$$= -16\frac{\Delta_{10}}{z_{12}^3}\left[\frac{\theta^{\bar{A}}}{f}\right]_0 - 2\frac{\Delta_{10}}{z_{12}^3}\partial\partial\left[\frac{\theta^{\bar{A}}}{f}\right]_0 z_{12}^2 + O(1) =$$

$$= -\frac{4}{3}\bar{\theta}_0^A + \frac{1}{6}(\partial\partial\bar{\theta}^A)_0\, z_{12}^2 + \dots.$$

$$\Delta h^*\, \partial\bar{\theta}_0^A = -32\,\frac{\Delta_{10}}{z_{12}^4}\left[\frac{\bar{\theta}_1^A}{f_1} - \frac{\bar{\theta}_z^A}{f_2}\right] = +\frac{8}{3}\,\partial\bar{\theta}_0^A + \dots$$

$$\Delta h^*\, \partial\partial\bar{\theta}_0^A = -12.16\,\frac{\Delta_{10}}{z_{12}^5}\left[\frac{\bar{\theta}_1^A}{f_1} + \frac{\bar{\theta}_2^A}{f_2}\right] + \frac{32}{z_{12}^2}\,\bar{\theta}_0^A + 4\partial\partial\bar{\theta}_0^A + \dots$$

$$\Delta h^*\, \partial x_0^I = -32\,\frac{\Delta_{10}}{z_{12}^4}\left(\frac{\partial x_1{}^I}{f_1} - \frac{\partial x_2{}^I}{f_2}\right) = -32\,\frac{\Delta_{10}}{z_{12}^3}\,\partial\left(\frac{\partial x^I}{f}\right)_0 + \dots = \frac{8}{3}\partial\partial x_0 + \dots$$

$$\Delta h^*\, \partial\partial x_0^I = -12.16\,\frac{\Delta_{10}}{z_{12}^5}\left(\frac{\partial x_1{}^I}{f_1} + \frac{\partial x_2{}^I}{f_2}\right) = \frac{32}{z_{12}^2}\,\partial x_0 + 4\partial\partial\partial x_0 + \dots. \quad (7.113)$$

where the first expression in each term is exact, while the right cones are the leading order contribution. From (7.113) we may now obtain Δh^* acting on each of the terms in (7.109). We obtain

$$\Delta h^*\, \partial x^R \theta^4 = \frac{20}{3}\,\partial x^R\theta^4 + \left[\frac{1}{6}\,\partial\partial\partial x^R\theta^4 + \frac{2}{3}\,(\partial x\partial x)\partial x^R\theta^4 + \frac{2}{3}\,\partial x\partial\partial\theta\theta^3\right] z_{12}^2$$

$$\Delta h^* \; \partial\partial\partial x^R \theta^4 = 32 \frac{\partial x^R \theta^4}{z_{12}^{\;2}} + {}^{28}/_3 \; \partial\partial\partial x^R \theta^4$$

$$\Delta h^* \; (\partial x \partial x)\partial x^R \theta^4 = {}^{40}/_3 \frac{\partial x^R \theta^4}{z_{12}^{\;2}} + {}^{28}/_3 (\partial x \partial x)\partial x^R \theta^4$$

$$\Delta h^* \; \partial x(\partial\partial\theta)\theta^3 = 32 \frac{\partial x^R \theta^4}{z_{12}^{\;2}} + {}^{28}/_3 \; \partial x(\partial\partial\theta)\theta^3 \tag{7.114}$$

and

$$\Delta h^* \; \partial x^R \theta^4 \; (\partial\theta\lambda) = + \frac{28}{3} \partial x^R \theta^4 \; (\partial\theta\lambda) + \frac{\partial x^R \theta^4}{z_{12}^{\;2}}$$

$$\Delta h^* \; \partial x^L = {}^{4}/_3 \; \partial x^L$$

$$\Delta h^* \; \partial x^i \theta 2^i \; (\theta\lambda) = {}^{16}/_3 \; \partial x^i \; \theta^{2\,i} \; (\theta\lambda)$$

$$\Delta h^* \; \partial\partial^R \partial\theta \; \theta^3 = {}^{28}/_3 \; \partial\partial x^R \; \partial\theta \; \theta^3$$

$$\Delta h^* \; \partial x \partial\theta\partial\theta\theta^2 = {}^{28}/_3 \; \partial x \; \partial\theta \; \partial\theta \; \theta^2$$

$$\Delta h^* \; \partial x^i (\theta\partial\theta)^i = {}^{16}/_3 \; \partial x^i \; (\theta\partial\partial)^i$$

$$\Delta h^* \; \partial\partial x^i \; \theta^{2\,i} = {}^{16}/_3 \; \partial\partial x^i \; \theta^{2\,i} \tag{7.115}$$

where all terms in (7.114) and (7.115) are evaluated at z_0. These are only the leading order contributions. Let us rewrite (7.106) explicitly to explain now the limit $1\to 0$, $2\to 0$. The equation is

$$[H,J^{R-}]_* + a_0[\Delta h, [H,J^{R-}]_*]_* + a_1[\Delta h,[\Delta h,[H,J^{R-}]_*]_*\Delta_*+\ldots$$

$$.+ a_n\underbrace{[\Delta h[\Delta h[\ldots.[\Delta h}_{n}[\dot{H},J^{R-}]_*]_* = 0 \qquad (7.116)$$

The structure of the equation is like an expansion in which Δh acts as a discretized derivative shifting fields from point $1\to 2$, and viceversa. The structure of the equation obtained by using $\Delta_F h$ defined in equation (7.80) is different. In that case the first two terms are the same but beyond the third term the expansion is in terms of $\Delta_F h$ and Δh. The solution in this case becomes more involved; it will not be considered further.

We now take the limit in (7.116) in the following sense. We evaluate $[H,J^{R-}]_*$ and expand $z_0 = \dfrac{z_1+z_2}{2}$, to obtain the eleven terms (7.109) in the form:

$$\frac{O(1)}{z_{12}^{\,2}} + O(1) + O(z_{12}^{\,2}), \qquad (7.117)$$

with only $\partial x^R\theta^4$ having non-zero coefficient in $\dfrac{1}{z_{12}^2}$.

We now annihilate terms $O(z_{12}^2)$, as if we were taking partial limits, and apply Δh,

$$[\Delta h\ ,\ \frac{O(1)}{z_{12}^{\,2}} + O(1)]_* \qquad (7.118)$$

After so doing we again have the structure (7.117) where now the O(1) terms have contributed to the $\frac{1}{z_{12}^2}$ term and vice-versa according to (7.114), (7.115). We proceed in this way n times to evaluate the term of highest degree. In that case there are a lot of combinations between different contributions of the eleven terms (7.109).

However the system may be explicitly described in the following way. Let us denote by V the eleven dimensional vector obtained from the eleven coefficients of the independent terms (7.109) in the first terms of (7.116). Let us denote by M the eleven dimensional matrix obtained from (7.114), (7.115)

$$M=\begin{bmatrix} 20/3 & 32 & 40/3 & 32 & & & & & & & \\ 1/6 & 28/3 & & & & & & & & & \\ 2/3 & & 28/3 & & & & & & & & \\ 2/3 & & & 28/3 & & & & & & & \\ & & & & 28/3 & & & & & & \\ & & & & & 4/3 & & & & & \\ & & & & & & 16/3 & & & & \\ & & & & & & & 28/3 & & & \\ & & & & & & & & 28/3 & & \\ & & & & & & & & & 16/3 & \\ & & & & & & & & & & 16/3 \end{bmatrix}$$

with the other entries zero.

It can be shown that the eleven independent equations contained in (7.116) may be expressed as

$$V + a_0\, MV + a_1\, M^2 V + \ldots. + a_n\, M^{n+1} V = 0 \qquad (7.119)$$

The necessary and sufficient condition for the existence of a solution to (7.119) is that the matrices N_1 and N_2 with columns

$$N_1 = (V, M_V, \ldots, M^{n+1}V)$$

$$N_2 = (M V, \ldots, M^{n+1}V)$$

have the same rank (since then V must be some linear combination of the columns of N_2). It is enough to have n = 11 to solve (7.119). In fact since det $M \neq 0$, any regular minor from N_2 may be obtained from N_2 by extracting a factor M from it.. We have thus proven the existence of $J_4{}^{R-}$ which closes (7.103) to quartic order . There is also a contribution to the previous analysis of a term of order O(1), with the structure $\partial X^R \theta^4$, whose coefficients depends on the derivatives of f(z) at the points 1 and 2, where

$$F'(z) = (z-z_1)(z-z_2)f(z)$$

This term is directly cancelled by expanding a_0, $a_1, \ldots a_n$ around

$z_0 = \frac{z_1+z_2}{z}$. It is enough to expand only one of the a_i, i=1,...,n. This term is different from the previous ones since it depends on the structure of the Mandelstam map F(z).

The cancellation of this term can be shown directly or it can be also incorporated in to the previous analysis. In that case the matrix M has an additional diagonal non zero term. We are now going to prove the closure of (7.104).

Let us apply to (7.103) the symbol j^{i-} to obtain

$$[\hbar,[j^{i-},J_4{}^{R-}]_* + [J^{i-},J^{R-}]_* + [J_4{}^{i-},j^{R-}]_*]_* = 0 \qquad (7.120)$$

We now have the symbol $\hbar$ acting on the commutation relation. It acts in a very different way than Δh. In fact it has the structure

$$\oint_C dz \frac{1}{(z-z_1)(z-z_2)} \quad , \qquad \text{C around 1 and 2,}$$

which implies that no matter how many contraction one performs it always preserves the order in z_{12} of the term on which acts. That is,

$$\oint dz \frac{G}{(z-z_1)(z-z_2)} \cdot \frac{1}{(z-z_1)^m(z-z_2)^n} = O(1)$$

if G is differentiable.

We now use this property in (7.120). The orders for $[J^{i-},J^{R-}]$ are

$$\frac{O(1)}{z_{12}^{\,3}} + \frac{O(1)}{z_{12}} + O(z_{12}),$$

the leading order term being $x^i x^R \theta^4$.

Since there are no contributions from $\frac{O(1)}{z_{12}}$ to $\frac{O(1)}{z_{12}^2}$ after applying $[\tilde{h},]_*$, and $[\tilde{h},]_*$ never annihilates a term, but only adds several derivatives to the factors, we conclude that $\frac{O(1)}{z_{12}^3} = 0$.

It can also be seen from the structure of h that it does not mix the $\frac{O(1)}{z_{12}}$ terms when applied to them. We thus obtain $\frac{O(1)}{z_{12}} = 0$. The closure of (7.104) to quartic order has thus been proved. This implies from the argument of section 7.2, the complete closure of sP_{10} to quartic order.

We wish now to prove that there are no higher order contribution to the algebra. We have then to show that

$$[J^{R-},J_4^{i-}]_* + [J_4^{R-},J^{i-}]_* = 0 \qquad (7.121)$$

and

$$[J_4^{R-},J_4^{i-}] = 0 \qquad (7.122)$$

From power counting, as was explained in [38], (7.122) is satisfied, while (7.121) is of order $\frac{1}{z}$. We are now going to use (7.92) in order to satisfy it.

Let us consider the first term in the expension of J_4^{i-}. We rewrite it in the form

$$G_{12}^{\ R} = [\Delta H, J^{R-}]_* \equiv \frac{1}{2}\ \Delta_{12}\ H_1 * J_2^{\ R-} - \frac{1}{2}\ \Delta_{21} J_1^{\ R-} * H_2$$

Its contribution to (7.121) is

$$J^{R-} * [\Delta H, J^{i-}]_* - [\Delta H, J^{i-}]_* * J^{R-} +$$

$$+[\Delta H, J^{R-}]_* * J^{i-} - J^{i-} * [\Delta H, J^{R-}]_* =$$

$$= J_1^{\ R-} * \frac{1}{2}(\Delta_{23} H_2 * J_3^{\ i-} - \Delta_{32}\ H_3 * J_2^{\ i-}) - \frac{1}{2}(\Delta_{12} H_1 * J_2^{i-} - \Delta_{21} H_2 * J_1^{\ R-}) * J_3^{R-} +$$

$$+\frac{1}{2}(\Delta_{12} H_1 * J_2^{\ R-} - \Delta_{21} H_2 * J_1^{\ R-}) * J_3^{i-} - J_1^{i-} * \frac{1}{2}(\Delta_{23} H_2 * J_3^{R-} - \Delta_{32} H_3 * J_2^{R-}) \qquad (7.123)$$

The calculations are done as in (7.117),(7.118) where the first limit corresponds to the $[\Delta H, J^{i-}]_*$ terms. For example, for the first term in (7.123) we evaluate $[\Delta H, J^{i-}]_*$, take limit $z_2 \to z_0$, $z_3 \to z_0$, where $z_0 = \frac{z_2+z_3}{2}$, then do the * product with J_1^{R-}, and take the lim $z_1 \to z_0$. In each limit the measure d Δ_{23} and d Δ_{10} respectively are included. The contribution of $[\Delta H, J^{-i}]$ is only $x_0^i \theta_0^4$ and is of order $O(1)$ in z_{23}. We then do the * product with J_1^{R-} and again obtain a term $x_0^i \theta_0^4$, but now of order $\frac{1}{z_{10}}$. The term arises from the following contraction

$$\overline{X_0^i \theta_0^4 \qquad F_1^R}$$

in this case.

The contributions from first and second terms in (7.123) cancel between each other and the same happens between third and fourth terms. It is easy to see that the same cancellations occur for all the terms in the expansions of J_4^{R-} and J_4^{i-}. (7.121) is then satisfied, as was already established in [38]. We have now completed the proof of the closure of the sP_{10} algebra in the light cone gauge.

We wish now to analyse the insertion factors associated to the non-linearly realized SUSY generators and to the Hamiltonian. We have for the SUSY generators

$$iQ_4^{-A} = [J_4^{L-}, Q^{+A}] \quad ,$$

$$i\bar{Q}_4^{-A} = [J_4^{R-}, Q^{+\bar{A}}] \quad , \tag{7.124}$$

where

$$Q^{+A} = - i\sqrt{2} \int dz\ \lambda^A F',$$

$$Q^{+\bar{A}} = = \frac{-i\sqrt{2}}{\pi} \int dz\ Q^{\bar{A}}.$$

We obtain

$$G^{A}_{12} = \frac{1}{2}\ \Delta_{12}\ (H*Q^{-A} + Q^{-A}*H)\ ,\ Q^{-A}_4 = K(G^{A}_{12})$$

$$G^{\bar{A}}_{12} = \frac{1}{2}\ \Delta_{12}\ (H*Q^{-\bar{A}} + Q^{-\bar{A}}*H)\ ,\ Q^{-\bar{A}}_4 = K(G^{\bar{A}}_{12}) \qquad (7.125)$$

We now show that there is no leading order contribution to the Hamiltonian from Q_4^{-A}, $Q_4^{-\bar{A}}$ and obtain the explicit expression of H_4. We must consider

$$[q^{-A}, Q_4^{-\bar{B}}]_* + [Q_4^{-A}, q^{-\bar{B}}]_* + [Q^{-A}, Q^{-\bar{B}}]_* = 2\delta^{A\bar{B}} H_4 \qquad (7.126)$$

This is according to the analysis of section 7.2 for the definition of H_4. From it we are going to obtain, as stated before,

$$H_4 = K(G^{H}_{12})$$

$$G^{H}_{12} = 2\Delta_{12}\ H_1 * H_2\ . \qquad (7.127)$$

where q^{-A} and $q^{-\bar{B}}$ are given by, (from section 2.2),

$$q^{-A} = \frac{-2^{\frac{1}{2}}}{\pi} \int dz \left[\sqrt{2}\, \rho^{i}_{\bar{A}B} \frac{\partial_z X^i}{F'} \theta^{\bar{B}} + 2\pi\, \partial_z X^L\, \lambda^A \right]$$

$$q^{-\bar{B}} = \frac{-2^{\frac{1}{2}}}{\pi} \int dz \left[2 \frac{\partial_z X^R}{F'} \theta^{\bar{B}} - \sqrt{2\pi}\, \partial_z X^i \rho^{i}_{BA} \lambda^A \right] \tag{7.128}$$

while

$$Q^{-A} = \frac{2}{3} Y^A \ , \ Q^{-\bar{A}} = Y^{\bar{A}} \tag{7.129}$$

Eg. (7.126) may be rewritten as

$$[Q^{-A}, Q^{-\bar{B}}]_* + k([\Delta h, [Q^{-A}, Q^{-\bar{B}}]_*]_*) + 4\Delta_{12}\ H{*}H\ \delta^{A\bar{B}} = 2\delta^{A\bar{B}} H_4 \tag{7.130}$$

The first two terms of the left member cancel using (7.106), and we obtain (7.127). Also the highest order contribution from the second and third terms cancel and hence we are left to leading order, O(1) terms, with

$$[Q^{-A}, Q^{-\bar{B}}]_* = 2\delta^{A\bar{B}} H_4 \tag{7.131}$$

which was the solution found in [38]. Since leading order terms are the only relevant ones in (7.131) we have thus shown that there are no contribution from Q_4^{-A}, $Q_4^{-\bar{B}}$ to the Hamiltonian and neither to the SUSY subalgebra. The proof of closure of sP_{10} is complete for the heterotic string.

We note finally as discussed briefly in appendix XIII in the special case of the open superstring tree amplitude for $p_1^+ = -p_4^+$, that $Q_4 \neq 0$ in that case. The situations for both the open and closed superstrings are still to be clarified as to closure of the sP_{10} algebra by a finite number of contact terms.

CHAPTER 8: FUTURE DIRECTIONS

Section 8.1: Introduction

An enormous amount of effort has been put into developing explicitly covariant approaches to superstrings, at both the first and the second quantised (field theoretic) levels. The covariant first quantised method of Polyakov for the bosonic and NRS string [22] has attracted much attention since it explicitly possesses world sheet co-ordinate invariance when the partition function (in the bosonic case) is expressed as

$$Z = \sum_h \int D_h\, g\, D_h\, X \quad e^{S^1(x,g)} \tag{8.1}$$

$$S^1(x,g) = \int d^2\sigma g^{\alpha\beta} \partial_\alpha X^\mu \partial_\beta X_\mu . \tag{8.2}$$

The integration is over both string embeddings X^μ (a flat background) metric is being used here) and the two-dimensional world-sheet metric $g^{\alpha\beta}$. The summation Σ in (8.1) is over all possible equivalence classes of metrics, so corresponding to a sum over Riemann surfaces with increasing genus. It is possible to quantise (8.1) by introducing ghosts, without loss of conformal invariance (which is a part of the symmetry of (8.2)). The resulting ghost structure is well-known, as it is for the corresponding NRS case (GSW, Chapter 3). There are , however, the two questions

(a) can (8.1) be justified?

(b) can (8.1) be used to evaluate non-perturbative properties?

These questions also arise in attempts that have been made to set up an explicitly covariant superstring field theory. We will consider that problem after considering the relation to (8.1) in the next section, and a discussion of the problem of non-trivial backgrounds in the following one.

Section 8.2: Relations to the covariant Polyakov approach

A particular limiting process, the short-string limit (SSL), was used in sections 5.4 to 5.7 to reduce the general multi-loop amplitude (for
$g \geq 0$) to a covariant form. After the limit has been taken, the only quantities which are not Lorentz covariant involve products $\underline{p}_r . \underline{p}_s$ (treating only the 0-modes for bosons). The positions $\tilde{\rho}$ of the interacting points were originally defined by means of (5.23), and are clearly not Lorentz invariant. At tree and one loop level the SSL removed such lack of covariance by replacing the $\tilde{\rho}$'s by the external source points z_r. For $g \geq 2$ it was shown in section 5.6 that under the SSL, N of the interaction positions may still be replaced by z_r's, whilst the remainder are defined by a suitable modification of the Ahlfors co-ordinates on Teichmüller space. On replacing the quantities $(\underline{p}_r . \underline{p}_s)$ or $(\underline{\zeta}_r . \underline{p}_s)$ by the corresponding lorentz invariants $(p_r . p_s)$ or $(\zeta_r . p_s)$ (which is valid for N=4, and even for $N \leq 10$), the expressions are thereby made explicitly Lorentz invariant.

Identical expressions are ultimately expected to arise from the covariant approach to superstrings using the two-dimensionnally supersymmetric extension of (8.1). A proof of unitarity of the covariant approach might also be given, in a similar manner to that of the purely bosonic string [27)]. It would appear useful, therefore, to indicate in what manner the various factors defined by the superstring Feynman rules in section 5.2 might be expected to arise in the covariant construction.

This problem appears to be resoluble into two parts
(a) in what manner the effects of the insertion factor V_3^H of (3.22c) arise from the covariant approach. This means finding the source of the

remnants of these insertion factors in the multi-loop amplitude expressions (5.63) and (5.87).

(b) in what manner the summation over spin structure produces the relatively simple expressions (5.63) and (5.87), which contain no hint of such summation.

It is to be noted that (a) and (b) are independent. Thus the NRS string in the L.C. gauge also requires insertion factors [55]. These have recently [107] been expressed more compactly by means of the introduction of suitable Grassmann-valued variables. That may also be done in the L.C. gauge for the superstring. This may be seen by writing

$$V_3^H(P) = \int_\Sigma \underline{X}.\underline{N} \tag{8.3a}$$

where

$$\underline{N} = (N^R, N^i, N^L) = -[2F''(\tilde{z})]^{-\frac{1}{2}}\{2^{-\frac{1}{2}}, Y^{i^2}(\tilde{z}), \frac{\sqrt{2}}{3} Y^4(\tilde{z})\}\partial_z \delta^2(z-\tilde{z}) \tag{8.3b}$$

and Y is given by (3.23b). The insertion factor may be elevated into an exponent by use of Berezhin integration, with $a = \int dc d\bar{c} e^{ca\bar{c}}$. Subsequent Gaussian integration over $\underline{X}$, as considered in section 5.2, will lead to an expression for type II multi-loop amplitudes, for example, as

$$\int \Pi d^2\tilde{\rho}_\alpha d\alpha_i d\beta_i\ d^4\theta_i d^4\tilde{\theta}_i \prod_\alpha dc_\alpha d\bar{c}_\alpha \exp[\underline{J}.G^*\underline{J}+M+N](\det \mathrm{Im}\Pi)^{-4} \tag{8.4}$$

where α denotes the interaction points, with

$$M = \sum_\alpha M_\alpha c_\alpha \bar{c}_\alpha, N = \sum_{\alpha,\beta} \underline{N}_\alpha^T G^* \underline{N}_\beta c_\alpha \bar{c}_\alpha c_\beta \bar{c}_\beta \tag{8.5a}$$

$$M_\alpha = (2F''_\alpha)^{-\frac{1}{2}} \sum_r G'(\tilde{z}_\alpha, z_r) \underset{\sim}{V}_3 (\underset{\sim}{p}_r, \underset{\sim}{Y}_\alpha) \tag{8.5b}$$

$$\underline{N}^T_\alpha G^* \underline{N}_\beta = (4F''_\alpha F''_\beta)^{-\frac{1}{2}} [\tfrac{1}{3}(Y^4_\beta + Y^4_\beta) + Y^{2i}_\alpha Y^{2i}_\beta] \partial_\alpha \partial_\beta G(\tilde{z}_\alpha, z_\beta) \tag{8.5c}$$

It is necessary to include L-movers $\tilde{\lambda}, \tilde{\theta}$ in (8.5) for type II, and also to fold in the external string zero-mode superfields, together with integration over the external θ's in those superfields with the usual measure, as discussed in earlier sub-sections.

It is interesting to note the similarity of (8.5) to the corresponding exponentiated version of the NRS in [107]. The latter may be written in terms of an integration over a super-Riemann surface (SRS), on which are the interacting points $(\tilde{z}, \theta)$. These are determined as super-variables, so with the additional Grassmann-valued companion θ to the variable $\tilde{z}$ considered here. The 1-loop extension of this analysis [108] appears to lead to the ambiguities in the definition of integration over super-moduli already discussed in the NRS case in [26]. Such ambiguities do not seem to arise in (8.5), where the variables c, $\bar{c}$ play an auxiliary role, and do not enter in a supersymmetric manner along with $\tilde{z}$'s. A similar feature should occur for the NRS string in the L.C. gauge, provided that the insertion $\underline{X}'.\underline{\psi}$, in that case, had been correctly justified from a field theory construction, as was done in Chapter 3 for the superstring in the L.C. gauge. Similar removal of the super-moduli integration ambiguity does not seem possible for the fully covariant first quantised multi-loop construction.

A detailed multi-loop construction has already been given in the covariant approach in [109] to within the specific question of the ambiguity of integration over super-moduli. It is this latter integration process which clearly produces the relics of the insertion factors V_3^H, as observed in [74] and [75]. That is clear from the covariant multi-loop integrand

$$\Pi V . \exp S \tag{8.6a}$$

$$S = \iint [\partial X \bar{\partial} X + \psi \bar{\partial} \psi + \bar{\psi} \partial \bar{\psi} + \chi \psi \bar{\partial} X + \bar{\chi} \bar{\psi} \partial X + \frac{1}{2} \bar{\psi} \psi \chi \bar{\chi}] \tag{8.6b}$$

where V are suitable vertex factors for the external states. For the gravitini with δ-function support, with χ given, for example, by

$$\chi = \sum_{a=1}^{2g-2} \zeta^a \delta^{(2)} (v - v_a) \tag{8.7}$$

(and the support of $\bar{\chi}$ being distinct from that of χ), Integration over the super moduli leads to the insertion factor

$$V_3^{NSR} = \Pi(\psi \partial X)(\bar{\psi} \bar{\partial} X) \tag{8.8}$$

in addition to the external V, where Π in (8.8) is taken over the positions of the δ-function supports in χ and $\bar{\chi}$. The insertion factors (8.8) are the covariant form of the NSR insertion factors ($\underline{\psi} . \underline{X}'$) remarked on earlier. Thus the question of the source of the insertion factors V_3^H in the amplitudes (5.63) and (5.87) from the covariant spinning string is reduced to that (to within ghost factors) in the L.C. gauge version of that string: these latter factors ($\underline{\psi} . \underline{X}'$) arise from integration out over super-moduli in the L.C. gauge-fixed version [107),108)] in an identical manner to the covariant factors (8.8) in the covariant NRS string.

It is thus effectively necessary to resolve problem (b) above - as to the manner in which summation over spin structures produces the effects of V_3^H in the L.C. gauged version of the NSR-string. The details of this have not been worked out at the multi-loop level, but the general picture seems clear. In particular the insertion factor V_3^{NSR} of (8.8)is not supersymmetric in 10 dimensions; summation over spin structures is equivalent to use of the GSO projection operator. Such projection must therefore lead to the same results as use of the explicitly 10-dimensional insertion V_3^H of (3.22c). This is clear at tree and 1-loop level, where the relic of the insertion factors is minimal; further study will need to be done to follow through the details of the spin structure to obtain the ultimate expressions (5.63) or (5.87).

It is useful also to point out the presence of non-factorisable terms arising from $\partial X \bar{\partial} X$ contractions in the type II case. These were already noted as being present in the covariant case [110], so agreeing with their presence in the final amplitudes (5.63) or (5.87).

In addition, we should note that there has been recent interesting work [111] attempting to use only the L.C. gauge for θ, not for X. It is not clear in this case whether or not insertion factors are present, and this approach deserves further study.

Finally we conclude that the answer to question (a) of section 8.1 is that whilst the answer is yes for the bosonic string the case of the superstring is uncertain; there are very difficult problems to resolve on supermoduli spaces, etc, before that answer can become clear. We consider question (b) of section 8.1 in the last section of this Chapter.

Section 8.3: General Backgrounds

Much work has been done recently on the problem of constructing string theories in a general background metric [112], especially from the point of view of obtaining conditions on the background values of the zero modes (gravitation, antisymmetric tensor and dilaton) for consistent propagation of the string [113), 114)]. There has also been analysis of the spectrum and first quantisation in special backgrounds [115)]. In order to better understand the features that may occur when a general background is present, it is the purpose of this section to consider how one may choose a light-cone gauge for the string propagation. This may allow a more direct analysis of the physical features, and in particular, permit construction of a complete L.C. field theory of strings and, hence that of superstrings. The resulting multi-loop amplitudes can then be investigated for finiteness.

The classical action for a string in a general background metric (where the other zero modes will not be considered here specifically for the sake of simplicity) is

$$S[G_{\mu\nu}] = \iint_{\Sigma} d\sigma d\tau \sqrt{(-\det g_{\alpha\beta})}\, g^{\alpha\beta}\partial_\alpha X^\mu \partial_\beta X^\nu G_{\mu\nu}(X(\sigma,\tau)) \qquad (8.9)$$

This action possesses both two-dimensional reparametrisation invariance on the world sheet Σ (with metric $g^{\alpha\beta}$) and general co-ordinate invariance $X^\mu \rightarrow X^{\nu\prime}(X^\nu)$. The former of these is a local invariance for the two-dimensional field $X^\mu(\sigma,\tau)$, the latter is a global one. The constraints associated with the local invariance when $g^{\alpha\beta} = \eta^{\alpha\beta}$ are [113)], in Hamiltonian form,

$$\phi_1 \equiv X^{\mu\prime} P_\mu = 0 \qquad (8.10a)$$

$$\phi_2 \equiv G^{\mu\nu} P_\mu P_\nu + X^{\mu\prime} X^{\nu\prime} G_{\mu\nu} = 0 \qquad (8.10b)$$

where, as usual, $P_\mu = \partial L/\partial \dot{X}^\mu$, L being the density of S of (8.9). The equations of motion from (8.9) are

$$\partial_+\partial_- X^\mu + \Gamma^\mu_{\nu\lambda}\partial_+ X^\nu \partial_- X^\lambda = 0 \qquad (8.11)$$

where L.C. variables $\sigma_\pm = \sigma \pm \tau$ are used in (8.11), and $\Gamma^\mu_{\nu\lambda}$ is the

Christoffel connection for the metric $G^{\mu\upsilon}$. In the L.C. variables $\sigma_\pm$ the constraints (8.10), in a lagrangian approach, become

$$T_{\pm\pm} = \partial_\pm X^\mu \partial_\pm X^\upsilon G_{\mu\upsilon} = 0 \tag{8.12}$$

where ++ and -- go together, and +- or -+ are not considered).

Under a variation $\delta\sigma^{\pm}$ of the co-ordinates, the constraints (8.12) vary as

$$\delta T_{++} = \partial_+ \delta\sigma^- . \partial_- X^\mu \partial_+ X^\upsilon G_{\mu\upsilon} + \delta\sigma^a \partial_a X^\lambda \partial_+ X^\mu \partial_+ X^\upsilon \Gamma_{\mu\upsilon,\lambda} \tag{8.13}$$

(with opposite signs for δT_{--}). Thus the residual reparametrisation invariance, after choosing the orthonormal metric $g_{\alpha\beta} = \eta_{\alpha\beta}$, is given by (8.13). For given initial values of $\delta\sigma^a$, the homogeneous first order equations (8.13) will have solutions with arbitrariness given by the null-space of the two-dimensional differential operator $\underline{D}$ on two-vectors $(\delta\sigma^+, \delta\sigma^-)^T$, with

$$\underline{D} = \begin{bmatrix} \partial_- + \Lambda^{-1}\partial_+ X^\lambda \partial_+ X^\mu \partial_+ X^\upsilon \Gamma_{\mu\upsilon,\lambda} & \Lambda^{-1}\partial_- X^\lambda \partial_+ X^\mu \partial_+ X^\upsilon \Gamma_{\mu\upsilon,\lambda} \\ \Lambda^{-1}\partial_+ X^\lambda \partial_- X^\mu \partial_- X^\upsilon \Gamma_{\mu\upsilon,\lambda} & \partial_+ + \Lambda^{-1}\partial_- X^\lambda \partial_- X^\mu \partial_- X^\upsilon \Gamma_{\mu\upsilon,\lambda} \end{bmatrix} \tag{8.14}$$

where $\Lambda = \partial_- X^\mu \partial_+ X^\upsilon G_{\mu\upsilon}$. In the flat case with $\Gamma_{\mu\upsilon,\lambda} \equiv 0$, the null-space of $\underline{D}$ is the space of vectors $(f(\sigma^+), g(\sigma^-))$, for arbitrary (smooth) f,g. In the general case, the $\sigma_\mp$ developments of $f_\pm$, respectively, are determined for the null-vector $(f_+(\sigma^+,\sigma^-), f_-(\sigma^+,\sigma^-))$ of $\underline{D}$ (8.14). Thus $f_+(\sigma^+,\sigma_0^-)$ and $f_-(\sigma_0^+,\sigma^-)$ can be specified arbitrarily for such a vector where $\sigma_0^\pm$ are particular co-ordinates. Therefore there is as much freedom left in gauge fixing the (σ,τ) parametrisation on Σ as in the flat case. There can be no special metrics on which this is no longer true; for the above analysis only fails if Λ is identically zero (and this occurs only when $G^{\mu\upsilon}$ is zero). Λ may indeed vanish on special choices of $X^\mu(\sigma,\tau)$, say, if $X^\mu = X^\mu(\sigma^+)$ or $X^\mu = X^\mu(\sigma^-)$ which also satisfy the string equation (8.11)). However the construction of a field theory of strings in an arbitrary background is that for general world sheets, and not restricted to those only satisfying the field equations, say. It would be disturbing if the background metric introduced a change in the physical degrees of freedom as the world-sheet Σ was changed; a consistent theory would appear difficult to construct under such a situation.

From the conclusion that there are two co-ordinate degrees of freedom which can still be fixed, let us consider the L.C. gauge fixing choices, in the Hamiltonian formalism (in which we will now remain)

$$X^+ = P_-\tau \tag{8.15a}$$

$$P'_- = 0 \tag{8.15b}$$

It is not, in general possible to choose $\dot{P}_- = 0$ in (8.15a), so then the $\mu = +$ equation of motion in (8.11) reads as the constraint

$$\Gamma^+_{\upsilon\lambda}\partial_+X^\upsilon\partial_-X^\lambda = 0 \tag{8.16}$$

This may be reduced, say in the L.C. gauge for $G^{\mu\upsilon}$ with $G^{++} = G^{+i} = 0$ and $G^{-+} = 1$, to the condition

$$\partial G^{\mu\upsilon}/\partial X^- = 0 \tag{8.17}$$

The condition that $\partial/\partial X^-$ is a Killing vector of $G^{\mu\upsilon}$ will become of relevance very shortly, with its associated simplifications.

The constraints (8.15a and 8.15b) must be preserved in time τ, so that

$$\chi_1 \equiv \dot{X}_+ - \dot{P}_- \tau - P_- = 0 \tag{8.18a}$$

$$\chi_2 \equiv \dot{P}'_- = 0 \tag{8.18b}$$

These conditions may be used to determine the parameters α,β in the general Hamiltonian

$$H = \int d\sigma(\alpha\phi_1 + \tfrac{1}{2}\beta\phi_2) \tag{8.19}$$

(8.18a,b) lead to the equations, respectively,

$$[\alpha' P_- + (\beta X^{\nu'} G_{-\nu})' - \tfrac{1}{2}\beta\phi_3]\tau - \beta p_\nu G^{+\nu} = 0 \tag{8.20a}$$

$$\alpha'' P_- + (\beta X^{\nu'} G_{-\nu})'' - \tfrac{1}{2}(\beta\phi_3)' = 0 \tag{8.20b}$$

where

$$\phi_3 \equiv P_\mu P_\nu \partial G^{\mu\nu}/\partial X^- + X^{\mu'} X^{\nu'} \partial G_{\mu\nu}/\partial X^- \tag{8.20c}$$

If $\partial/\partial X^-$ is a Killing vector then $\phi_3 = 0$, and the equations (8.20), in the L.C. gauge for $G^{\mu\nu}$, reduce to the usual values in flat space,

$$\alpha = 0, \ \beta = 1 \tag{8.21}$$

Moreover, from

$$\dot{P}_- = \alpha' P_- - \frac{1}{2}\beta\phi_3 + \beta X^{\upsilon} G_{-\upsilon} \qquad (8.22)$$

results that $\phi_3 = 0$ and when the L.C. gauge is chosen for $G^{\mu\upsilon}$, P_- is conserved. These, then, are the promised simplifications when $\partial/\partial X^-$ is a Killing vector.

In the general case (8.20) may be solved for α and β as

$$\beta = (P_{\upsilon} G^{+\upsilon})^{-1}(P_- - \alpha_0 \tau) \qquad (8.23a)$$

$$\alpha = (P_-)^{-1}\{\frac{1}{2}\int_0^{\sigma} d\sigma' \beta(\sigma')\phi_3(\sigma') - \beta X^{\upsilon'} G_{-\upsilon} - \alpha_0 \sigma - \alpha_1\} \qquad (8.23b)$$

where α_0 must be chosen so that α is periodic in σ with period $P_-\pi$:

$$\alpha_0 = (\frac{1}{2P_-\pi})\int_0^{\pi P_-} d\sigma \beta(\sigma)\phi_3(\sigma) \qquad (8.23c)$$

and α_1 is a constant which may be set to zero. In this case it is to be noted from (8.22) that P_- is no longer conserved. The general consistency of the Hamiltonian process of gauge fixing guarantees that the difficulty noted in (8.16) no longer arises in the general case; the Hamiltonian equations of motion arising from (8.19) will be consistent though not easily reducible to lagrangian equations, as we will see shortly.

It is now possible to solve the constraints ϕ_1, ϕ_2 of (8.10) in the L.C. gauge for $G^{\mu\upsilon}$, as

$$X^- = x^- - (P_-)^{-1}\int_0^\sigma d\sigma' P_i X^{i\,\prime} P_i X^{i\,\prime}(\sigma') \qquad (8.24a)$$

$$P_+ = -(2P_-)^{-1}[P_i P_j G^{ij} + X_{I'} X^{j\,\prime} G_{ij}] - P_i G^{i-} - \frac{1}{2}P_- G^{--} \qquad (8.24b)$$

It is moreover possible to solve the constraint ϕ_2 without choosing the L.C. gauge for $G^{\mu\nu}$, but since the L.C. gauge conditions (8.18) destroy explicit general co-ordinate invariance, we may as well use the freedom in the choice of gauge for $G^{\mu\nu}$ (before fixing (8.18)). The canonical first quantisation of the string then leads to transition amplitudes from the string state $X_1(\sigma)$ at τ_1 to X_2 (σ) at τ_2 given by

$$<X_1(\sigma),\tau_1|X_2(\sigma),\tau_2> = \int D\underline{X}D\underline{P}\exp\{i\int_{\tau_1}^{\tau_2} d\tau[P_j\dot{X}^j + P_+\dot{X}^+ + P_-\dot{X}^-]\} \times$$

$$\times \prod_{i=1}^{2} \delta(\underline{X}(\sigma,\tau_i) - \underline{X}_i(\sigma)).\det\{\chi_r,\phi_s\} \qquad (8.25)$$

in which the gauge-fixing constraints χ_1,χ_2 and solutions (8.24) are substituted where necessary in (8.25). It is to be noted that integration over $\underline{P}$ will not be possible explicitly if $\partial/\partial X^-$ is not a Killing vector of $G^{\mu\nu}$, since $\underline{P}$ will enter in the $G^{\mu\nu}$ terms in P_+ in the exponent in (8.25) from the dependence of X^- on $\underline{P}$ given by (8.24a). This is an indication of the additional complexity of the lagrangian equations of motion beyond the Hamiltonian ones noted earlier.

The Hamiltonian may be read off from the exponent in (8.25) to be

$$H_1 = [P_+P_- + \dot{P}_-(P_+\tau - X^-)] \qquad (8.26)$$

where the first term on the r.h.s of (8.26) is the usual one, and the extra terms are written in terms of the dynamical variables $\underline{X},\underline{P},x^-$ and P_- by means of (8.22) and (8.24) (where (8.24) are also used in (8.22)).

There remains to discuss the preservation of the global symmetry, that of general co-ordinate transformations (G.C.T.)

$$X^\mu \to X^\mu + \zeta^\mu(x) \tag{8.27}$$

with generators, on the string fields, as

$$T_\xi = \int d\sigma\, \xi^\mu(X(\sigma))P_\mu(\sigma) \tag{8.28}$$

with algebra·

$$[T_\xi, T_\eta]_- = T_{[\xi,\eta]} \tag{8.29a}$$

where

$$[\xi,\eta]^\mu = \xi^\upsilon \partial_\upsilon \eta^\mu - \eta^\upsilon \partial_\upsilon \xi^\mu \tag{8.29b}$$

This algebra contains the global inhomogeneous Lorentz transformations as $\xi^\mu = \epsilon^\mu{}_\upsilon X^\upsilon + \epsilon^\mu$. Closure of that sub-algebra, in the flat case, was only possible in the critical dimension $d_{crit} = 26$ [44]. It is necessary to consider the question of closing (8.29) in this more general framework.

The problem of defining a normal ordering of the operators (8.28), using the solutions (8.24) of the constraints is very difficult in general. The case when the embedding space-time is MxG, where M is flat Minkowski space-time and G a group manifold, is known to lead to critical dimension $26-(1+(2k)^{-1}C_A)^{-1}d_G$ [115], where d_G is the dimension of the group, k is an integer and C_A is the eigenvalue of the second Casimir operator of G in the adjoint representation. This condition arises from the vanishing of $[M^{i-}, M^{-j}]_-$ for i,j transverse labels of M. The conditions for the more general case of an arbitrary curved space-time for G are unknown to the authors, as are those for the more general cases of an arbitrary curved manifold replacing MxG. The difficulty essentially arises from the non-linearity of the equations of motion

(8.11), leading to a poorly defined normal ordering. To this is compounded the question as to which co-ordinate frame the calculations are related.

It is possible to write down a general form of multi-loop amplitude at each order of perturbation theory. Only the case that $\partial/\partial X^-$ is a Killing vector., with $G^{\mu\upsilon}$ in its L.C. gauge will be considered explicitly. Then from (8.25) and (8.26) the general multi-loop amplitude will be

$$M = \int D\underline{X}\ \exp -\frac{1}{2}\iint\{[X^{i\prime}X^{j\prime}+(x^i-P_-G^{i-})(\dot{X}^j-P_-G^{j-})]G_{ij}+P_-^2G^{--}\}$$

$$\times\ \exp\iint \underline{J}.\underline{X} \tag{8.30}$$

which reduces to

$$M = \exp[-\underline{J}.G*J]x<\exp\{-\frac{1}{2}\iint[X^{i\prime}X^{j\prime}+\dot{X}^i\dot{X}^j)(G_{ij}-\eta_{ij})$$

$$-2P_-G^{i-}\dot{X}^jG_{ij}+P_-^2(G^{i-}G^{j-}G_{ij}+G^{--})]\}. \tag{8.31}$$

where < > is evaluated using $<X_1^iX_2^j> = G_{12}\delta^{ij}$. The exponent in (8.31) inside the bracket < > may be expanded in powers k to give a weak field approximation to the amplitude.

We have thus accomplished our purpose in this section. It is necsssary to extend these results to the superstring by solving the constraint in that case, in a suitable L.C. gauge. It is to be expected that this is a considerably more complicated exercise, but one which should, in principle, be possible to complete. It should then be possible to investigate the nature of backgrounds on which multi-loop amplitudes are finite, extending the work of the previous section. Such an approach is being developed in [116], with expressions for multi-loop superstring amplitudes being obtained on a class of spaces $M_4 \times G$, with M_4 Minkowski 4 space. The relevance of such constructions will be considered briefly in section 8.5.

Section 8.4: Covariant Superstring Field Theory

We have not considered covariant string or superstring field theory in any manner at all as yet, having based our development totally on a gauge-fixed approach to superstrings, using the light-cone gauge. In spite of this having led to finite covariant multi-loop amplitudes obtained from a string field theory possessing the global symmetries (closure of the full sP_{10} algebra, as shown in the previous Chapter), it has been felt by many researchers in the field that there is a lack of alegance and power in such an approach. In comparison, the strength of the approach of Einstein's general theory of relativity is its undeniable beauty as a complete description of the gravitational field; all its local symmetries are laid bare. Moreover the resulting framework allows the local symmetries to be used to write down elegant and simple covariant expressions for quantities which in a non-covariant formalism would appear very much more complicated. When quantum corrections - such as multi-loop amplitudes-are being considered then the complexity of terms arising in a non-covariant approach is very high. Moreover it has been possible to put such strong restrictions on possible quantum counter-terms arising in certain supersymmetric field theories (when expressed in an explicitly supersymmetric manner that no suitable terms are allowed. Hence the total finiteness of these theories have been proven [117), 118)]. It was hoped to construct a similar explicitly symmetric superstring field theory, and similarly prove the absence of any possible quantum corrections.

Such a hope has not yet been fulfilled. No universally appealing candidate has appeared for the full local symmetry of superstring field when closed superstrings are involved as independent fields. The best candidate so far is the open superstring field theory of Witten [119], which is a cubic action

$$\Phi\, Q\, \Phi + \Phi^3 \tag{8.32}$$

where Q is the BRST charge and the Φ^3 interaction term is obtained by sewing two string together along half of their lengths to make the third in the Φ^3 term. The action (8.32) is invariant under the local symmetry

$$\delta\Phi = \epsilon Q\Lambda \tag{8.33}$$

which may be shown to correspond to the invariance associated with the Virasoro algebra (8.32) has also been shown to lead to the expected multi-loop amplitudes [41].

There have been numerous attempts to construct an extension of this action to the closed string case, but none has avoided the difficulty of non-local interactions which arise when the open string of (8.32) are closed by joining their ends. A further difficulty for closed strings is that it is necessary to generate multi-loop amplitudes which give a single coverage of supermoduli space. This is an as yet unsolved problem since in particular this latter space is not yet well-defined (as noted earlier). The same problem arises in the construction of a covariant field theory for closed bosonic strings. A possible solution to this problem has been recently proposed [120], and involves an infinite power series in interactions, $a_n\phi^n$. These higher-order interaction terms ϕ^n, arise from the edges of moduli space where increasing numbers of string interaction points coincide. It is interesting to note that the heterotic superstring avoids such an infinite series. It was shown in the previous Chapter that it only needs ϕ^3 and ϕ^4 terms to achieve closure of the algebra. Of course that construction is only in a flat background space-time; in a curved space-time there may be higher order terms for the closure.

Another approach to covariant string field theory has been to try to develop a theory of co-ordinate invariance in loop space, the space of sets of co-ordinates $\{X\mu(\sigma)\}$, $0 \leq \sigma \leq 2\pi$ (together with ghost co-ordinates) [121),122)]. Co-ordinate transformations can be performed independently at each value of σ, and the generators of such transformations, together with σ-reparametrisations, used to develop a general relativity on such a space of transformations (in a similar manner to general relativity bein specified in a certain manner by transformations equal to the set of translations and Lorentz rotations. No natural action corresponding to Einsteins R/detg is apparent in this approach, although $\Phi\ Q\ \Phi$ does arise at quadratic level from some choices [121)]. Whilst this frame has a certain appeal as extending general covariance, it is not yet clear that it is based on a suitable set of local symmetries.

Section 8.5: Whither Superstrings?

We list a set of pressing questions which are now facing superstrings:

(1) What are the physical predictions of a given superstring action?
(2) How do superstrings handle the age-old question of quantum gravity-what cosmology do they imply, what is the origin of physical space-time, are space-time singularities avoided, etc, etc?
(3) What is the full symmetry of the theory?

More could be added to the list although the above seem the most pressing; we have already attempted to answer (3) in the previous section, so we will only consider (1) and (2).

Numerous attampts to obtain physics from superstrings have been, and are still being, made [123)]. Such approaches in general attempt to construct four-dimensional superstrings with an internal symmetry group G which contains the SU(3) X SU(2) X U(1) standard model symmetry. G is then broken by some unknown manner to the latter, which is then broken by the standard mechanisms to the low-energy theories. Such an approach assumes an expanding cosmology, with decreasing temparature, to have arisen from the superstring theory; and decoupled from the massive matter modes of the superstring. Mechanisms for such features are completely unknown; they may even not exist.

The only way presently to attempt to answer question (1), and hopefully at the same time the ancient questions in (2), is to begin a carefully non-perturbative analysis of the superstring theory. Since there is no explicitly covariant theory, the only way ahead seems to be to use the L.C. gauge-fixed theory constructed in this book. That itself is only on a flat background space-time. One may ask if there is a non-trivial ground-state or if different physics (and space-times, etc) would occur if a non-trivial background were used for the basic construction. One may follow John Schwarz [124)] in hoping that only a flat background construction is needed, and that the other consistent and finite superstrings, constructed on curved space-time, are other ground states of this flat background construction. There is no justification of this appealing hypothesis. However non-perturbative analysis of the L.C. superstring field theory presented here is, in principle, able to determine if the hypothesis is correct. It may be expected to be, however, some time before such a check is able to be achieved.

APPENDICES

Appendix I: Riemann surfaces

In the following appendices we give some definitions and results concerning the theory of Riemann surfaces which we have used in the text. We have extracted material from references [66),67),69),70)].

A manifold may be considered as a generalization of a surface in Euclidean 3-space. We may locally represent the surface by the parametrization

$$x = x(u,v),\ y = y(u,v),\ z = z(u,v),$$

where these functions are supposed to be sufficiently smooth. The derivative of these functions define two tangent vectors. If these vectors are independent at a point P, then the surface has at that point the topology of Euclidean 2-space.

A topological space T which is separable is called a Hausdorff space, that is, given any two elements t_1 and $t_2 \in T$ there exists two disjoint open sets T_1 and T_2 such that $t_1 \in T_1$ and $t_2 \in T_2$. A continuous mapping with continuous inverse between two topological spaces is called a homeomorphism. Properties which are preserved under homeomorphisms of one space onto another are called topological invariants.

A connected Hausdorff space T such that each point of T is contained in an open set homeomorphic to an open set in Euclidean n-dimensional space is called an n-dimensional manifold. An analytic manifold or a Riemann surface is a manifold M such that:

a) There is an atlas of M, that is a collection $\{U_i, \phi_i\}_{i\epsilon I}$ where $\{U_i\}_{i\epsilon I}$ is an open covering of M and ϕ_i a homeomorphism of U_i onto an open set in the complex z-plane C.

b) When $U_i \cap U_j \neq 0$ (the null or empty set), then $\phi_j(\phi_i^{-1})(z)$ is an analytic function of z in $\phi_i(U_i \cap U_j)$.
It follows from this definition that $\phi_j(\phi_i^{-1})$ is a conformal, sense preserving mapping $\phi_i(U_i \cup U_j)$ onto $\phi_j(U_i \cap U_j)$ and that $\frac{d}{dz}\phi_j(\phi_i\ (z)) \neq 0$. The atlas $\{U_i, \phi_i\}_{i \in I}$ defines the analytic structure of the Riemann surface. Another atlas $\{V_j, \psi_j\}_{j \in J}$ defines the same analytic structure with the new atlas $\{W_k, \lambda_k\}_{k \in K}$, where $\{W_k\}_{k \in K} = \{U_i\}_{i \in I} \cup \{V_j\}_{i \in J}$ and λ_k is the corresponding ϕ_i or ψ_j according to which U_i or V_j we are considering, and satisfies (a) and (b).

Given an atlas $\{U_i, \phi_i\}$, the mapping ϕ_i defines a local uniformization of the manifold M. $z \in \phi_i(U_i)$ are the local coordinates. The angle between two intersecting curves on M is defined as the angle between their tangents in any local coordinates. This is a well defined geometrical object since $\phi_j(\phi_i^{-1})$ is conformal.

The analytic structure of Riemann surfaces allows the introduction of analytic functions f with domain on one Riemann surface R and images on another Riemann surface S. If ϕ is any uniformizer of R around P and ψ any uniformizer in an open set containing f(P), we define f: R → S to be analytic if

$$w = \psi(f\phi^{-1}(z))) = w(z)$$

is an analytic function of z for all $P \in R$.

We say that two Riemann surfaces R and S are conformally equivalent if there is a one to one analytic mapping f of R onto S.

Appendix II: Uniformization

Let us consider a surface S, not necessarily a Riemann surface. Suppose $\tilde{S}$ is also a surface and π is a locally topological mapping of $\tilde{S}$ into S. The pair $(\tilde{S},\pi)$ is called a smooth covering surface of S, π is the projection mapping. The points $\pi^{-1}(q)$, $q \in S$ are said to lie over q, and $\pi^{-1}(q)$ is the fiber over q. The surface S is called unlimited when for each curve γ on S with initial point q there exists a curve $\tilde{\gamma}$ on $\tilde{S}$ with initial point on $\pi^{-1}(q)$ such that $\pi\tilde{\gamma} = \gamma$.

The homeomorphism of $\tilde{S}$ on itself which map points on a fiber into points in the same fiber are called covering transformations. From all unlimited smooth coverings of a surface S there is only one which is simply connected. It is called the universal covering, $\hat{S}$. The group of covering transformations is usually denoted $\hat{H}$.

Let us consider now a Riemann surface R with universal covering surface $\hat{R}$. In this case π is an analytic mapping of $\hat{R}$ onto R and $\hat{H}$ is a group of conformal automorphisms of $\hat{R}$.

Riemann's mapping Theorem states that every simply connected Riemann surface is conformally equivalent to the extended plane $\hat{\underline{C}}$, the finite plane $\underline{C}$ or the unit disk (which is conformally equivalent to the upper half plane U). Any simply connected plane region with at least two boundary points is conformally equivalent to the unit disk.

As we have already defined in Appendix I, conformally equivalent means that there exists an analytic one to one mapping $w = f(\phi^{-1}(z))$, where ϕ is the local homeomorphism of the Riemann surface $\hat{R}$ from $\phi(\hat{P})$, $\hat{P} \in \hat{R}$ into one of the three canonical regions G of the theorem.

We can define now a conformal mapping from G onto R.

$$P = \pi\, f^{-1}(W) \equiv F(W)$$

It is analytic but not one to one; F^{-1} is not single valued.

Let us consider now a multiple valued analytic function g(P) regular at every point of R. It can be shown (from the monodromy theorem) that the function g(F(W)) is a single valued function of $W \epsilon G$. We say that the mapping $F: G \to R$ uniformizes any function on R. This is the general uniformization theorem of Klein, Poincaré and Koebe.

It is clear, then, how to find the different uniformizations of R. We have to find the different ways in which $\hat{R}$ is mapped onto the canonical region G. Then we construct the different mappings F. It is enough, therefore, to determine the conformal mappings of G onto itself.

The general result can be summarized in the following proposition.
(a) The one to one conformal self mappings of $\hat{\underline{C}}$ are the Möbius transformations (linear fractional transformations)

$$w = \frac{az+b}{cz+b}, \qquad ad-bc = 1, \qquad a,b,c,d \text{ complex parameters}$$

This is a three complex parameter group.
(b) The one to one conformal self mappings of $\underline{C}$ are the Möbius transformations

$$w = az+b, \qquad a \neq 0.$$

This is a four real parameter group.

(c) The one to one conformal self mappings of $\underline{U}$ are the Möbius transformations.

$$w = \frac{az+b}{cz+d}, \qquad ad-bc = 1, \qquad a,b,c,d \text{ real parameters}$$

This is a three real parameter group.

In all cases the uniformizing function is determined uniquely up to a Möbius transformation.

Let us consider now the group of covering transformations $\hat{H}$. Let f be a mapping from $\hat{R}$ to G and $h \in \hat{H}$. Then fhf^{-1} is a one to one conformal mapping of G onto itself. There is then an isomorphism between the group $\hat{H}$ and the group Γ defined in (a),(b),(c). Consequently the transformations in Γ are fixed-point free since this is true for $\hat{H}$.

For the same reason Γ is discontinuous, that is, any set of equivalent points with respect to Γ has no point of accumulation in G.

A connected subset of G which contains one and only one point from each equivalent set of points with respect to Γ is called a fundamental domain. It is essentially the image of one "sheet" of $\hat{R}$ in G. Γ is thus a discontinuous group of fixed point free linear transformations of G. Two equivalent points with respect to Γ are mapped by f^{-1} into two points of $\hat{R}$ that have the same projection in R. In this way we may define a one to one conformal mapping from $R \to G/\Gamma$. We may take G/Γ as a normal form of R.

If $\hat{R}$, the universal covering of R, is conformally equivalent to $\hat{\underline{C}}$, the condition that Γ be fixed point free implies that its only element is the identity. Consequently R itself is conformal equivalent to $\hat{\underline{C}}$ (or to the sphere). If $\hat{R}$ is conformally equivalent to $\underline{C}$, the condition of being fixed point free for Γ implies that

$$w = z+b \tag{A2.1}$$

There are then three cases for Γ.

i) The only element of Γ is the identity. R is then conformally equivalent to $\underline{C}$. That is, to a one-punctured sphere.

ii) Γ is generated by $w = z+b$. The group Γ consists of all translations $w = z+nb$. R is conformally equivalent to a doubly punctured sphere.
iii) Γ is generated by $w_1 = z+a$, $w_2 = z+b$ where $\text{Im } a/b \neq 0$. (A2.2)
Γ consists of the mappings $w = z+ma+nb$. R is conformally equivalent to a torus. The converse is also true, every compact Riemann surface of genus 1 has $\underline{C}$ as a universal covering.

The Riemann surfaces which are not included in the above list have $\underline{U}$ as universal covering. This is the content of the general uniformization theorem. Before giving a precise statement of it we consider some further definitions.

Appendix III: Kleinian and Fuchsian groups and uniformizations

A Möbius transformation different from the identity is either parabolic (conjugate to $w = z+1$) or elliptic (conjugate to $w = \lambda z$, $|\lambda| = 1$) or loxodromic (conjugate to $w = \lambda z$, $|\lambda| \neq 1$). A loxodromic transformation is called hyperbolic of λ is greater than one and positive.

The limit set of Γ is defined as the set of accumulation points of orbits (equivalent points with respect to Γ). It can be shown that is is also the closure of the set of fixed points of non-elliptic elements of Γ, distinct from the identity.

A discrete group Γ of Möbius transformations is called Kleinian if its limit set $\Lambda(\Gamma)$ is not the whole $\hat{\underline{C}}$.

A Kleinian group Γ is called Fuchsian if all its loxodromic elements are hyperbolic and Γ leaves a disc or a half-plane fixed. By performing a conjugation one can always consider the upper half plane. The transformations of Γ

$$w = \frac{az+b}{cz+d}, \qquad ad-bc = 1 \tag{A3.3}$$

have real coefficients.

In this case the limit set either coincides with the real axis and Γ is called of the first kind or the limit set is a nowhere dense subset of the real axis and Γ is called of the second kind. The upper half plane $\underline{U}$ is taken with the Poincaré metric.

$$ds^2 = \frac{dz d\bar{z}}{y^2} \qquad x = x+iy \tag{A3.4}$$

as a model of non-Euclidean geometry.

The group defined in Appendix II(b) is an example of the Kleinian group. The group defined in Appendix II(c) is Fuchsian.

The Fuchsian uniformization theorem may now be expressed as:

Every Riemann surface R which is not conformally equivalent to $\hat{\underline{C}}$, $\underline{C}$ (the once punctured sphere), doubly punctured sphere or a compact Riemann surface of genus 1 can be represented as $\underline{U}/\Gamma$ where Γ is a Fuchsian group without elliptic elements. The elliptic elements are excluded as a consequence of the fixed point free restriction on Γ.

If the Riemann surface is compact the parabolic elements are also excluded. Interesting examples of Kleinian groups are the Schottky groups. Let $C_1, C_1', \ldots, C_p, C_p'$ be 2p disjoint Jordan curves on $\hat{\underline{C}}$ that bound a domain of connectivity 2p, and let $g_1, \ldots, g_p$ be Möbius transformations such that g_j maps the exterior of C_j onto the domain interior to C_j'. Then g_j, $j = 1, \ldots, p$, generate a Kleinian group called a Schottky group. The domain exterior to all C_j is a fundamental region. The region of discontinuity of a Kleinian group Γ is defined as

$$\Omega(\Gamma) = \hat{\underline{C}} - \Lambda(\Gamma),$$

where $\Lambda(\Gamma)$ is the limit set of Γ.

$\Omega(\Gamma)$ must be open and dense in $\hat{\underline{C}}$ and Γ acts discontinuously in Ω. That is to say, for every compact set $K \subset \Omega$, $K \neq \phi$, one has the intersection $\gamma(K) \cap K = \phi$ for all but finitely many $\gamma \in \Gamma$. For the Schottky groups we have the following uniformization theorem:

Every compact Riemann surface can be represented as Ω/Γ where Γ is a Schottky group.

Appendix IV: Differentials

We begin with some definitions which we are going to use to characterize the analytic structure of the Riemann surfaces.

If Ω is a 1-differential form on M the conjugation operator * is defined as

$$\Omega = \alpha(x,y)dx + \beta(x,y)dy,$$
$$*\Omega = -\beta(x,y)dx + \alpha(x,y)dy. \tag{A4.1}$$

It has the right transformation law of a 1-differential form. It has the properties

$$*(\Omega_1+\Omega_2) = *\Omega_1+*\Omega_2,$$
$$**\Omega = *(*\Omega) = -\Omega, \tag{A4.2}$$

If $\Omega_1 = \alpha_1 dx+\beta_1 dy$ and $\Omega_2 = \alpha_2 dx + \beta_2 dy$ then

$$\Omega_1 \wedge *\Omega_2 = (\alpha_1\alpha_2+\beta_1\beta_2)dx\wedge dy \tag{A4.3}$$

If $\Omega = df$, Ω is called exact. If $d\Omega = 0$, Ω is called closed and locally is always exact.

If $\Omega = *(df) \equiv *df$, Ω is called co-exact. If $d*\Omega = 0$, Ω is called co-closed. A co-closed differential form is always locally co-exact:

$$d*\Omega = 0 \Rightarrow *\Omega = df \text{ (locally)}$$
$$\Rightarrow \Omega = -*df$$

The Laplace operator Δ may be expressed as

$$\Delta f = d*df \tag{A4.4}$$

A differential Ω is called harmonic on the Riemann surface M if locally it is exact (a closed form)

$$\Omega = df,$$

and f is harmonic

$$\Delta f = 0. \tag{A4.5}$$

One may check directly that Ω is harmonic if and only if $d\Omega = 0$ and $d{*}\Omega = 0$.

Ω is called a holomorphic or analytic differential if locally it is an exact 1-form

$$\Omega = df, \tag{A4.6}$$

and f is a holomorphic function.

One may check directly that Ω is holomorphic if and only if $d\Omega = 0$ and ${*}\Omega = -i\Omega$. In the above definitions "locally" means that "there exists an open parametric disk".

The above differentials are all defined on the Riemann surface. However by using a local homeomorphism of the manifold one may express them (by pull back) in terms of the local coordinates.

We introduce now an inner product in the space of differentials. We consider measurable 1-differential forms Ω on the Riemann surface M which are Lebesgue integrable. That is to say, locally $\Omega = \alpha(x,y)dx+\beta(x,y)dy$, where α,β are measurable and $|\alpha|^2, |\beta|^2$ are locally integrable. The inner product of two differentials Ω_1 and Ω_2 is defined by

$$(\Omega_1,\Omega_2) = \int_M \Omega_1 \wedge {}^*\Omega_2 \tag{A4.7}$$

Using (3.3) we obtain locally

$$\Omega\, {}^*\overline{\Omega} = (|\alpha|^2 + |\beta|^2)\, dx \wedge dy.$$

This product satisfies

$$\begin{aligned}(\Omega_1,\Omega_2) &= \overline{(\Omega_2,\Omega_1)},\\ (a\Omega_1,\Omega_2) &= a(\Omega_1,\Omega_2),\\ (\Omega_1+\Omega_2,\Omega_3) &= (\Omega_1,\Omega_3)+(\Omega_2,\Omega_3)\\ (\Omega,\Omega) &\geq 0\end{aligned} \tag{A4.8a}$$

If we identify two differentials which differ in local coordinates on a set of measure zero we have in addition

$$(\Omega,\Omega) = 0 \Rightarrow \Omega = 0 \tag{A4.8b}$$

Ω now represents the equivalence class of differentials that in local coordinates agree up to a set of measure zero.

(A4.7) then defines an inner product in the vector space of differentials, which is called H. Moreover H is complete in the norm induced by this inner product

$$\|\Omega\|^2 = (\Omega,\Omega) \tag{A4.9}$$

The proof of the completeness of H is analogous to the proof of completeness of the L^2 spaces in the theory of functions. H is then a Hilbert space, usually denoted as $L^2(M)$ where M is the Riemann surface on which the differentials are defined.

We introduce now subspaces of $L^2(M)$ which are very useful in order to decompose $L^2(M)$ in terms of orthogonal subspaces. $f \in C^k$ denotes that the complex valued function f has k continuous partial derivatives.

We denote $\delta(M)$ the closure in $L^2(M)$ of the vector space of exact differentials df, $f \in C^2$ and f has compact carrier $\subset M$. We denote $*\delta(M)$ the closure in $L^2(M)$ of the vector space of co-exact differentials *df, $f \in C^2$ and f has compact carrier $\subset M$. This means that if $\Omega_1 \in \delta(M)$, then

$$\Omega_1 = \lim_{n\to\infty} df_n, \tag{A4.10}$$

in the norm of $L^2(M)$. Analogously

$$\Omega_2 \in *\delta(M), \quad \Omega_2 = \lim_{n\to\infty} *dg_n. \tag{A4.11}$$

We denote $\Delta(M)$ the subspace of $L^2(M)$ orthogonal to $\delta(M)$ and $*\delta(M)$

$$\Delta(M) = \delta(M)^{\perp} \cap (*\delta(M))^{\perp}.$$

We observe that $\delta(M)$ and $*\delta(M)$ are orthogonal. In fact from (A4.10) and (A4.11) it follows that

$$(df_n, *dg_n) = \int_M df_n \wedge **d\bar{g}_n.$$

Using (A4.2) and the hypothesis that f_n have compact carriers we obtain for all n

$$(df_n, *dg_n) = -\int_M d(f_n\, d\bar{g}_n) = 0$$

hence

$$(\Omega_1, \Omega_2) = 0.$$

The orthogonality of $\delta(M)$ and $*\delta(M)$ implies that

$$\Delta(M) = (\delta(M) \oplus *\delta(M))^{\perp} \tag{A4.12}$$

and from the properties of a Hilbert space we conclude

$$L^2(M) = \delta(M) \oplus *\delta(M) \oplus \Delta(M). \tag{A4.13}$$

This means that a differential $\Omega \in L^2(M)$ can be decomposed as the sum

$$\Omega = \Omega_1 + \Omega_2 + \omega \tag{A4.14}$$

$\Omega_1 \in \delta(M)$, $\Omega_2 \in *\delta(M)$ and $\omega \in \Delta(M)$, where Ω_1 and Ω_2 are only the limit of exact and co-exact differentials. The limit, being defined in terms of the $L^2(M)$ norm, is therefore a mean limit, in spite of the fact that f_n and g_n are C^2. Thus Ω_1 and Ω_2, in general, will be non-regular. However it is possible to prove that $\omega \in \Delta(M)$ is always C^1.

From the definition of $\Delta(M)$, we have in particular

$$(\omega, df) = 0,$$
$$(\omega, *dg) = 0. \tag{A4.15}$$

We thus obtain

$$0 = \int_M \omega \wedge *d\bar{f} = -\int_M *\omega \wedge d\bar{f} = -\int_M d*\omega . \bar{f}$$

$$0 = -\int_M \omega \wedge d\bar{g} = -\int_M d\omega . \bar{g}, \tag{A4.16}$$

which imply $d\omega = d*\omega = 0$.

This implies, from (A4.5), that ω is harmonic. Conversely, if Ω is harmonic then (A4.15) and (A4.16) are valid for df_n and $*dg_n$ of (A4.10) and (A4.11). Therefore in the limit

$$(\Omega,\Omega_1) = (\Omega,\Omega_2) = 0,$$

hence $\Omega \in \Delta(M)$.

We have thus shown that $\Omega \in \Delta(M)$ if and only if Ω is harmonic in M. Finally by imposing only regularity conditions on Ω one can obtain a stronger decomposition than (A4.14). In fact it can be shown that if $\Omega \in L^2(M)$ satisfies only the regularity condition $\Omega \in C^3$ then

$$\Omega = df + *dg + \omega, \qquad \text{(A4.17)}$$

where f and g are C^2-complex functions and $\omega \in \Delta(M)$.

The pairwise orthogonality of the subspaces $\delta(M)$, $*\delta(M)$ and $\Delta(M)$ asssures the uniqueness of this decomposition.

Appendix V: On the existence of harmonic and analytic differentials

The results of the previous section may be used directly to impose restrictions on the existence of harmonic and analytic differentials on a Riemann surface. We first observe that a harmonic function f, $f \in C^2$, on a compact Riemann surface must be constant. In fact df is an exact and harmonic differential form on M. The orthogonality of the subspaces $\delta(M)$ and $\Delta(M)$ then implies f = constant.

We can thus look for meromorphic functions over a compact Riemann surface. However there are also restrictions on the class of meromorphic functions we must consider. Let Ω be a meromorphic differential, that is, locally, $\Omega = df$, f having a finite number of poles on M. Denote P_i, $i = 1,\ldots,n$, the points on M at which f is singular and c_i the corresponding residue. Let us take non-intersecting parametric discs D_i, $P_i \in D_i$. If we apply Stoke's theorem to $R = S - \Sigma D_i$ we have

$$0 = \int_R d\Omega = \int_{\partial R} \Omega = -\sum_i \int_{\partial D_i} \Omega = -2\pi i \sum_i c_i .$$

The sum of the residues is thus zero.

These two 'non-existence' results are simple consequences of the discussion in the previous sections. The existence theorems on harmonic and analytic differentials are more difficult to prove. Several approaches have been used since Riemann considered the Dirichlet principle to approach this problem. In particular it may be proved [69),70)] that there exist on a arbitrary Riemann surface M exact harmonic differentials with singularity $d(1/z^n)$, $(n > 1)$, at an arbitrary point P $\in$ M and of an analytic differential with singularity $d(1/z^n)$, $n \geqq 1$, at P with exact real part. Since $n \geqq 1$, the residues are always zero and the non-existence theorem is avoided. For n = 0 we must consider functions which are singular at least in two points on compact Riemann surfaces. If P and Q are two arbitrary distinct points on a Riemann surface M then there exist a harmonic and an analytic differential on M with singularity $-dz/z$ at P and dz/z at Q. Moreover there exist a real harmonic function on M having singularity $-\log|z|$ at P and $\log|z|$ at Q. Finally given P_i, $i = 1,\ldots,n$, on M and arbitrary non-zero complex numbers c_i, $\Sigma c_i = 0$, then there exists a meromorphic differential on M with poles of order 1 at P_i, $i = 1,\ldots,n$ and residues c_i. This result shows that the only restriction to the existence of meromorphic differentials on a compact Riemann surface is the result we have established at the beginning of the section.

The restrictions are only for compact Riemann surfaces. It can be shown that on non compact surfaces there always exist exact harmonic and analytic differentials which do not vanish identically.

We now consider differentials which are not exact. It can be shown there exist harmonic differentials everywhere regular on M. Moreover for a compact Riemann surface of genus g the vector space of harmonic differentials $\Delta(M)$ over the complex number field has dimension 2g.

The analytic or holomorphic differentials on a compact Riemann surface are also called abelian differentials of the first kind. Given a harmonic differential Ω on a compact Riemann surface we may construct the differential

$$\phi = \Omega + i{*}\Omega.$$

It is closed and satisfies $\phi = i{*}\phi$. Hence it is an analytic differential. The existence of abelian differentials of the first kind is then guaranteed by the existence of harmonic differentials. Moreover if we define

$$\tilde{\phi} = \bar{\Omega} + i{*}\bar{\Omega},$$

where $\bar{\Omega}$ is the complex conjugate to Ω, we observe

i) $\tilde{\phi}$ is analytic (with respect to $\bar{z}$ in a local parametrization),

ii) $\Omega = \frac{1}{2}(\phi+\tilde{\phi})$,

iii) $(\phi,\bar{\tilde{\phi}}) = 0$

We conclude that the vector space of analytic differentials on a compact Riemann surface of genus g has dimension g.

Appendix VI: Period Matrix

We consider in this section a compact Riemann surface M of genus g. The normal form of M is a polygon Π with 4g sides denoted by

$$a_1 b_1 a_1^{-1} b_1^{-1} \ \dots \ a_g b_g a_g^{-1} b_g^{-1} \ ,$$

where a_i and b_i are piecewise analytic curves and they form a homology basis for M. Any cycle c on M is thus homologous to an integral linear combination of a_i and b_i:

$$\sim \sum (n_i a_i + m_i b_i).$$

The period of a 1-differential closed form on a cycle c

$$\int_c \Omega$$

is then equal to

$$\Omega = \sum_{i=1}^{g} (m_i \int_{a_i} \Omega + n_i \int_{b_i} \Omega)$$

$$= \sum_{i=1}^{g} (m_i \ A_i + n_i \ B_i), \tag{A6.1}$$

where we have denoted

$$A_i = \int_{a_i} \Omega \ , \ b_i = \int_{b_i} \Omega \tag{A6.2}$$

Consequently A_i, B_i, $i = 1,\ldots,g$, determine all the periods of a closed differential. Let us assume that Ω is closed, $\epsilon\ C^1$ and $\epsilon\ L^2(M)$. We then have for any complex function g

$$0 = \int_M d\Omega \cdot \bar{g} = \int_M \Omega \wedge d\bar{g} = -(\Omega, {*dg}). \qquad \text{(A6.3)}$$

Ω is thus orthonogal to $*\delta(M)$, so that from (A4.13)

$$\Omega = \Omega_1 + \omega. \qquad \text{(A6.4)}$$

Moreover, Ω and ω are C^1, hence Ω_1 is C^1. This is enough to guarantee that Ω_1 is exact. We thus have

$$\Omega = df + \omega, \qquad \text{(A6.5)}$$

ω harmonic. The decomposition is unique. That is if Ω is closed there exists one and only one harmonic differential differing from Ω by an exact differential. This is the Hodge theorem for compact Riemann surfaces.

The period on any cycle c of Ω is then equal to the period of ω, and they only depend on the 2g periods A_i, B_i, $i = 1,\ldots,g$. This property is directly related to the fact that the vector space of harmonic differentials has dimension 2g. The basis of the harmonic differentials ω_i, $i = 1,\ldots,2g$ may be constructed in such a way as to satisfy

$$\int_{a_j} \omega_i = \delta_{ij} \quad , \quad \int_{b_j} \omega_i = 0,$$

$$\int_{a_j} \omega_{g+i} = 0 \quad , \quad \int_{b_j} \omega_{g+i} = \delta_{ij}, \quad i = 1,\ldots,g \qquad \text{(A6.6)}$$

Let us consider now the vector space of abelian differentials of the first kind. It has dimension g. Denote ϕ_i , i = 1,...,g a basis in that space and

$$A_{ij} = \int_{a_j} \phi_i \, .$$

The matrix $\{A_{ij}\}$ is non singular. Otherwise there would be non trivial $\bar{\lambda}_i$ which solve the system of linear equations

$$\bar{\lambda}_i A_{ij} = 0, \quad j = 1,\ldots g$$

and the differentials $\sum_i \bar{\lambda}_i \phi_i$ would have vanishing periods for all cycle c, being then exact. But an exact and harmonic differential on a compact Riemann surface necessarily vanishes, contradicting the assumption of independence of ϕ_i, i = 1,...,g.

Since $\{A_{ij}\}$ is non singular, there exists a non trivial solution to the system of linear equations.

$$\sum_{i=1}^{g} \lambda_{ik} A_{ij} = \delta_{jk} \, .$$

We can then define the basis $\{\phi_i\}$ of the vector space normalized by the condition

$$A_{ij} = \delta_{ij} \, . \tag{A6.7}$$

We denote B_{ij} the period of ϕ_i over the cycle b_j. We call $\{\phi_i\}$ with this normalization a canonical basis for the abelian differentials of the first kind. Consider how two abelian differentials of the first kind ϕ_1 and ϕ_2. It can be shown that

$$i(\phi_1,\bar{\phi}_2)=\sum_{j=1}^{g}({}^1A_j\ {}^2B_j - {}^1B_j\ {}^2A_j) = 0. \tag{A6.8a}$$

where ${}^1A_i, {}^1B_i$ and ${}^2A_i, {}^2B_i$ are the periods of ϕ_1 and ϕ_2 respectively over a canonical basis a_i, b_i, $i = 1,\ldots,g$.

Additionally by direct evaluation we obtain

$$(\phi_1,\phi_2) = i\sum_{j=1}^{g}({}^1A_j{}^2\bar{B}_j - {}^1B_j{}^2\bar{A}_j). \tag{A6.8b}$$

In particular

$$0 \leq (\phi,\phi) = i\sum_{i=1}^{g}(A_j\bar{B}_j - B_j\bar{A}_j) \tag{A6.8c}$$

(A6.8) are called the bilinear relations of Riemann.

(A6.8c) implies that if A_i or B_j are zero, $j = 1,\ldots,g$, then $\phi = 0$ and also if A_j and B_j are real, $j = 1,\ldots,g$, then $\phi = 0$. If we apply (6.8a) to the elements of a canonical basis ϕ_i and ϕ_j we conclude

$$B_{ij} - B_{ji} = 0, \tag{A6.9}$$

the matrix $\{B_{ij}\}$ is symmetric. In addition, if we apply (6.8c) to the abelian differential of the first kind

$$\phi = \sum \lambda_i\phi_i$$

using that

$$A_i = \lambda_i ,$$

$$B_j = \sum_i \lambda_i B_{ij} ,$$

we obtain

$$\lambda^T \operatorname{Im} B\lambda \geq 0 \qquad \text{(A6.10)}$$

where $\lambda^T = (\lambda_i, \ldots, \lambda_g)$ and $B = \{B_{ij}\}$.
The matrix B is then symmetric and its imaginary part is positive definite.

Appendix VII: Analytical structure of Teichmüller and Torelli spaces

All the Riemann surfaces we consider in this section are compact. Following Rauch [67] and Bers [125] we consider triplets (S,S',α) where α is a homeomorphism of S onto S'. Two triplets are Teichmüller equivalent

$$(S,S_1,\alpha_1) \sim (S,S_2,\alpha_2), \tag{A7.1}$$

if there exists f: $S_1 \to S_2$, conformal,

$$\begin{array}{ccc} S & \xrightarrow{\alpha_1} & S_1 \\ & & \Big\downarrow f \\ S & \xrightarrow{\alpha_2} & S_2 \end{array} \tag{A7.2}$$

such that f is homotopic to $\alpha_2\alpha_1^{-1}$.

A class of Teichmüller equivalent triplets is a Teichmüller surface denoted by $\langle S,S',\alpha\rangle$. It contains (S,S',α) as a representative. $T^g(S)$ is the totality of Teichmüller surfaces. $\langle S,S,1\rangle$ is the origin of $T^g(S)$. g is the genus of the reference surfaces. Two triplets are called Torelli equivalent

$$(S,S_1,\alpha_1) \approx (S,S_2,\alpha_2)$$

if there exists f: $S_1 \to S_2$, conformal, such that f induces the same authomorphism of $H_1(S)$, the first homology group, as $\alpha_2\alpha_1^{-1}$. A class of Torelli equivalent triplets is a Torelli surface denoted by $\{S,S',\alpha\}$. $J^g(S)$ is the totality of Torelli surfaces. $\{S,S,1\}$ is the origin of $J^g(S)$.

The triplets (A7.1) are conformally equivalent if there exists f: $S_1 \to S_2$ conformal. The conformal equivalence class is denoted $[S_1]$. M^g is the totality of conformal equivalence classes. A homeomorphism β of S onto itself induces a homeomorphism of $T^g(S)$ ($J^g(s)$) into itself:

$$\langle S,S',\alpha\rangle \to \langle S,S',\beta\alpha\rangle$$
$$\{S,S',\alpha\} \to \{S,S',\beta\alpha\}$$

The set of all these self homeomorphisms of $T^g(S)$ ($J^g(s)$) is denoted by $\Gamma^g(S)(\Lambda^g(S))$ the modular group of Teichmüller (Torelli) space. $\Delta^g(S) \subset T^g(S)$ denotes the subset consisting of those mappings $\epsilon\ \Gamma^g(S)$ for which the self-map of S induces the same automorphism of $H_1(S)$ as the identity.

It can be shown [67)] that $\Gamma^g(S)$ and $\Lambda^g(S)$ can be given structures as groups of one to one self maps. With these structures one has

$$J^g(S) = T^g(S)/\Delta^g(s),$$
$$M^g = T^g(S)/\Gamma^g(S) = J^g(S)/\Lambda^g(S)$$

We know from the uniformization theorem that every compact Riemann surface (CRS) of genus o is conformally equivalent to a sphere. The conformal type of a CRS of genus 1 depends on one complex parameter. It can always be mapped conformally (by Schottky uniformization) onto the Riemann surface obtained by identifying the inner and outer boundaries C and C′ of an annulus by the mapping $W = \lambda z$, λ being the complex parameter.

Already Riemann knew that the conformal type of a CRS of genus g > 1 depends on 3g-3 complex parameters (moduli). The relevance of the Teichmüller and Torelli approaches is to make explicit the dependence on these complex parameters. The problem becomes more transparent if one generalizes the conformal mappings to quasiconformal ones.

Let S and S′ be two homeomorphic Riemann surfaces, and take f: S → S′ to be an orientation preserving diffeomorphism. If ϕ and ϕ' are local homeomorphisms on S and S′ we may rewrite

$$z' = \phi' f \phi^{-1}(z) = z'(z)$$

The map f is called quasiconformal if there is a constant $K \geq 1$ such that for every local homeomorphism ϕ and ϕ' and at every point of S

$$\left|\frac{\partial \bar{z}}{\partial z}\right| \leq \frac{K-1}{K+1}\left|\frac{\partial z'}{\partial z}\right|,$$

where $\bar{z}$ is the complex conjugate of z. Locally f maps every circle into an ellipse whose major axis is at most K times the minor axis. The smallest K for which f is K-quasiconformal is called the dilation of f and is denoted K(f).

When K(f) = 1, f is conformal. Given α: S → S′, there exists β: S → S′ homotopic to α and quasi-conformal. The topology in $T^g(S)$ is given by the Teichmüller distance between two surfaces $<S,S_1,\alpha_1>$ and $<S,S_2,\alpha_2>$

$$d(<S,S_1,\alpha_1>, <S,S_2,\alpha_2>) = \tfrac{1}{2} \log \inf K(\alpha)$$

where α ranges over all quasiconformal mappings $\alpha: S_1 \to S_2$ in the homotopy class of $\alpha_2\alpha_1^{-1}$. The Teichmüller space T^g is a complete metric space and the modular group M^g is a group of isometries. The Teichmüller space T^g is homeomorphic to R^{6g-6} for $g \geq 2$, (to R^2 for g= 1), and the modular group is properly discontinuous group of homeomorphisms.

A Beltrami differential μ is a tensor such that $\mu\, d\bar{z}/dz$ is invariant under change of parameters. Let B(S) be the space of all Beltrami differentials on S, and A(S) the space of all quadratic differentials on S. A(S) has dimension 3g-3 over the field of complex numbers. Define the functional

$$(\phi,\mu) \equiv \frac{i}{2}\int_S \phi\mu \, dz \wedge d\bar{z},$$

where $\phi \in A(s)$ and $\mu \in A(S)$.

The subspace $N(S) \subset B(S)$ is defined by $(\phi,\mu) = 0$. Then the space $B(S')/N(S')$ has dimension 3g-3 $(g \geq 2)$. Let $\mu_1,\ldots,\mu_{3g-3}$ be a basis of that space, and $\underline{c} = (c_1,\ldots,c_{3g-3})$ a vector $\underline{C}^{3g-3}$. For $\|\underline{c}\|^2 = |c_1|^2 + \ldots + |c_{3g-3}|^2$ sufficiently small then $\mu = \Sigma\, \mu_i c_i$ is a Beltrami differential and the map $\underline{c} \to \langle S,S',\alpha\rangle$ is a homeomorphism. We can introduce local coordinates for $T^g(S)$ in this way.

$T^g(S)$ with this local coordinate becomes a complex analytic manifold whose complex dimension is 3g-3 for $g \geq 2$, 1 for $g = 1$ and 0 for $g = 0$. $\Gamma^g(s)$, in addition to the properties already mentioned, becomes complex analytic. By means of the projection $T^g(S) \to J^g(S)$, $J^g(S)$ also becomes a complex analytic manifold. $\Lambda^g(S)$ is also realized by complex analytic maps.

Appendix VIII: Expansions around the interaction point

In this appendix the nature of objects associated with $\theta^{\bar{A}}$ and $\partial_z X^I$ at an interaction point will be evaluated using functional techniques rather than by mode expansions, as in [47] (to which the expressions derived below will be related). We will concentrate on $\theta^{\bar{A}}$, since $\partial_z X^I$ (and λ^A) can be treated in an identical fashion.

At next order, (5.23a) is

$$\epsilon = (\rho-\tilde{\rho}) = \frac{1}{2}(z-\tilde{z})^2 F''(\tilde{z}) + \frac{1}{6}(z-\tilde{z})^3 F^{(3)}(\tilde{z}) + O((z-\tilde{z})^4) \tag{A8.1}$$

Then to this order

$$(z-\tilde{z})^2 = 2[F''(\tilde{z})]^{-1}\epsilon + c\epsilon^{3/2} \tag{A8.2}$$

$$c = -(2/F'')^{3/2}(F^{(3)}/3F'') \tag{A8.3}$$

where all derivatives F", $F^{(3)}$, etc. in (A8.2), (A8.3), and so on, are evaluated at $z = \tilde{z}$. From the expression

$$\theta(\rho) = [\theta(\tilde{z})+(z-\tilde{z})\theta'(\tilde{z})][(z-\tilde{z})F'']^{-1}[1+\frac{1}{2}(z-\tilde{z})(F^{(3)}/F'']^{-1} \tag{A8.4}$$

and use of (A8.2), (A8.3), it follows that (5.22) is extended to

$$\theta(\rho) = (2\epsilon F'')^{-\frac{1}{2}}\{\theta(\tilde{z})-\epsilon^{\frac{1}{2}}[2^{\frac{1}{2}}F^{(3)}/3(F'')^{3/2}]\theta(\tilde{z}) +$$

$$+ (2\epsilon/F'')^{\frac{1}{2}}\theta'(\tilde{z}) + O(\epsilon)\} \tag{A8.5}$$

This is to be related to (3.60) of [47], with the identification, beyond (5.22), that

$$i2^{-\frac{1}{2}}\bar{G}^{A} = (F'')^{-1}\bar{\theta}^{A\prime}(\tilde{z}). \quad (A8.6)$$

Similarly for ∂X^I,

$$i2^{-\frac{1}{2}}G^{I} = (F'')^{-1}\partial_z^2 X^I(\tilde{z}) \quad (A8.7)$$

It is possible to extend (A8.5) to higher powers, with the next order being

$$\theta(\rho) = (2F''\epsilon)^{-\frac{1}{2}}\{\theta(\tilde{z})(1+A\epsilon^{\frac{1}{2}}+B\epsilon)+\theta'(\tilde{z})(C\epsilon^{\frac{1}{2}}+D\epsilon)$$

$$+\ \epsilon(F'')^{-1}\theta''(\tilde{z})+O(\epsilon^{3/2})\}$$

with

$$A = -2^{\frac{1}{2}}F^{(3)}/(3(F'')^{3/2})$$

$$B = \frac{7}{12}\cdot(F^{(3)})^2/(F'')^3 - \frac{1}{4}\cdot F^{(4)}/(F'')^2$$

$$C = (2/F'')^{\frac{1}{2}}$$

$$D = -F^{(3)}/(F'')^2 \quad (A8.8)$$

This allows the identification

$$W^{\bar{A}} = \frac{1}{2}[(F'')^{-1}\tilde{\theta}''(z)-F^{(3)}(F'')^{-2}\tilde{\theta}'(z)] \tag{A8.9}$$

in (3.42) of [47].

Appendix IX: Non-zero Super-Poincaré algebra

The non-zero (anti) commutators are

$$[J^{IJ},J^{KL}] = i(\delta^{IK}J^{JL}-\delta^{IL}J^{JK}J^{IL}+\delta^{JL}J^{IK})$$

$$[J^{IJ},J^{K-}] = i(\delta^{IK}J^{J-}-\delta^{JK}J^{I-})$$

$$[J^{IJ},J^{K+}] = -i(\delta^{IJ}J^{+-}+J^{IJ})$$

$$[J^{+-},J^{I-}] = -iJ^{I-}$$

$$[J^{+-},J^{I+}] = iJ^{I+}$$

$$[J^{IJ},P^{K}] = i(\delta^{IK}P^{J}-\delta^{JK}P^{I})$$

$$[J^{I-},P^{K}] = i\delta^{IK}H$$

$$[J^{I+},P^{K}] = i\delta^{IK}P^{+}$$

$$[J^{J-},P^{+}] = iP^{I}$$

$$[J^{+-},P^{+}] = iP^{+}$$

$$[J^{+-},H] = -iH$$

$$\{Q^{+A},Q^{-\bar{B}}\} = 2\delta^{A\bar{B}}P^{+} \qquad \{Q^{+\bar{A}},Q^{-B}\} = 2\delta^{\bar{A}B}P^{L}$$

$$\{Q^{+A},Q^{-B}\} = -\sqrt{2}\rho^{jAB}P^{j} \qquad \{Q^{+\bar{A}},Q^{-\bar{B}}\} = \sqrt{2}\rho^{j\overline{AB}}P^{j}$$

$$\{Q^{+A},Q^{-\bar{B}}\} = 2\delta^{A\bar{B}}P^{R} \qquad \{Q^{-A},Q^{-\bar{B}}\} = 2\delta^{A\bar{B}}H$$

$$\{J^{ij},Q^{+A}] = \frac{1}{2}i\rho^{ij\bar{B}A}Q^{+B} \qquad [J^{Li},Q^{+A}]_{-} = \sqrt{\frac{1}{2}}i\rho^{iAB}Q^{+\bar{B}}$$

$$[J^{ij},Q^{+\bar{A}}] = -\frac{1}{2}i\rho^{ij\bar{A}B}Q^{+\bar{B}} \qquad [J^{ij},Q^{-A}] = \frac{1}{2}i\rho^{ij\bar{B}A}Q^{-B}$$

$$[J^{ij},Q^{-\bar{A}}] = -\frac{1}{2}i\rho^{ij\bar{A}B}Q^{-\bar{B}} \qquad [J^{Li},Q^{-\bar{A}}]_{-} = \sqrt{\frac{1}{2}}i\rho^{i\overline{AB}}Q^{-B}$$

$$[J^{Ri},Q^{+\bar{A}}] = -\sqrt{\frac{1}{2}}i\rho^{i\overline{AB}}Q^{+B} \qquad [J^{+-},Q^{+A}] = \frac{1}{2}Q^{+A}$$

$$[J^{Ri},Q^{-A}] = -\sqrt{\frac{1}{2}}i\rho^{iAB}Q^{-\bar{B}} \qquad [J^{+-},Q^{+\bar{A}}] = \frac{1}{2}iQ^{+\bar{A}}$$

$$[J^{+-},Q^{-\bar{A}}] = -\frac{1}{2}iQ^{-\bar{A}} \qquad [J^{+-},Q^{-A}] = -\frac{1}{2}iQ^{-A}$$

$$[J^{LR},Q^{+A}] = \frac{1}{2}iQ^{+A} \qquad [J^{LR},Q^{-A}] = -\frac{1}{2}iQ^{-A}$$

$$[J^{LR},Q^{+\bar{A}}] = -\frac{1}{2}iQ^{+\bar{A}} \qquad [J^{LR},Q^{-\bar{A}}] = \frac{1}{2}iQ^{-\bar{A}}$$

$$[J^{i-},Q^{+A}] = -\sqrt{\frac{1}{2}}i\rho^{iAB}Q^{-\bar{B}} \qquad [J^{i+},Q^{-A}] = i\sqrt{2}\rho^{iAB}Q^{+\bar{B}}$$

$$[J^{i-},Q^{+\bar{A}}] = \sqrt{\frac{1}{2}}i\rho^{i\overline{AB}}Q^{-B} \qquad [J^{i+},Q^{-\bar{A}}] = -i\sqrt{2}\rho^{i\overline{AB}}Q^{+B}$$

$$[J^{L-},Q^{+A}] = iQ^{A} \qquad [J^{R+},Q^{-A}] = 2iQ^{+A}$$

$$[J^{R-},Q^{+\bar{A}}] = iQ^{-\bar{A}} \qquad [J^{L+}, Q^{-\bar{A}}] = 2iQ^{+\bar{A}}$$

Appendix X: SU(4) Formulae

Define

$$\theta^4 = \epsilon^{ABCD}\theta^{\bar{A}}\theta^{\bar{B}}\theta^{\bar{C}}\theta^{\bar{D}} \tag{A10.1}$$

$$\theta^{3A} = \epsilon^{ABCD}\theta^{\bar{B}}\theta^{\bar{C}}\theta^{\bar{D}} \tag{A10.2}$$

Using the SU(4) matrices ρ^i_{AB}, $\rho^i_{\bar{A}\bar{B}}$, with

$$\rho^i_{AB} = \frac{1}{2}\,\epsilon_{ABCD}\rho^i_{\bar{C}\bar{D}}$$

then

$$\theta^{\bar{A}}\theta^{3B} = \frac{1}{4}\delta^{\bar{A}\bar{B}}\theta^4 \tag{A10.3}$$

$$\theta^{\bar{B}}\theta^{\bar{C}}\theta^{\bar{D}} = \left(\frac{1}{3!}\right)\overline{\epsilon^{ABCD}}\theta^{3A} \tag{A10.4}$$

and

$$\epsilon_{\overline{ACDE}}\theta^{\bar{B}}\theta^{\bar{D}}\theta^{\bar{E}} = \frac{1}{3}(\theta^{3A}\delta^{B}_{\bar{C}} - \delta^{\bar{B}}_{\bar{A}}\theta^{3C}) \tag{A10.5}$$

$$\epsilon_{\overline{ACDE}}\rho^i_{BC}\theta^{\bar{D}}\theta^{\bar{E}} = \delta^{A}_{\bar{B}}\rho^i_{\overline{DE}}\theta^{\bar{D}}\theta^{\bar{E}} - 2\rho^i_{\overline{AC}}\theta^{\bar{B}}\theta^{\bar{C}} \tag{A10.6}$$

Appendix XI: Coalescences

At a coalescence of two zeros of F', say $\tilde{z}_1 \sim \tilde{z}_2$, we may expand various derivatives of F around $\tilde{z}_1$ in powers of $\tilde{z}_2-\tilde{z}_1 = z$. Thus using

$$F'(\tilde{z}_2) = 0 = zF''_1 + \frac{1}{2}z^2F_1^{(3)} + \frac{1}{6}z^3F^{(4)} + \frac{1}{24}z^4F^{(5)}$$

we obtain

$$F''_1 = -\frac{1}{2}zF_1^{(3)} - \frac{1}{6}z^2F_1^{(4)} - \frac{1}{24}z^3F_1^{(5)} \tag{A11.1}$$

whilst similarly

$$F''_2 = \frac{1}{2}zF_1^{(3)} + \frac{1}{3}z^2F_1^{(4)} + \frac{1}{8}z^3F_1^{(5)} \tag{A11.2}$$

Also

$$F_2^{(3)} = F_1^{(3)} + zF_1^{(4)} + \frac{1}{2}z^2F_1^{(5)}$$

and

$$\rho_{21} = \frac{1}{2}z^2F''_1 + \frac{1}{6}z^3F_1^{(3)} + 0(z^4)$$

$$= -\frac{1}{12}z^3F^{(3)} + 0(z^4) \tag{A11.3}$$

If $g = -(F''_1/F''_2)$, then from (A11.1) and (A11.2) we have

$$g = 1 - \frac{1}{3}z(F^{(4)}/F^{(3)}) + 0\ (z^2) \qquad \text{(A11.4)}$$

At a coalescence of three zeros of F′

$$\rho_{31} = z_{31}^2 z_{21} z_{32} F^{(4)} + 0(z^5) \qquad \text{(A11.5)}$$

Appendix XII: Special Case $p_1^+ = -p_4^+$

In this case the Mandelstam map, at the tree level, is

$$\rho = p_4^+ \ell n[(z-x)/(1-z)] + p_3^+ \ell n z \equiv F(z) \tag{A12.1}$$

and the solutions $\tilde{z}_{1,2}$ of $F'(z) = 0$ thus satisfy

$$z^2+[m(x-1)-(1+x)]z+x = 0 \tag{A12.2}$$

where $m = p_4^+/p_3^+$. If we let $\epsilon = (x-1)$ then

$$\tilde{z}_{1,2} = \frac{1}{2}[2+\epsilon(1-m)\pm((1-m)^2\epsilon^2-4m\epsilon)^{\frac{1}{2}}] \tag{A12.3}$$

and so

$$\tilde{z}_{1,2}-1 = \frac{1}{2}(1-m)\epsilon\pm\epsilon^{\frac{1}{2}}d \tag{A12.4}$$

$$\tilde{z}_{1,2}-x = -\frac{1}{2}(1+m)\epsilon\pm\epsilon^{\frac{1}{2}}d \tag{A12.5}$$

where $d^2 = (1-m)^2\epsilon-4m$.

We may write

$$F''(\tilde{z}_1) = A\tilde{z}_{12}[\tilde{z}_1(\tilde{z}_1-x)(\tilde{z}_1-1)]^{-1} \tag{A12.6}$$

$$F''(\tilde{z}_2) = A\tilde{z}_{21}[\tilde{z}_2(\tilde{z}_2-x)(\tilde{z}_2-1)]^{-1} \tag{A12.7}$$

From (A12.4) -(A12.7), as $\epsilon \sim 0$

$$(\tilde{z}_{1,2}-1),(\tilde{z}_{1,2}-x) = O(\epsilon^{\frac{1}{2}}) \tag{A12.8}$$

so

$$F''(\tilde{z}_2) = O(\epsilon^{-\frac{1}{2}}) \tag{A12.9}$$

For general z, but only for zero modes θ's

$$\theta(z) = p_1^+(z-1)^{-1}\theta_{14}+p_3^+z^{-1}\theta_{34}+\theta_4F'(z) \tag{A12.10}$$

so that

$$\theta'(\tilde{z}_1) = p_1^+(\tilde{z}_i-1)^{-1}\theta_{14} + p_3^+\tilde{z}^{-1}\theta_{34} \tag{A12.11}$$

and from (A12.8) as $\epsilon \sim 0$

$$\theta'(\tilde{z}_i) = O(\epsilon^{-\frac{1}{2}}) \tag{A12.12}$$

The behaviour (A12.8), (A12.9), (A12.12) will be valid on a general Riemann surface when 4-string matrix elements are taken with $p_1^+ = -p_4^+$, though next-to-leading-orders will be modified.

Appendix XIII: Contributions to [10] SUSY at $p_1^+ = -p_4^+$

At tree level, in the configuration $p_1^+ = -p_4^+$, then, from (A12.10)

$$\theta(\tilde{z}_i) = \frac{p_1^+\theta_{14}}{(\tilde{z}_i-1)} + \frac{p_3^+\theta_{34}}{\tilde{z}_i}$$

so that, combined with $F''(\tilde{z}_i)$ (from (A12.6), (A12.7))

$$\psi_\ell = [F(\tilde{z}_\ell)'']^{-1}\theta(\tilde{z}_\ell) = (-1)^\ell(\theta+\psi\delta_\ell)\tilde{z}_\ell(\tilde{z}_\ell-x)p_1^+(p_3^+\tilde{z}_{21})^{-1} \qquad \text{(A13.1)}$$

where $\theta = \theta_{14}$, $\psi = \theta_{34}$, $\delta_i = \frac{p_3^+}{p_1^+}\left(\frac{\tilde{z}_i-1}{\tilde{z}_i}\right)$. We may express the contracted term on the right hand side of (7.74) as proportional to

$$(F''_1)^{3/2}(F_2)^{\frac{1}{2}}[\psi_1^{3B}\ \bar{\psi}_2^{A}-g^{-1}\bar{\psi}_1^{A}\psi_2^{3B}] \qquad \text{(A13.2)}$$

with $g = -F_1''/F_2''$, so using (A13.1) and App.XII, (A13.2) reduces to

$$\left(\frac{p_1^{+4}}{p_3^{+2}\tilde{z}_{21}^2}\right)\left[\frac{\tilde{z}_1(\tilde{z}_1-x)}{(\tilde{z}_1-1)}\right]^{3/2}\left[\frac{\tilde{z}_2(\tilde{z}_2-x)}{(\tilde{z}_2-1)}\right]^{\frac{1}{2}}\{(\theta+\psi.\delta_1)^{3B}(\theta+\psi\delta_2)^{\bar{A}}-h(\theta+\psi\delta_2)^{3B}\} \qquad \text{(A13.3)}$$

where

$$h = \frac{\tilde{z}_2(\tilde{z}_2-x)(\tilde{z}_1-1)}{\tilde{z}_1(\tilde{z}_1-x)(\tilde{z}_2-1)} \qquad \text{(A13.4)}$$

From Appendix XII, h in (A13.4) behaves, as $x-1 = \epsilon \sim 0$, as

$$h = 1 + O(\epsilon^{\frac{1}{2}}) \tag{A13.5}$$

as do the second and third factors in (A13.3); $\delta_1 = O(\epsilon^{\frac{1}{2}})$. The last bracket in (A13.3) can be written, without approximation, as

$$\begin{aligned}
&\delta^{\bar{A}B}[-\tfrac{1}{4}\theta^4(1+h)+\theta^{3C}\psi^{\bar{C}}(\delta_1+h\delta_2)-\psi^{3C}\theta^{\bar{C}}(\delta_1^2\delta_2+h\delta_1\delta_2^2)\\
&-\tfrac{1}{4}\psi^4(\delta_1^3\delta_2+h\delta_2^3\delta_1)]+\theta^{3B}\psi^{\bar{A}}(\delta_2-\delta_1)(1-h)+3(\theta\psi^2)^B\theta^{\bar{A}}(\delta_1^2+h\delta_2^2)\\
&+3(\theta^2\psi)^B\psi^{\bar{A}}\delta_1\delta_2(1+h)+(\delta_1+\delta_2)\psi^3_{3B}\theta^{\bar{A}}(\delta_1^2+h\delta_2)
\end{aligned} \tag{A13.6}$$

From Appendix XII,

$$(1-h) = \tilde{z}_{21}(1-x)(1+\frac{p_4^+}{p_3^+})/\tilde{z}_1(\tilde{z}_1-x)(\tilde{z}_2-1) \tag{A13.7}$$

$$(\delta_2-\delta_1) = -\,p_3^+\tilde{z}_{21}/(p_1^+x) \tag{A13.8}$$

Then there is a term in (A13.3) which is non-zero as $\epsilon \sim 0$, with value proportional to

$$\frac{p_4^{+3}}{p_3^+}(1+\frac{p_4^+}{p_3^+})\left[\frac{\tilde{z}_1(\tilde{z}_1-x)}{(\tilde{z}_1-1)}\right]^{3/2}\left[\frac{\tilde{z}_2(\tilde{z}_2-x)}{(\tilde{z}_2-1)}\right]^{\frac{1}{2}}\frac{(x-1)}{\tilde{z}_1(\tilde{z}_1-x)(\tilde{z}_2-1)}\theta^{3B}\psi^{\bar{A}} \tag{A13.9}$$

From (A12.4), (A12.5), in the limit $\epsilon \sim 0$, (A13.9) becomes

$$\frac{p_4^+}{p_3^+}\left(1 + \frac{p_4^+}{p_3^+}\right) d^{-2} \theta^{3B} \psi^{\bar{A}} + O(\epsilon^{\frac{1}{2}}) \tag{A13.10}$$

This is the only non-zero term in (A13.2) in this limit, which is not proportional to $\delta^{\bar{A}B}$. In order to close the supersymmetry algebra for the open super-string it therefore appears necessary to have a non-zero Q_4 (as discussed in detail in [38]).

Appendix XIV: On the solution $\partial_\rho \tilde{\theta} = 0$

We discuss in this appendix the general solution to the following linear differential equation on odd Grassmannian fields

$$\partial_{\bar\rho} \tilde{\theta} = 0 \tag{A14.1}$$

with $\tilde{\theta} \to \theta_r$ when $\rho \to \rho_r$, $r = 1,\ldots,N$ and where all objects are defined in the L.C. diagram, and ρ_r represents the sources. We introduce a (0,1) form in the z-plane

$$\bar{\theta}(z,\bar{z}) = \bar{F}'(\bar{z})\tilde{\theta}(\rho(z),\bar{\rho}(\bar{z})) \tag{A14.2}$$

where

$$\rho = F(z) = \sum_r \alpha_r \, \underline{G}(z,z_r) \tag{A14.3}$$

is the usual Mandelstam map (5.21), with $\alpha_r = \pi p_r^+$ and Re$\underline{G}$ is the Green function of the closed Riemann surface under consideration. Under an automorphic
transformation in the z-plane, $z \to \gamma(z)$, we have $\bar{\theta} \to [\bar{\gamma}'(\bar{z})]^{-1}\bar{\theta}$. We denote

$$E_z = F'(z), \qquad E_{\bar z} = \bar{F}'(\bar{z})$$

$$E^z = E_z^{-1}, \qquad E^{\bar z} = E_{\bar z}^{-1} \tag{A14.4}$$

The metric in the z-coordinates is given by

$$d\rho d\bar{\rho} = E_z E_{\bar{z}} dz d\bar{z} \qquad (A14.5)$$

We may rewrite (A14.1) as

$$E^{\bar{z}} \partial_z (E^z \bar{\theta}) = E^{\bar{z}} E^z (\partial_z \bar{\theta} + (E_{\bar{z}} \partial_z E^{\bar{z}}) \bar{\theta}) = 0 \qquad (A14.6)$$

and $\bar{\theta} \to \frac{\alpha_r \theta_r}{z - z_r}(\bar{z}-\bar{z}_r)$ when $\bar{z} \to \bar{z}_r$, $\bar{z}_r$ $r = 1,\ldots,N$,
in one fundamental region. We notice that the metric has isolated zeros and singularities at the interaction points and sources respectively and is
otherwise regular. We can thus drop the term $E_{\bar{z}} \partial_z E^{\bar{z}}$ in (A14.6). We are left with the problem of solving

$$E^{\bar{z}} E^z \partial_z \bar{\theta} = 0 \qquad (A14.7)$$

with $\bar{\theta} \to \dfrac{\alpha_r \theta_r}{\bar{z} - \bar{z}_r}$ when $\bar{z} \to \bar{z}_r$, $\bar{z}_r \in$ one fundamental region (A14.7).

The problem is invariant under automorphic transformations.

A solution is given by

$$\bar{\theta} = \sum_r \alpha_r \bar{\theta}_r \partial_{\bar{z}} \bar{G}(\bar{z}, \bar{z}_r) + \sum_{r=1}^{g} \bar{\theta}_i \partial_{\bar{z}} u_i(z) \qquad (A14.8)$$

where $\partial_{\bar{z}} u_i d\bar{z}$ is a basis of the first abelian differential on the Riemann surface .$\partial_{\bar{z}} u_i$ is anti holomorphic, hence (A14.8) has the right behaviour when $\bar{z} \to \bar{z}_r$. (A14.8) is the most general solution of (A14.7). In fact given any solution θ_p of (A14.7), the difference

$$\Delta = \theta_p - \sum \alpha_r \theta_r \partial_{\bar{z}} \bar{G}$$

is also a solution of (A14.7) and it is antiholomorphic. Hence $\Delta d\bar{z}$ is the first abelian differential expressed in local coordinates, and may always be expressed in the form (A14.8) but now with θ_r's being zero. The geometrical objects $\bar{\theta}_i$ $i = 1,\ldots,g$, defined in (A14.8) are odd elements of a Grassmann-algebra. They are independent of $(z,\bar{z})$, invariant under automorphic transformations, but have a non-trivial transformation under the modular groups as discussed in the text. The result that (A14.8) is the most general solution of (A14.1) is thus proven.

Appendix XV: On the fermionic measure

In this appendix we present another way of obtaining the functional integral measure for the odd G-valued fields, in complete agreement with the usual function with fields inserted at points P. To simplify the notation we consider only one θ field and not a 4 of SU(4) as in the superstring theory. The action is

$$S = \int_{\Sigma} d\rho d\tilde{\rho}(\lambda\partial_{\tilde{\rho}}\theta+\tilde{\lambda}\partial_{\rho}\tilde{\theta}) = S_L+S_R \tag{A15.1}$$

The usual result [23] is

$$\left|\int D\lambda D\theta \prod_{i=1}^{g} \theta(P_i)\lambda(Q)e^{-S_L}\right|^2 = \frac{\det u_i(P_j).\det \bar{u}_i(P_j)}{\det \mathrm{Im}\Pi} \cdot \frac{(\det'-\nabla^2)}{\iint\sqrt{g}} \tag{A15.2}$$

The functional integral we consider is

$$I = \int D\lambda D\theta D\tilde{\lambda}D\tilde{\theta}\; e^{-(S_l+S_R)} \prod_{i=1}^{N-2+2g} V(X(P_i)\; \theta(P_i)). \prod_{i=1}^{N-2+2g} \tilde{}\; V(X(P_i),\tilde{\theta}(P_i)) \tag{A15.3}$$

The zero mode sector of the λ and $\tilde{\lambda}$ fields, being (0,0) tensors, are constants on the Riemann surface. As in (A15.2), they are integrated away without consequences. We may consider directly in (A15.3) the integration on non-zero modes only. The zero mode section of θ is the vector space of first abelian differentials $\sum_{i=1}^{g} \theta_i u_i'$. The zero mode integration in (A15.3) may be performed as in (A15.2) by extracting from the vertices all possible factors $\prod_i^g A(P_i)$ and replacing them by their zero mode contributions $\prod_i^g \theta_i \det\theta_i'(P_j)$ and integrating the θ_i variables. The result of the zero mode integration is therefore

$$I = \int D\lambda' D\theta' d\tilde{\lambda}' D\tilde{\theta}' \; e^{-(S_L+S_R)} \sum_{\substack{\text{all}\\ \text{ways}}} \det u_i'(P_j).\det \bar{u}_i'(P_j)$$

$$[\prod_j^g \frac{\partial V}{\partial \theta}(X(P_j),\theta(P_j)) \prod_k^{N-2+g} V(X(P_k),\theta(P_k))] \prod \tilde{V}\frac{\tilde{\partial}}{\partial\tilde{\theta}}\tilde{V} \qquad (A15.4)$$

where the square bracket is evaluated only with $\theta,\tilde{\theta}$ having non-zero modes. The integration is now on non-zero modes only, and is performed in the standard way,

$$I = \frac{\det'-\nabla^2}{\iint\sqrt{g}}.\sum \frac{\det u_i'(P_j)\det \bar{u}_i'(P_j)}{\det \operatorname{Im}\Pi} (\prod^g \partial_\theta V \prod^{N-2+g} V)_{\theta=\Sigma\alpha_r\theta_r\bar{G}'}$$

$$.(\prod^g \partial_{\tilde{\theta}}V. \prod^{\overline{N-2+g}} V)_{\tilde{\theta}=\Sigma\alpha_r\tilde{\theta}_r\bar{G}'} \qquad (A15.5)$$

There is a det $\operatorname{Im}\Pi^{-1}$ coming from the measure of the zero modes in agreement with (A15.2). (A15.4) or (A15.5) may now be written, by direct evaluation, as

$$I = \int D\lambda' D\theta' D\tilde{\lambda}' D\tilde{\theta}'_{\;i} \prod d\theta_i \prod_j d\tilde{\theta} \frac{1}{\det \operatorname{Im}\Pi}$$

$$x(\det\operatorname{Im}\Pi)[\prod_i V(X(P_i),\sum_r \alpha_r\theta_r\bar{G}' + \sum_j \theta_j u_j'(P_i)).\prod_j \bar{V}] \times e^{-(S_L+S_R)} \qquad (A15.6)$$

This is the expression obtained in the text.

Appendix XVI: Chiral determinants

As ground work for analysing chiral determinants in the L.C. gauge it is useful to consider the covariant case, associated with the Polyakov string. We shall follow the notation of Alvarez [26]. Let Σ be a Riemann surface with metric g. Locally, conformal coordinates may be introduced such that the metric takes the form

$$d^2 s = g_{z\bar{z}}\ dzd\bar{z} \tag{A16.1}$$

Under analytic coordinate transformation, $z \to z'$, $\partial z'/\partial\bar{z} = 0$, the metric transforms as $g_{z\bar{z}} \to e^{\phi}\ g_{z\bar{z}}$; therefore complex structure is equivalent to conformal structure for a Riemann surface. Define τ^n to be the space of tensors with conformal weight n, (strictly speaking weight (n,0)). that is, τ^n consists of tensors T, with only z indices, which transform under an analytic change of coordinates as

$$T \to (\partial z'/\partial z)^n\ T \tag{A16.2}$$

The index n equals $n_- - n_+$ where n_+ and n_- are the number of upper and lower z indices respectively. Thus $T_{zz}{}^z$ and T_z both belong to τ^1. A tensor
with mixed indices z, $\bar{z}$ can always be expressed as one with just z indices by using the metric $g_{z\bar{z}}$. For example, $T_{zz\bar{z}} = g_{z\bar{z}}\ T_{zz}{}^z$

Define the covariant derivatives $\nabla^z_n : \tau^n \to \tau^{n-1}$ by

$$\nabla^z_n = g^{z\bar{z}}\ \partial_{\bar{z}} T \tag{A16.3}$$

where $g^{z\bar{z}} = (g_{z\bar{z}})^{-1}$. The adjoint operator is

$$(\nabla^z_n)\dagger = -\nabla^{n-1}_z \qquad \text{(A16.4)}$$

where $\nabla^n_z : \mathcal{T}^n \rightarrow \mathcal{T}^{n+1}$ has the explicit form

$$\nabla^n_z = (g_{z\bar{z}})^n \, \partial_z[(g^{z\bar{z}})^n \, T] \qquad \text{(A16.5)}$$

The adjoint in (A16.4) is defined with respect to the inner product on tensor fields given by

$$(\psi,\psi') = i \int_\Sigma (g^{z\bar{z}})^{n-1} \, \bar{\psi}\psi' \, dz \wedge d\bar{z} \qquad \text{(A16.6)}$$

where ψ,ψ' have weight n.

Using this notation define the Laplacian $\Delta_n : \mathcal{T}^n \rightarrow \mathcal{T}^n$ by

$$\Delta_n = (\nabla^n_z)\dagger \, \nabla^n_z = \nabla^z_{n+1} \nabla^n_z \qquad \text{(A16.7)}$$

The determinant of this Laplacian may be evaluated by analysing the change of Δ_n under a conformal transformation of the metric using a heat kernel regularisation [126]. However, the chiral operators (A16.3) and (A16.5) are mappings between non-isomorphic infinite-dimensional vector spaces and their associated determinants are not easily defined. One possible approach is to use the result of Belavin and Knizhnik [83] who Knizhnik considered families of operators in moduli space. They showed that Δ_n essentially factorises into a holomorphic and an anti-holomorphic function of the moduli Σ, except for a holomorphic anomaly term involving the Liouville action S_L. More explicitly,

$$\det'[(\nabla_z^n)^\dagger \nabla_z^n]/[\det(\phi_i,\phi_j)\det(\psi_i,\psi_j)]$$

$$= \exp\{\{3(n+1)^2 - 1\}S_L[\phi,\hat{g}]/24\pi\}\ |Q(\underline{\mu})|^2 \quad \text{(A16.8)}$$

where $\{\phi_i\}$ and $(\psi_i\}$ span the kernels of ∇_z^n and $(\nabla_z^n)^\dagger$ and where

$$S_L[\phi,\hat{g}] = \int_\Sigma d^2\sigma\sqrt{\hat{g}}\ (\hat{g}^{\alpha\beta}\partial_\alpha\phi\ \partial_\beta\phi - \hat{R}\phi) \quad \text{(A16.9)}$$

The metric $g_{\alpha\beta}$, $(\alpha,\beta = 0,1)$, of Σ has been decomposed into the product $g = \hat{g}e^{2\phi}$ with ϕ a conformal mode and $\hat{g}$ a fiducial metric with constant curvature $\hat{R}$. The moduli of Σ are denoted by $\underline{\mu}$; $Q(\underline{\mu})$ is a holomorphic function of $\underline{\mu}$. (For more details see ref.[23]). Equation (A16.8) is a particular example of Quillen's theorem [128] and one may identify $\overline{Q(\underline{\mu})}$ as the chiral determinant $\det\nabla_z$ and $Q(\mu)$ as $\det(\nabla_z)^\dagger$. Note, however, that there is a phase ambiguity in these definitions of chiral determinants since we are taking the square root of a modulus squared.

How does the above covariant formalism translate to the L.C. gauge?. Firstly recall that multi-loop amplitudes are evaluated by mapping the appropriate L.C. diagram Σ_ρ to a region in the z-plane Σ_z which has a flat metric i.e. $g_{z\bar{z}} = 1/2$. This implies that the Liouville action in (A16.8) vanishes. Furthermore, setting $g_{z\bar{z}} = 1/2$ in equations (A16.3) and (A16.5), ∇_z and ∇^z simply become ∂_z and $\partial_{\bar{z}}$ respectively. Therefore we shall denote the chiral operators in the L.C. gauge by $\partial_z^{(n)}$ and $\partial_{\bar{z}}^{(n)}$ with n indicating the associated conformal weight. The Laplacian Δ_z and Σ_z operates on bosonic modes X which are conformal scalars (n = 0). Therefore, we may identify Δ_z with Δ_0. In other words, $\Delta_z = \partial_{\bar{z}}^{(1)}\ \partial_z^{(0)}$. Setting n = 0 and $S_L = 0$ in equation (A16.8) leads to

$$\det{}'[\partial_{\bar{z}}^{(1)}\partial_z^{(0)}]/[\det(u_i,u_j)\det(1,1)] = |Q(\mu)|^2 \tag{A16.10}$$

where the bases for $\mathrm{Ker}\partial_z^{(0)}$ are constant functions and the bases for $\ker\partial_{\bar{z}}^{(1)}$ are the first Abelian integrals u_i. In particular, $\det(u_i,u_j) = \det\mathrm{Im}\Pi$ which follows from (A16.6) and a Riemann bilinear relation. So using methods from algebraic geometry leads to the L.C. gauge result

$$\det{}'\Delta_z = \det\mathrm{Im}\Pi|Q(\mu)|^2 \tag{A16.11}$$

Equation (A16.11) was used in (5.74) under the Schottky representation of the moduli $\mu = (z_{1p}, z_{2p}, w_p)$. Using (A16.10) the chiral determinant $\det\partial_{\bar{z}}^{(1)}$ is then identified with the holomorphic function $Q(\mu)$,

$$\det\partial_{\bar{z}}^{(1)} = Q(\mu) \tag{A16.12}$$

This is an expression for the chiral determinant in the z-plane corresponding to the fermions $\bar{\theta}^A$, which have weight 1. We then expect that the chiral determinant $\det\partial_z^{(1)}$ arising from the fermions $\tilde{\theta}^A$ of weight -1 is equal to $\overline{Q(\mu)}$. To show this note that for flat metrics an equivalent definition of the Laplacian Δ_n is

$$\Delta_n = \partial_z^{(1)}\partial_{\bar{z}}^{(0)} \tag{A16.13}$$

Repeating the above analysis then leads to the required result. Hence

$$\det\partial_z^{(1)} = \overline{Q(\mu)} \tag{A16.14}$$

However, there are problems with the above approach. Firstly, as we have already mentioned, there is a phase ambiguity in the definition of the chiral determinants (A16.12) and (A16.14). Secondly, the analysis has been performed in the z-plane, whereas the chiral determinants appearing in the amplitudes (5.29 and (5.87) are defined in the ρ-plane. In particular, the ρ-plane version of (A16.10) is

$$\det{}'[\partial_{\bar{\rho}}^{(1)}\partial_{\rho}^{(0)}]/\det\mathrm{Im}\Omega = |A|^2\,|Q(\mu)|^2 \tag{A16.15}$$

where A is the closed string conformal anomaly of (5.77),

$$A = (\prod_P F")^{1/24}\ (\Pi\alpha_r)^{1/12}(\Pi\epsilon_r)^{1/12} \tag{A16.16}$$

One may then define of the foolowing product of chiral determinants,

$$\det\partial_{\bar{\rho}}^{(1)}\ \det\partial_{\rho}^{(0)} = |A|^2\,|Q(\mu)|^2 \tag{A16.17}$$

However, one cannot take a holomorphic square root of (A16.17) as A is itself non-holomorphic. [Note that equations (A16.15) and (A16.16) imply that the ratio of determinants in (5.29) is equal to unity. A solution to the second problem has recently been suggested by Taylor and Restuccia [38]. It involves splitting up the conformal transformation between the z-plane and the ρ-plane into two stages, which we now briefly describe. The metric in the L.C. diagram Σ_ρ is given by

$$d\rho d\bar{\rho} = F'(z)\bar{F}'(z)\ dzd\bar{z} \tag{A16.18}$$

which has curvature singularities at the interaction points. The evaluation of $\det'\Delta_\rho$ under the conformal change of the metric described by

$$g_{z\bar{z}}=e^{2\phi}/2 = F\bar{F}'/2 \rightarrow g_{z\bar{z}} = 1/2 \tag{A16.19}$$

This relates $\det'\Delta_\rho$ to $\det'\Delta_z$ and leads to the result for closed strings, used in (A16.15), that

$$\det\Delta_\rho = |A'|^2 \det\Delta_z \tag{A16.20}$$

Then $\det\Delta_z$ is itself evaluated. Such an approach may be extended to the case of chiral determinants by performing (A16.19) in two steps:

$$\bar{F}F' \to F' \to 1 \tag{A16.21}$$

Each step introduces a chiral change in the metric, the first with factor $\bar{F}$, inducing changes in $\det\partial_\rho$ and the second with factor F', inducing changes in $\det\partial_{\bar{\rho}}$. Using a careful regularisation of the determinants it is shown that [38] (up to a constant phase)

$$\det\partial_{\bar{\rho}} = A \det\partial_{\bar{z}}, \qquad \det\partial_\rho = \bar{A} \det\partial_z \tag{A16.22}$$

There is an alternative way to derive (A16.22). Our starting point is the canonical quatisation of systems with second-class constraints developed in ref.[129] as a generalisation of the Fadeev-Popov procedure. Applying this to the 16 chiral components of the heterotic string with second-class constraints (4.88) leads to the partition function

$$Z = \int D\underline{X}(\sigma) D\underline{P}(\sigma) D\underline{W} D\underline{V} \exp[i\int d\tau\ (d\sigma \int \underline{P}.\partial_\tau \underline{X} + \underline{W}.\partial_\tau \underline{V} - H)]$$

$$\delta(\underline{x}_R) \prod_\sigma \delta(\Lambda^I(\sigma)) \det\{\Lambda^I,\Lambda^J\}^{\frac{1}{2}} \tag{A16.23}$$

with $\det\{\Lambda^I,\Lambda^J\}^{\frac{1}{2}} = (\det\partial_\sigma)^8$ = and $\underline{x}_R = \underline{X} - \underline{V}$, (equation (4.53)). We are assuming for the moment that R^{16}/Γ has been lifted to R^{16}. On integrating out the momenta, Z becomes

$$Z = \int D'\underline{X}(\sigma)\exp[i\int d\tau \int_{-\pi}^{\pi} d\sigma\ \partial_\sigma \underline{X}(\partial_\tau - \partial_\sigma)\underline{X}/2\pi](\det\partial_\sigma)^8$$

$$\int D\underline{x}_L\ \exp[i\int d\tau \dot{\underline{x}}_L^2/2] \tag{A16.24}$$

where $D'\underline{X}$ denotes exclusion of the zero mode $\underline{x}_L$. The contribution of $\underline{x}_L$ to (A16.24) gives a term proportional to the area of the world sheet and may be subtracted out. After Euclideanisation, this leads to an expression of the form

$$Z = \int D'\underline{X}(\sigma)\ \exp[i\int d^2\rho\ \partial_{\bar{\rho}}\ X\ (\partial_{\bar{\rho}}^{(0)} - \partial_{\rho}^{(0)})\underline{X}/2\text{Å}]$$

$$[\det\ (\partial_{\bar{\rho}} - \partial_\rho)]^8$$

$$\equiv [\det\ \partial_{\bar{\rho}}^{(0)}]^{-8} \tag{A16.25}$$

Note that the Hamiltonian approach is only applicable to the free bosonic modes with the partition function Z (with appropriate boundary conditions), corresponding to a first-quantised propagator. In other words, the integration region of the action in (A16.25) is a cylinder. The extension to the multi-loop interacting case is non-trivial. In particular, the Lagrangian cannot be rewritten in terms of global one-forms allowing an analysis of inexact modes. [It is crucial to take into account such modes in the case of compactified bosonic components of the string]. However, we shall now show how Z can be rewritten in such a way that its Euclideanised functional form has a well-defined extension to arbitrary Riemann surfaces. The resulting expression then determines the chiral determinant $[\det\ \partial_{\bar{\rho}}^{(0)}\ ]^{-8}$

Return to equation (A16.23) and perform the canonical transformation of $(\underline{X}(\sigma), \underline{P}(s), \underline{W}, \underline{V}) \rightarrow (\underline{\Phi}_L(\sigma), \underline{\Pi}_L(\sigma), \underline{\Phi}_R(\sigma), \underline{\Pi}_R(\sigma))$. This gives

$$Z = \int D\underline{\Phi}_L(\sigma) D\underline{\Phi}_R(\sigma) D\underline{\Pi}_L(\sigma) D\underline{\Pi}_R(\sigma) \prod_{\sigma=0}^{\pi} [\delta(\underline{\Pi}_R(\sigma))\delta(\partial_\sigma \underline{\Phi}_R(\sigma))]\delta(\underline{x}_R)$$

$$\exp[i\int d\tau \,(\int_0^\pi d\sigma\, \underline{\Pi}_L \partial_\tau \underline{\Phi}_L + \int_0^\pi d\sigma \underline{\Pi}_R \partial_\tau \underline{\Phi}_R - H_L - H_R)]$$

$$\det(\partial_\sigma)^8 \qquad \text{(A16.26)}$$

Note that $\det(\partial_\sigma)^8$ is defined over the range $-\pi \le \sigma \le \pi$. The delta-functions in (A16.26) set all right-moving modes to zero and yield a factor of $\det(\partial_\sigma)^{-16}|_{\sigma \le 0} = \det(\partial_\sigma)^{-8}$. Hence, on integrating out $\underline{\Phi}_R$, $\underline{\Pi}_R$, $\underline{\Pi}_L$ we obtain

$$Z = \int D\underline{\Phi}_L \;(\sigma) \exp[i\int d\tau \int_0^\pi d\sigma \;(\dot{\underline{\Phi}}_L^2 - \underline{\Phi}_L'^2)/4\pi] \qquad \text{(A16.27)}$$

Z is now in a form which, on Euclideanisation, may be extended to an arbitrary L.C. diagram Σ_ρ. Thus

$$Z = \int D\underline{\Phi}_L(\sigma) \exp(-\int_{\Sigma_\rho} d\underline{\Phi}_L \wedge {}^*d\underline{\Phi}_L/8\pi) \qquad \text{(A16.28)}$$

We recognise equation (A16.28) as the left-moving contribution to the scattering amplitude, $f_L(\tilde{\bar{\underline{\rho}}}, \underline{\alpha}, \underline{\beta})$, of equation (5.87), but with all external sources $\underline{J}_L$ set to zero. Therefore, settingall momenta $\tilde{\underline{p}}_r$ to zero in (5.87) we find

$$Z = [(\Pi F")^{1/24}(\Pi\alpha_r)^{1/12}\ (\Pi\epsilon_r)^{1/12}Q(\omega_p\cdot z_{1p}\cdot z_{2p})]^{-8} \qquad (A16.29)$$

We must include in (A16.29) the term ϵ_r arising from the conformal anomaly A equation (A16.16) (such terms could be neglected in multi-loop amplitudes due to cancellation with the singular parts of $\exp\{-\Sigma\ \tilde{\underline{p}}_r^{\ 2}\ H(z_{r'},z_r)\}$ etc.). Combining equations (A16.25) and (A16.29) we conclude that

$$\det\partial_{\bar{\rho}}^{(0)} = (\Pi F")^{1/24}(\Pi\ \alpha_r)^{1/12}(\Pi\ \epsilon_r)^{1/12}Q(\omega_p\cdot z_{1p}\cdot z_{2p}) \qquad (A16.30)$$

From equation (A16.16) this may be rewritten as

$$\det\partial_{\bar{\rho}}^{(0)} = AQ(\underline{\mu}) \qquad (A16.31)$$

Similarly the corresponding right-moving determinant satisfies

$$\det\partial_{\rho}^{(0)} = \bar{A}\ \overline{Q(\underline{\mu})} \qquad (A16.32)$$

Equations (A16.17), (A16.31) and (A16.32) then determine the expressions for $\det\partial_{\bar{\rho}}^{(1)}$ and $\det_{\rho}^{(1)}$ as given in (A16.22). In fact

$$\det\partial_{\bar{\rho}}^{(1)} = \det\partial_{\bar{\rho}}^{(0)},\quad \det\partial_{\rho}^{(1)} = \det\partial_{\rho}^{(0)} \qquad (A16.33)$$

We end this appendix by briefly mentioning another approach to chiral determinants in the context of the Polyakov string [130]. It is based on the Siegel action [131),132]

$$S = \int d\tau \int d\sigma\ (-\partial_+ X^I \partial_- X^I\ +\ \lambda\partial_- X^I \partial_- X^I) \qquad (A16.34)$$

In (A16.34) the chiral constraints (4.88) have been squared to give (classically) a first-class constraint which has been added to the kinetic action using a Lagrange multiplier λ. The action has a local gauge symmetry

$$\delta X^I = \epsilon \partial_+ X^I$$

$$\delta \lambda = 2\partial_- \epsilon + \epsilon \partial_+ \lambda - \lambda \partial_+ \epsilon \quad \text{(A16.35)}$$

The Lagrange multiplier may be considered as an auxilary metric and the transformation law (A16.35) as follows from reparametrisation invariance. The presence of a local gauge symmetry allows λ to be gauged to zero, at least locally, so that it is a pure gauge degree of freedom and does not propagate.

In ref [130] the Siegel action is applied to the bosonic Polyakov string by the introduction of two metrics, the world-sheet metric and the Lagrange multiplier λ. Extra non-propagating fields are also introduced to cancel a Virasoro anomaly arising from the normal-ordering of the squared constraints. The two sets of moduli are then used to define chiral determinants. It may be possible to gauge fix this formalism to the L.C. gauge. However, difficulties may occur associated with the results of Teitelboim for systems with constriants quadratic in momenta [133].

Appendix XVII: Multi-loop amplitude reduction

Two steps are needed to justify use of perturbation rules of subsections 5.1 and 5.2 applied to the mappings (5.43) and (5.54). These are that (1) the change of insertion factors to the new interaction points and loop string widths and twists is correctly given on change of the Mandelstam map (5.21) to the new maps (5.43) and (5.54) of the same Riemann surface. (2) the extension of step (1) to account for [10]-SUSY can also be obtained. Step (1) has already been proven for bosonic strings by Mandelstam [103], where he showed that interaction vertices are shifted along with the interaction positions, on performing suitable transformations under M^{i-} and M^{+-} together with the associated conformal transformations to correct string widths. This argument can be seen to arise from the commutator $[M^{i-},H]$ at cubic order:

$$[M_3^{i-},H_3]_- + [M_3^{i-},P^-]_- = 0 \qquad \text{(A17.1)}$$

Since the insertion factor for M_3^{i-} is $X^i V_3^H(---)$ where $(---)$ denotes associated string field products, then (A17.1)) may be rewritten as

$$\delta V_3^H = [M_2^{i-},H_3]_- - -[X^i V_3^H,P_2^-]_- \qquad \text{(A17.2)}$$

which is identical to equation (2.2) of [103] when reduced to first quantised form. Similar expressions occur for M^{+-}. Following Mandelstam it then results that open string L.C.D. integrands may be changed from those determined by the Mandelstam map (5.21) to those by the Arfaei map (5.54). The insertion factors will be the original ones at the interaction points of the new diagram fig.11 at one loop, and of similar diagrams at higher loops. It should be noted that the proof of [103] was originally for the open string; the canonical transformations of section 4 (to two open strings) will allow that proof to be adapted to the closed string case.

Feature (2) of the Arfaei map may be obtained by extending equations (A17.1, (A17.2) to the superstring case, using the same vanishing commutator $[J^{i-},H]_-$. The same equation (A17.1) results, but now (A17.2) has the additional fermionic contribution

$$[QS,P_2^-]. \tag{A17.3}$$

where Q denotes the supersymmetry generators $Q^{-\bar{A}}$, Q^{-A} and S a set of Grassmann valued variables involving θ^A,λ^A etc. at the interaction point. Thus (A17.3)) must be regarded as an additional [10]-SUSY transformation. The [10]-super-Poincaré invariant volume elements formed from a given L.C.D. will thus have no contribution from this term (this method avoids the difficulty of [103], in which the fermionic part was not able to be discussed). Thus our result extends that on p. 417 of [103] to be that "the fractional change of the integrand of the S-matrix under a [10]-super-Poincaré transformation and the associated conformal transformation is equal to the Jacobian factor necessary to change the variables of integration to the new interaction terms".

To determine how the integration range can be determined, let us first consider the one-loop case. The map (5.43) is

$$\hat{\rho} - \rho = \gamma \ell nz \tag{A17.4}$$

and the string length and twist parameters $\hat{\alpha},\hat{\beta}$ are given as

$$\hat{\alpha} = \alpha+\gamma$$

$$\hat{\beta} = \beta+\gamma\tau \tag{A17.5}$$

where α and β are those from the map (5.21). The identification of opposite sides of the rectangle of fig.12 implies that $[\mathrm{Re}(\gamma\tau)+i\mathrm{Re}\gamma]$, $[\mathrm{Im}\gamma+i\mathrm{Im}(\gamma(\tau))]$ are the complex time variable $\hat{T}$ and width variable α

associated with fig.12. We may sum over conformally inequivalent diagrams by fixing Υ. In particular we may take $\Upsilon = 1$, which corresponds to the choice of (5.43). In general we have that

$$\partial(\hat{\alpha},\hat{\beta})/\partial(\mathrm{Re}\tau,\mathrm{Im}\tau) = |\gamma|^2/\mathrm{Im}\tau \qquad (A17.6)$$

The factor $|\gamma|^2$ in (A17.6) is exactly cancelled by other parts of the measure, as seen in sub-section 5.5, so giving γ-independence, and the result as given at 1-loop in equation (3.42). This cancellation occurs with equal factors γ and $\bar{\gamma}$ arising from the measure in the type II case. For the heterotic there is cancellation with powers of γ arising from the right-moving SUSY vertices and in $\bar{\gamma}$ from the factors $\prod_{\tilde{P}}[\bar{F}''(\tilde{P})]^{\frac{1}{2}-}$ arising from $(\det\partial_\rho)^{-12}$.

We may generalize the above one loop analyses to the multi-loop case

$$\hat{\rho} = \hat{F}(z) = F(z) + \sum_i \gamma_i u_i(z) \qquad (A17.6b)$$

Again $d\hat{\rho} = dp + \sum_i \gamma_i u_i' dz$ is a meromorphic one form defined on an unmarked R-S. We have that (A17.5) is now modified to

$$\hat{\alpha}_i = \alpha_i + \gamma_i$$

$$\hat{\beta}_i = \beta_i + \Pi_{ij}\gamma_j \qquad (A17.7)$$

In the ρ-plane the points ρ_{2i} and ρ_{1i}, $\tilde{\rho}_{2i}$ and $\tilde{\rho}_{1i}$, with

$$\rho_{2i} - \rho_{1i} = \alpha_i + \gamma_i$$

$$\tilde{\rho}_{2i} - \tilde{\rho}_{1i} = \beta_i + \Pi_{ij}\gamma_j \tag{A17.8}$$

have to be identified, so generalising (A17.5). As in the one-loop case we introduce the complex time variables

$$\tilde{T}_i = \mathrm{Re}\Pi_{ij}\gamma_j + i\mathrm{Re}\,\gamma_i \tag{A17.9}$$

and the complex widths

$$\mathrm{Im}\gamma_i + i\mathrm{Im}\Pi_{ij}\gamma_i \tag{A17.10}$$

both associated for the ρ-diagram of fig 23a. We now write the amplitude in terms of the $\hat{\tilde{\rho}}_j$ interacting points and $\hat{\alpha}_i$ and $\hat{\beta}_i$ variables. We sum over all conformal inequivalent classes, for fixed $\tilde{T}_i$. The amplitude turns out to be independent of $\tilde{T}_i$. There are various remarks to be made on the above:

(1) If we take $\tilde{T}_i = 1$, then

$$\mathrm{Re}\gamma_i = 0,$$

$$\mathrm{Re}\Pi_{ij}\gamma_j = 1.$$

We thus have

$$\mathrm{Im}\gamma_i = -(\mathrm{Im}\Pi)^{-1}_{ij}e_j \tag{A17.11}$$

where $e_j = 1$, $\forall_j$, as used in sub-section 5.6, eqn.(5.62). The $\hat{\rho}$-diagram is now as in fig .23b.

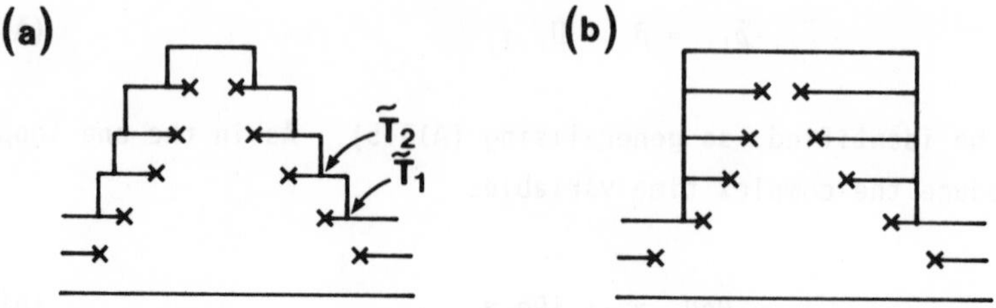

Fig. 23 (a) The new L.C. diagram in the $\hat{\rho}$-variable resulting from using the mapping (J.6) to map from the RS Σ. (b) The same as (a), but for the special choice of complex times $\tilde{T}_i$=1

(2) The measure in terms of the $\hat{\rho}$ variables has the same modular invariant properties as the one in terms of the original light cone diagram.

(3) Decoupling of the non-holomorphic factors may be achieved in the amplitude. To do that, we introduce new integration variable u_{oi}, i=1,...,g by

$$\gamma_i = u_{0j}(\mathrm{Im}\Pi)^{-1}_{ji} \tag{A17.12}$$

Under a modular transformation u_{0i} tranforms as a first Abelain integral, so $\gamma_i u_i$ is then modular invariant. In terms of the fixed variables $\tilde{T}_i$ we obtain

$$\mathrm{Re}\vec{\gamma} = (\mathrm{Im}\Pi)^{-1}\mathrm{Re}\,\vec{u}_0 = \mathrm{Im}\,\vec{\tilde{T}}$$

$$\mathrm{Re}(\Pi\vec{\gamma}) = \mathrm{Re}\Pi.\mathrm{Im}\vec{\tilde{T}} - \mathrm{Im}\,\vec{u}_0 = \mathrm{Re}\,\vec{\tilde{T}} \tag{A17.13}$$

We thus have

$$\vec{u}_0 = i\Pi\,\mathrm{Im}\,\vec{\tilde{T}} - i\,\mathrm{Re}\,\vec{\tilde{T}} \tag{A17.14}$$

We change variables from $(\hat{\alpha}_i, \hat{\beta}_i)$ to u_{0i} leaving fixed the other moduli of the surface. The jacobian is now

$$\det(\mathrm{Im}\Pi)^{-1}.\det[\delta_{ij} + \frac{\partial \mathrm{Re}\epsilon\Pi_{i\rho}}{\partial \mathrm{Imu}_{0m}} \mathrm{Im}\hat{\alpha}_{\ell} \frac{\partial(\mathrm{Im}\Pi)_{mk}\,\mathrm{Im}\hat{\alpha}}{\partial\, \mathrm{Re}\, u_{0j}} k] \tag{A17.15}$$

The dependence on $\vec{u}_0$ of Π, hence the factors $\frac{\partial \mathrm{Re}\Pi}{\partial \mathrm{Imu}_0}, \frac{\partial \mathrm{Im}\Pi}{\partial \mathrm{Reu}_0}$ are only function of $\vec{\tilde{T}}$. Furthermore $\mathrm{Im}\alpha_i = \gamma_\ell - \mathrm{Im}T_\ell$. The Jacobian may thus be expressed as

$$\det(\mathrm{Im}\Pi)^{-1}.Q(\gamma) \tag{A17.16}$$

where $Q(\gamma) = \vec{\gamma}^T M \vec{\gamma}$, with the restriction

$$\frac{\partial \mathrm{Re}\Pi_{i\ell}}{\partial \mathrm{Imu}_{0k}} = \frac{\partial \mathrm{Im}\Pi_{i\ell}}{\partial \mathrm{Reu}_{0k}} \tag{A17.17}$$

which can always be satisfied.

If we like we may now change variables for example to the Schottky uniformization. The change is now completely holomorphic in the moduli since Π depends holomorphically on multipliers and fixed points, while $\vec{u}_0$ depends holomorphically on $\bar{\Pi}$. The non-holomorphic dependence on $\mathrm{Im}\Pi$ has been decoupled into the factor $\det(\mathrm{Im}\Pi)^{-1}$ and the implicit dependence in γ_i on $\mathrm{Im}\Pi$.

We may now attempt to use analytic dependence on γ_i of the amplitude to prove independence of γ_i as in one loop. But it is simpler to appeal to the equivalence of the amplitude in the $\hat{\rho}$-plane to the amplitude in the ρ-plane (see section 5.6 and earlier in this appendix). This implies independence on $\tilde{T}$ and hence on $\vec{\gamma}$.

Appendix XVIII: Heterotic graviton amplitudes

We may extend the scattering amplitudes for external massless states in the heterotic string containing coloured gauge bosons, as contructed in section 5.7, to those containing gravitons by using the formulae of section 5.9 developed for non-zero mode scattering. In particular the change from section 5.7 is that L-going modes are described by external states $\tilde{\zeta}_r . \underline{\tilde{a}}_{r-1}|0\rangle$, in the notation of (5.150), where $\tilde{\zeta}_r$ is the polarisation vector associated with the rth L-going mode. Then the relevant factor for these latter modes may be read off directly from (5.151b) as $A_{-1r,-1s}$, which is proportional to $\partial_{z_r}\partial_{z_s}G(z_r,z_s)$. There will also be contributions from $A_{or',-1s}$ and $A_{-1r,0s}$, which are proportional to $\partial_{z_s}G(z_r,z_s)$ and $\partial_{z_r}G(z_r,z_s)$ respectively. These terms lead to the usual factor $\exp[\frac{1}{2}\zeta_r\zeta_s\partial_{z_r}\partial_{z_s}G(z_r,z_s) + \zeta_r.\underline{p}_s\partial_{z_r}G(z_r,z_s)]$ for vectors meson scattering in the bosonic string obtained by operator methods [65], and to amplitudes identical to those discussed in [127] at the one-loop level. The use of the SSL of section 5.7 leads to an evident extension of the latter to the multi-loop level.

References

[1] Kaluza, T., Sitzuhgsber-Preuss. Acad. Wiss, 966 (1921); Klein O, Z. Phys. 37 895 (1926).

[2] Piran, T. and Weinberg, S., Physics in Higher Dimensions, Vol 2, World Scientific, Singapore, (1986).

[3] Wess, J. and Zumino, B., Nucl. Phys. B70, 39 (1974).

[4] Ferrara, S., Freedman, D.Z. and van Niewenhuizen, Phys. Rev. D13, 3214 (1976); Deser, S. and Zumino, B., Phys. lett. B62 335 (1976).

[5] Green, M.B. and Schwarz, J.H., Nucl. Phys. B181 502 (1981) and Phys. lett 109B 444 (1982).

[6] Green, M.B. and Schwarz, J.H., Phys. lett. 136B 367 (1984).

[7] Christensen, S.M., (ed) Quantum Theory of Gravity, Adam Hilger Ltd., Bristol, (1984).

[8] Brezin, E., Itzykson, C., Parisi, G. and Zuber, J.B., Comm. Math. Phys. 59 35 (1978); Kazakov, V. and Migdal, A., Nucl. Phys. B311 171 (1989).

[9] Becchi, C., Rouet, A. and Stora, R., Comm. Math. Phys. 42 127 (1975); Tyutin, S., unpublished report.

[10] Henneaux, M., Phys. Rep. 126 ,1(1985).

[11] Dirac, P.A.M., Rev. Mod. Phys. 21 392 (1949); a recent review is Leibbrandt, G., Rev. Mod. Phys 59 1067 (1987).

[12] Kogut, J.B. and Soper, D.E., Phys. Rev. D1 2901 (1970); Bjorken, J.D., Kogut, J.B. and Soper, D.E., Phys. Rev. D3 1382 (1971).

[13] Leibbrandt, G., Phys. Rev. D29 1699 (1984); Tang, A.C., Phys. Rev. D37 3014 (1988)

[14] Rivelles, V.O., and Taylor, J.G., Phys. lett. 104B 131 (1981); ibid 121B 37 (1983); Siegel, W., and Rocek, M., Phys. Lett. 150B 275 (1981); Taylor, J.G., J. Phys. A15 867 (1982).

[15] Mandelstam, S., Nucl. Phys. B213 149 (1983).

[16] Taylor, J.G., P. 211 in Supersymmetry and Supergravity 1983, ed. B. Milewski (World Scientific Publ. co. Singapore); Bengtsson, A.K.H., Nucl. Phys. B228 190 (1983); Hori, T., and Kamimura, K., Phys. Lett. 132B 65 (1983); Mondaini, R.P., Restuccia, A., and Taylor, J.G., Phys Lett. 145B 201 (1984).

[17] Scherk, J. and Schwarz, J.H., Gen. Rel. and Grav. 6 537 (1975); Kaku, M., Nucl. Phys. B91 99 (1975); Aragone, C., and Chela-Flores, J., Nuov. Cim. 25B 225 (1975); Aragone, C., and Restuccia, A., Phys. Rev. D13 207 (1976).

[18] Green, M.B. and Schwarz, J.H., Phys. Lett. 149B 117 (1984).

[19] Green, M.B. and Schwarz, J.H., Phys. Lett. 151B 1 (1985).

[20] Gross, D.G., Harvey, J.A., Martinec, E. and Rohm, R., Nucl. Phys. B267 75 (1986); Yahikozawa, S., Phys. Lett. 166B 135 (1966).

[21] Gross, D.G., Harvey, J.A., Martinec, E. and Rohm, R., Phys. Rev. Lett. 54 502 (1985); ibid Nucl. Phys. B256 253 (1985).

[22] Polyakov, A., Phys. Lett. B103 207, 211(1981).

[23] Nelson, P., Phys. Rep. 149 337 (1987); Alvarez-Gaumé, Moore, G., and Vafa, C., Comm. Math. Phys. 106 1 (1986).

[24] Ramond, P., Phys. D1 2415 (1971); Neveu, A. and Schwarz, J.H., Nucl. Phys. B 31 86 (1971).

[25] Friedan, D., "Notes on string theory and two dimensional conformal field theory" in Workshop on Unified String Theories (ITP, Santa Barbara 1985) ed. Green, M. and Gross, D., World Scientific, 1986.

[26] Atick, J.J., Moore, G. and Sen, A., Nucl. Phys B308 1 (1988); Atick, J.J., Rabin, J.M. and Sen, A., Nucl. Phys. B299 279 (1988).

[27] D'Hoker, E., and Giddings, S.B., Nucl. Phys. B291 90 (1987).

[28] Kaku, M. and Kikkawa, K., Phys. Rev. D10 110, 1832 (1974).

[29] Brink, L., Green, M.B., and Schwarz, J.H., Nucl. Phys. B219 417 (1981); Green, M.B. and Schwarz, J.H., Nucl. Phys. B243 475 (1984); Brink, L., p244 in Unified String Theories, ed. Green, M. and Gross, D., World Scientific, (1986).

[30] Gross, D., and Periwal, V., Nucl. Phys. B287 1 (1987).

[31] Bressloff, P.C., Restuccia, A. and Taylor, J.G., Int. J. Mod. Phys. 3 45 (1988); Bressloff, P.C., Restuccia, A. and Taylor, J.G., Phys. Lett. 190B 69 (1987).

[32] Restuccia, A. and Taylor, J.G., Phys. Lett. 174B 56 (1986); ibid 177 B39 (1986) ; ibid Phys. Lett. 192B 89 (1987); ibid Phys. Rev. D36 489 (1987); Taylor, J.G., "Quantisation and Divergences of Superstrings in the Light-Cone Gauge" in Workshop on Constraint's Theory and Relativistic Dynamics, ed. G. Longhi and Lusanna, World Scientific, 1987.

[33] Restuccia, A. and Taylor, J.G., Phys. Lett. 187B 267,273 (1987); ibid Comm. Math. Phys. 112 447 (1987).

[34] Greensite, J. and Klinkhamer, F.R., Nucl. Phys. B281 269 (1987); ibid B291 557 (1987).

[35] Greensite, J. and Klinkhamer, F.R., Nucl. Phys. B304 108 (1988).

[36] Restuccia, A. and Taylor, J.G., "Closure of the Light-Cone Gauge Closed Superstring Algebra", Phys. Lett. B 213 11 (1988); ibid "Constructing the Superstring Space-Time SUSY Algebra in the L.C. Gauge", Int. J. Mod. Phys. A 3 2855 (1988)

[37] Green, M.B. and Seiberg, N., Nucl. Phys. B299 559 (1988).

[38] Restuccia, A. and Taylor, J.G., Phys. Rep. 174, 285 (1989).

[39] Restuccia, A. and Taylor, J.G., "Closure of the Superstring algebra in the Lightcone Gauge", King's College preprint, Oct 1991.

[40] Witten, E., Nucl. Phys. B276 291 (1986).

[41] Wendt, C., Nucl. Phys. B314 209 (1989).

[42] Siegel, W. and Zwiebach, Nucl. Phys. B299 206 (1988), and earlier references therein.

[43] Hata, H., Itoh, K., Kugo, T., Kunitomo, H. and Ogawa, K., Phys. Rev. D34 2360 (1986); ibid 35 1318 (1987).

[44] Goddard, P., Goldstone, J., Rebbi, C., and Thorn, C.B., Nucl. Phys. B56 109 (1973).

[45] Nambu, Y., Lectures at the Copenhagen Summer Symposium, 1970.

[46] Kastrup, H.A., Phys. Rev. 101 3 (1983).

[47] Linden, N., Nucl. Phys. B286 429 (1987).

[48] Green, M.B., Schwarz, J., and Witten E, Superstring Theory, Vols I and II, Camb. Univ. Press, 1987.

[49] Brink, L, Green M.B. and Schwarz, J.H. Nucl. Phys. B219 437 (1983).

[50] Green, M.B. and Schwarz, J.H., Nucl. Phys. B243 475 (1984); Brink, L., p 244 in Unified String Theories, ed. Green, M. and Gross, D. World Scientific (1986).

[51] Osterwalder, R and Schroder, R. Comm. Math. Phys. 31 83 (1973).

[52] Mandelstam, S., Phys. Rep. C13 259 (1974).

[53] Giddings, S. and Wolpert, S., Comm. Math. Phys. 109 177 (1987).

[54] Cremmer, E. and Gervais, J-L., Nucl. Phys. B90 410 (1975).

[55] Mandelstam, S., Phys. Rep. C13 259 (1974).

[56] Dowker, J.S., J. Phys. 5A 936 (1972).

[57] Isham, C.J. and Linden, N., J.C. Class. and Quant. Grav. 5 71 (1988)

[58] Ezawa, Z.F., Nakamura, S., and Tezuka, A., Tohoku preprint 1986.

[59] Frenkel and Kac, Inv. Math. 62, 23, (1980).

[60] Gross, D.J. and Percival, V., Phys. Rev. Lett. 60 2105 (1988).

[61] Taylor, J.G., Phys. Lett. 193B 23 (1987).

[62] Restuccia, A., and Taylor, J.G., Phys. Lett. 192B 89 (1987).

[63] Bressloff, P.C., Restuccia, A., and Taylor, J.G., Class and Quant Grav. 2 685 (1990).

[64] Arfaei, H., Nucl. Phys. B112 256 (1986).

[65] Schwarz, J.H., Phys. Rep. 89 223 (1982).

[66] Ahlfors, L., "The Complex Analytic Structure of the Space of Closed Riemann Surfaces" in Analytic Functions, Princeton Univ. Press, Princeton, N.J. (1960).

[67] Rauch, H.E., Bull. Amer. Math. Soc. 71 1 (1965).

[68] Weinberg, S., "Radiative Corrections in String Theories", Texas preprint UTTG-22-85 (1985).

[69] Fricke, R. and Klein, F., Vorlendsugen uber die Theorie der Automorphen Funktionen, Teubner, Leipzig, (1926).

[70] Ford, L.R., Automorphic Functions, Chelsea, NY, (1951).

[71] Alessadrini, V., Nuovo Cim, 2A 321 (1971).

[72] Mandelstam, S., in Unified String Theory, ed. Green, M. and Gross, D. World Scientific Pub. Co. p 577 (1986)

[73] Blau, S.K., Clemens, M. and de la Pietra, S., Nucl. Phys. B301 285 (1988)

[74] Mumford, D., Tata lectures on Theta, Birkhauser, Boston, (1983).

[75] Jacobs, M., Phys. Reports 1, North Holland, (1974).

[76] Siegel, C.L., Topics in Complex Function Theory, Wiley, (1971).

[77] Bressloff, P.C., Restuccia, A. and Taylor, J.G. Int. J. Mod. Phys. A3 451 (1988).

[78] Jetzer, P. and Mizrachi, L., Phys. Rev. Lett. 58 89 (1987); Jetzer, P., Lacki, J., and Mizrachi, L., Int. J. Mod. Phys. A3 243 (1988) ; Ellis, J., Jetzer, P., and Mizrachi, L., Nucl. Phys. B303 1 (1988).

[79] Glashow, S.L., Nucl. Phys. 22 579 (1961); Salam, A., Proc. 8th Nobel Symp., Aspenasgarden, 1968, ed. N. Svartholm, Almqvist and Wiksell, Stockholm, 1968, p 367; Weinberg, S., Phys. Rev. Lett. 9 1264 (1967).

[80] Shapiro, J.A., Phys. Rev. D11 2937 (1975); Ademollo, M., D'Adda, A., d'Auria, R., Gliozzi, F., Napolitano, E., and Scuto, S., Nucl. Phys. B94 221 (1975); Kaku, M., and Scherk, J., Phys. Rev. D3 (1975) 430; Alessandrini, V., and Amati, D., N. Cim,., 4A 793 (1971).

[81] Scherk, J., Rev. Mod. Phys. 47 123 (1975).

[82] Gava, E., Iengo, R., Jayaraman, T., and Ramachandran, R., Phys. Lett. 168B 201 (1986).

[83] Belavin, A.A. and Knizhnik, V.G., Phys. Lett. 168B 201 (1986); Catenacci, R., Cornalba, M., Martellini, M., and Reina, C., Phys. Lett. 172B 328 (1986); Bost, J.B. and Jolicoeur, T., Phys. Lett. 174B 2713 (1986); Gomez, C., Phys. Lett. 175B 32 (1986).

[84] Belavin, A.A. and Knizhnik, V.G., Phys. Lett. 168B 201 (1986).

[85] Atick, J.J., Moore, G., and Sen, A., Nucl. Phys. B307 221 (1988).

[86] Hodgkin, L.M., Letters in Math. Physics, 14 47, 177 (1987).

[87] Martellini, M., and Teofilatto, P., Phys. Lett. 211B 293 (1988).

[88] Rothstein, M.J., Trans. Amer. Math. Soc. 229 387 (1987).

[89] Schiffer, M., and Spencer, D., Functionals of Finite Riemann Surfaces, Princeton Univ. Press, (1954).

[90] Alessandrini, A., Nuovo Cim. "A , 321 (1971).

[91] Restuccia, A. and Taylor, J.G., "On the Infinities of Closed Superstring Amplitudes", Mod. Phys. Lett. A3 883 (1988).

[92] Lebowitz, "On the Degeneration of Riemann Surfaces", p.265 in Advances in the Theory of Riemann Surfaces, ed. L.V. Ahlfors et al., Annals of Math. Studies, Princeton Univ. Press (1971).

[93] Fay, J. Theta Functions on Riemann Surfaces, Lecture Notes in Mathematics 352, Springer-Verlag, (1973)

[94] Masur, H., Duke Math. Journal 43 623 (1976).

[95] Wolpert, L.S., "Asymptotics of length spectra and the Selberg Zeta-Function", College Park (Md), unpublished (1986).

[96] Bateman, H., Higher Transcendental Functions, vol.II (McGraw-Hill, 1953).

[97] Jahnke-Emde-Losch, Tables of Higher Functions, (McGraw-Hill, 1960).

[98] Restuccia, A. and Taylor, J.G., "On the Infinites of Closed Superstring Amplitudes", King's College preprint, Feb 1988.

[99] Bers, L., Bull. Amer. Math. Soc. 67 206 (1961).

[100] Yahikozawa, S., Phys.Lett. 166B 135 (1986).

[101] Atick, J.J., Rabin, J.M. and Sen, A., Nucl. Phys. B299 279 (1988).

[102] Restuccia, A. and Taylor, J.G., Phys. Lett. B213, 16 (1988).

[103] Mandelstam, S., Nucl. Phys. B83 413 (1974).

[104] Berezin, F. and Shubin, M., Schrodinger Equation (Moscow Univ. Press, Moscow, 1983).

[105] Batalin, I. and Fradkin, E., Ann. Inst. Henri Poincare 49 (1988) 145.

[106] Buchbinder, I., Fradkin, E., Lyakhorich, S. and Pershin, V., Int. J. of Math. Phys. 3 1211 (1991).

[107] Berkovits, N., Nucl. Phys. B276 650 (1986).

[108] Hamada, K., Phys. Lett. 201B 440 (1988).

[109] d'Hoker, E. and Phong, D.H., Nucl. Phys. B278 225 (1986).

[110] Knizhnik, V.G., Phys. Lett. 196B (1987) 473.

[111] Kallosh, R. and Morozov, A., Phys. Lett. 207B 164 (1988).

[112] Lovelace, C., Phys. Lett. 135B 75 (1984); ibid Nucl. Phys. B273 413 (1986); Fradkin, E.S., and Tseytlin, A.A., Nucl. Phys. B261 1 (1985).

[113] Callan, C.G., Friedan, D., Matinex, E.J. and Perry, M.J., Nucl. Phys. B262 593 (1985); Jain, S., Shankar, R. and Wadia, S.R., Phys. Rev. 32D 2713 (1985); Nemeschansky, D. and Yankielowicz, Phys. Rev. Lett. 54 620 (1985); Maharana, J. and Veneziano, G., Nucl. Phys. B283 126 (1987); Sen, A., Nucl. Phys. 284B 423 (1987); de Alwis, S.P., Phys. Rev. D34 3760 (1986); Callan, C.G., Klebanov and Perry, M.J., Nucl. Phys. B278 78 (1986); Akhoury, R. and Okada, Y., Phys. Rev. D35 1917 (1987).

[114] Bowick, M.J. and Rajeev, S.G., Phys. Rev. Lett. 58 (1987) 535; ibid Nucl. Phys. B293 (1987) 348; Saito, T. and Wu, K., Phys. Lett. 200B 31 (1988).

[115] Bergshoeff, E., Randjbar-Daemi, S., Salam, A., Sarmadi, H., Sezgin, E., Nucl. Phys. B269 77 (1986).

[116] Mattos, O, PhD Thesis, University of London.

[117] Mandelstam, S., Nucl. Phys. B213, 149 (1983).

[118] Howe, P., Stelle, K., and Townsend, P., Nucl. Phys. B214, 519, (1982).

[119] Witten, E., Nucl. Phys. B268, 253, 1986; ibid B276, 2, (1987).

[120] Kaku, M., Phys. Lett. B250, 64, (1990).

[121] Restuccia, A. and Taylor, J.G., Phys. Lett. B214, 527, (1988).

[122] Bardakci, K, Nucl. Phys. B271 561 (1986); Bars, I. and Yankielowiez, S., Phys. Rev. D35 3878 (1987); Kaku, M., Int. J. Mod. Phys. A2 1 (1987).

[123] del Aguila, F. and Coughlan, G.D., Nucl Phys. B351 90 (1991); Font, A., Ibanez, L.E., Quevedo, F. and Sierra, A., Nucl. Phys. B331 421 (1990); Casas, J.A., and Munoz, C., Phys. Lett. B214 543 (1988); Greene, B., Kirklin, K.H., Miron, P.J. and Ross, G.G., Phys. Lett. B292 606 (1987); Antoniadis, I., Ellis, J., Kelley S. and Nanopoulos, D.V., "The Prize of Deriving the Standard Model from String", CERN preprint CERN-TH6169/91 (1991); Casas, J.A., Gomez, F. and Minoz, C., "Complete Structure of the Z_n Yukawa Couplings", CERN preprint CERN-TH6194/91 (1991).

[124] Schwarz, J., private communication, (1989).

[125] Bers, L., Bull. Amer. Math. Soc.(NS) 5 131 (1981)

[126] Alvarez, O., Nucl. Phys. B216 125 (1983).

[127] Jetzer, P. and Mizrachi, L., Phys. Rev. Lett. 58 89 (1987); Jetzer, P., Lacki, J. and Mizrachi, L., Int. J. Mod. Phys. A3 243 (1988); Ellis, J., Jetzer, P. and Mizrachi, L., Nucl. Phys. B303 1 (1988).

[128] Quillen, D., Funk. anal. i. Prilozen, 19, 37, (1985).

[129] Senjanovitch, P., Ann. of Phys. 100 273 (1976).

[130] Hull, C.M., Phys. Lett. 178B 357 (1986).

[131] Siegel, W., Nucl. Phys. B238 307 (1984).

[132] Labastida, J.M.F. and Pernici, M., Nucl. Phys. B297 557 (1988).

[133] Teitelboim, Phys. Rev. D25 3159 (1982).